Risk Analysis for Process Plant, Pipelines and Transport

Risk Analysis for Process Plant, Pipelines and Transport

J.R. TAYLOR

Taylor Associates Aps
Glumsø
Denmark

Taylor & Francis
Taylor & Francis Group
LONDON AND NEW YORK

Published 1994 by Taylor & Francis
2 Park Square, Milton Park, Abingdon, Oxon, OX14 4RN
52 Vanderbilt Avenue, New York, NY 10017, USA

First issued in paperback 2020

Taylor & Francis is an imprint of the Taylor & Francis Group, an informa business

First edition 1994

A catalogue record for this book is available from the British Library

Library of Congress Cataloging-in-Publication data available

ISBN 13: 978-0-367-57984-5 (pbk)
ISBN 13: 978-0-419-19090-5 (hbk)

Contents

Acknowledgements

The material in this book is for the most part based on material collected during practical risk analysis projects. Nevertheless, no book covering a subject as wide as risk analysis could be written without information, help, and support from others.

The debt due to Professor Jens Rasmussen in the area of human reliability analysis will be obvious to anyone who reads chapter 6 of this book. Less apparent is the support he has given through many years of research into methods and data, only part of which is reflected here.

Thanks are also due to my many friends in Venezuela, particularly Nelson Garcia, Beatriz Tosta, and Enrique Garcia who provided the encouragement for writing the first version of this book. And thanks to Henning Styhr Pedersen for the diligent criticism of concepts and presentation.

Finally, thanks to my wife, Hanne Taylor, for many hours of typing and editing and patience.

Any errors or omissions are entirely the responsibility of the author.

J.R. Taylor
Glumsø

1 Major Hazards Analysis, Motivation and Management

This book is intended as a practical self study guide, and as a support for teaching courses, in risk analysis of process plant. The techniques described here are for the most part fairly standard, but some advances in techniques, developed by the author and his colleagues, have been included. All the methods and approaches described here are tried and tested, having been applied in many risk analyses of process plant, offshore platforms, pipelines, and hazardous goods transportation.

Risk analysis is the systematic application of methods to identify and evaluate hazards. As such it is a very wide subject, stretching from financial investment analyses, to analysis of continuing health hazards, insurance questions, as well as the problems of safety. Even within the field of safety there is a wide variation in technique, depending on what is to be analysed. Risk analysis is routinely applied for weapons systems, space systems, communications systems, aircraft, road and transport, and bridge design as well as to process plant. The techniques described her cover therefore just a small part of the total field of risk analysis, but an important one.

Major Accidents and Accident phenomena

Accidents in chemical plants can be grouped into a few dominant classes:

- Mechanical crushing, falls, dropped objects.
- Traffic accidents.
- Corrosive burns, from spray and splashes.
- Leaks of toxic substances, causing poisoning.
- Asphyxiation after tank entry.
- Fire.
- Explosions.
- Major toxic releases.

Of these, the last three can give rise to major hazards with large consequences. (Multiple fatalities and major environmental incidents). Fires and explosions can at the same time be starting points for major toxic and environmental accidents.

Since the 1970's, attention has focussed increasingly on environmental accidents. Many of these have been the result of continuing pollution or deliberate dumping, but increasingly it is accidental release which gives the largest contribution to consequences, as routine environmental protection measures become increasingly effective. Here leakage and spills can lead to

- ground pollution.
- groundwater pollution.
- waterway and sea area pollution.

Developments over the last 20 to 30 years have led to a steady fall in the number of fatalities on average. However, at the same time there has been a continuing increase in the size of individual toxic release accidents, and in the average losses from chemical and petrochemical plants. (W.G.Garrison 1989). In some areas, the size of the largest losses has made it difficult, as well as expensive, to organise insurance for chemical plant. These trends can be explained in terms of improving safety techniques, coupled with a continuing increase in size of production units, and overall production volume.

Fires

Chemical and petrochemical plant have always been subject to fires, since the substances produced are often very flammable, and because many processes involve highly flammable solvents.

Liquid fires in process equipment typically involve leaks from flanges, pump packings, broken low gage pipes, corrosion holes, burst vessels, etc. or overflow from tanks and drains. Sometimes fires arise from deliberate releases, such as draining of pipes or removal of sludge.If the leaking liquid is then above the autoignition temperature, ignition is immediate. Otherwise an ignition source is necessary.

Some of the largest liquid fires involve storage tanks, particularly as a result of an initial explosion (cone roof tanks) or rim fires (floating roof tanks or after a small explosion in a cone roof tank). A typical course of events in such a fire involves an initial spill, an ignition at the tank, and a resulting local fire. This local fire may then be extinguished if response is rapid and the emergency facilities have sufficient capacity. Delays, failure of emergency equipment, and inadequate capacity of equipment compared with the size of the release result in gradual increase in the size of the fire. If the fire is large enough, it can spread through collapse of the tank, or through boil over and frothing, or through radiation to nearby buildings and other tanks.

It is rare for a simple pool fire due to a spill of flammable liquid to present a threat for persons outside the vicinity of the plant. Pool fires under transport, and pipelines, by contrast, very often cause a hazard to buildings and persons in them. Additionally, fires, and firefighting actions represent one of the most significant acute threats to the environment.

Pool fires have long been a major topic in process plant accident analysis. In practice, such fires often take the form of tank top fires, or "running fires". In these fires, liquids run down slopes, run into channels and spread in that way. Running fires of this kind require special analytical treatment.

For releases of liquids at high pressure from piping, a different mechanism applies. The liquid forms a spray or jet, which then leads to a jet fire. Such fires can be very intense, and the flames surprisingly long. Additionally they can impinge on other equipment and cause damage, leading to still larger releases.

Toxic and environmental releases from fires have recently been recognized as a major problem. Fires on pesticide mixtures will often release pesticide vapour. In most situations, toxic smoke will rise above the level at which it can cause acute toxic effects. However, fires in the neighborhood of high rise buildings present a risk, as do fires in narrow valleys and under unusual weather conditions.

Fire fighting generally involves the use of large amounts of water. In a well designed process plant, such water is impounded in bund areas, or in a hold up basin. Failure or incorrect design of these facilities gives major problems. for example, a fire in a pesticide storage unit at the pesticide storage plant in Basle, 1986, caused pollution of the Rhine along most of it length to the sea.

An example of ground area pollution as a result of a fire is that of reactor II at Tjernobyl. The fire generated smoke polluted with radioactive strontium and caesium, which were then deposited as a result of rainfall over large parts of Europe. The net result was that roughly 1000.000 square kilometers of productive land was rendered unusable for a period of several months, and sheep and reindeer livestock in areas as far away as Scotland and Sweden were rendered unsuitable for human consumption.

BLEVES

One of the most violent types of accidents is a so called "boiling liquid expanding vapour explosion" or BLEVE. The typical pattern of a BLEVE starts with a fire under or around a pressurised vessel containing liquefied petroleum gas or a similar flammable liquid. As a result, the pressure rises, and the safety valves open. If the unwetted part of the vessel, above the level of the liquefied gas, is not cooled, it overheats, and as a result the vessel is weakened. When the vessel fails, a large part of the contents of the vessel are thrown out, and the result is a large cloud of flashing liquid and gas, mixing violently with air, and burning. The whole contents of a large vessel may burn in under 40 sec, causing intense heat radiation. (Note: in principle the term BLEVE also covers explosion such as steam explosions, in which no fire occurs. Such usage is unusual however.)

A case of this kind occurred at San Juan Ixhuatepec, near Mexico City in 1984, as a result of a large flash fire following a pipe break. In all four spheres with a capacity of 1600 m^3, two of 2400 m^3 and 48 horizontal cylinders were involved, as well as a household gas packing depot, with small cylinders. Large cylinders were thrown up to 1200 m. and over 500 people were killed and 7000 injured.(C.M. Pietersen 1986)

Gas releases

Releases of flammable gas and vapours behave quite differently from releases of liquids below their boiling point. Initially, the gas of vapour spreads as a jet, with air moved into the jet by the turbulence of the gas flow. This mixing may dilute the jet to a concentration below that of the lower explosion limit. (In the case of a boiling liquid or liquefied gas however, this is unlikely).

Once the jet is expended (jet velocity below wind velocity), the gas will disperse with the wind. For natural gas, for ammonia, and similar light gases, the gas tends to rise and spread with the wind turbulence. For higher hydrocarbons (propane and above) and most liquefied gases, the mixture of gas and air will be heavier than air, and the cloud spreads as a flat pancake on the ground. Dispersion is then much slower.

For liquefied gases, the mixture of gas and air will generally be heavy, because the air is cold, even if the gas itself is light.

Flammable gases spreading as a cloud may ignite. Ignition is generally either at the source (in which case a cloud will not have time to form) or at the edge of the cloud.

An example of this was the release of ammonia due to the collapse of a large cryogenic storage tank, at Jonava, Lithuania in 1990. Overpressuring, possibly caused by rollover (that is, rapid mixing of hot and cold layers of liquefied gas) caused tank collapse. The resulting vapour cloud ignited and destroyed a considerable part of the ammonia plant, as well as causing some seven deaths.

The form of fire depends very much on the amount of air which has mixed with the gas. If very little mixing has taken place (releases at low pressure, or into a confined space, or releases of cryogenic liquids which then evaporate) then the fire will take place at the edge of the cloud, as air diffuses into the cloud. Such a cloud can in some cases be stable, if there is a continuing release of gas.

In the more usual case the fire proceeds through the mass of gas, with increasing turbulence and mixing with air. If the burning mass is large enough it may form a fireball which rises from the ground. The rapid rate of burning gives rise to high heat radiation levels which can be especially dangerous for persons in the area of the fire.

Toxic liquid releases

Toxic liquid releases can arise at pipelines, in plants, or during transport. Almost always they present an environmental threat, for example to water courses, agricultural ground, dwelling areas, and to ground water resources. Often too such releases represent a threat from evaporates (vapour) which can penetrate houses, requiring decontamination of the affected soil.

Toxic gas releases

Releases of toxic gases generally fall into one of two categories. Light gases such as ammonia spread initially upward (unless they are mixed with air and are cold, for example as a result of evaporation of the liquid), later diffusing neutrally. These rarely present a serious threat to life unless they occur in a narrow valley, between buildings, or indoors.

Jet releases of liquefies gases however will often be cold, and by mixing with air form a cold heavy mixture.

Heavy gases and mixtures tend to spread horizontally, forming a dense low lying cloud of gas. This may travel with the wind across a populated area or collect in narrow valleys, so that the hazard is much greater. The cloud will most often disperse, but in some few cases will exist as a cloud in valleys for a longer period, if the weather is very still.

Internal equipment explosions

Chemical plant equipment can explode as a result of runaway reactions, or as a result of ingress of air to the equipment, followed by burning. Such explosions can be very violent, and affect persons outside the plant. An example is nitration reactions. An explosion in a dinitrotoluene transfer line occurred at Institute , West Virginia in 1972 as a result of overheating of a steam jacket, and subsequent decomposition. 300 ft. of 2 inch pipeline were affected and caused severe damage to surrounding equipment. (Bateman, Small, and Snyder, 1973).

Distillation of nitrocellulose paint led to a similar type of accident in Germany, (Ernst, 1989) apparently as a result of a stirrer scraping a tank wall. 900 l. of paint were involved, and the explosion destroyed the building in which the explosion took place.

Gas explosions

Flammable gas clouds, and gas in buildings, can explode, sometimes violently, if the conditions are right. Inside buildings, the only requirement is that gas be mixed with air in a flammable mixture (i.e. between the lower and upper explosion limits). Outdoors, much more complex requirements must be fulfilled (see Ch 12), but the most important, for petroleum gases, is that the amount of gas in the cloud must be large.

Unconfined gas explosions have in many incidents completely destroyed chemical plants e.g. the accident at Flixborough, England in 1974, in which some 60 T of hexane exploded, demolishing the plant, and killing 26 people.

Domino effects

When large fires and explosions occur in a plant, it is possible for these to affect other plants, and for the accident to spread from unit to unit. This is termed a domino effect. An example of this is the explosion of a LPG storage sphere at Texas City refinery in 1986. The explosion spread to half of the refinery and destroyed it. Initially , one sphere burst due to a weak weld, and the release ignited. As a result, other fires were started at neighbouring tanks, and several of these underwent BLEVE type explosions.

The risk analysis process

The overall process of risk analysis involves (fig. 1.1):

- First describing the scope and objectives of the analysis
- Identifying the potential hazards
- Quantifying the probability or frequency of accidents
- Quantifying the consequences of accidents,e.g. damage, injury, and fatality.
- Integrating the information derived into an overall picture of risks.
- Assessing whether the risks are acceptable or tolerable.
- Generally, revising or improving plant designs and operations.
- Following up, to ensure that the assumptions made remain valid.

Defining the scope of the analysis is important. It is easy to leave out important contributors to risk, because of the choice of method for analysis or because of preconceptions about what is most hazardous. A preliminary hazards analysis (PHA) can help to avoid these problems (see chapter 3).

Unfortunately any risk analysis is very dependent on the assumptions made. If the situation changes, the analysis becomes invalid, and may serve only as a false assurance. An analysis will become misleading for example if:

- storage and transport patterns in a plant change, as new plant units are built, or the level of activity in the company increases.
- the types of production in the plant unit changes, or production levels change.
- the manning levels are changed, e.g. due to improved automation or the type of staff changes e.g. due to recruiting difficulties or a new generation of employees.
- maintenance or repair activities on the analysed plant, or nearby.

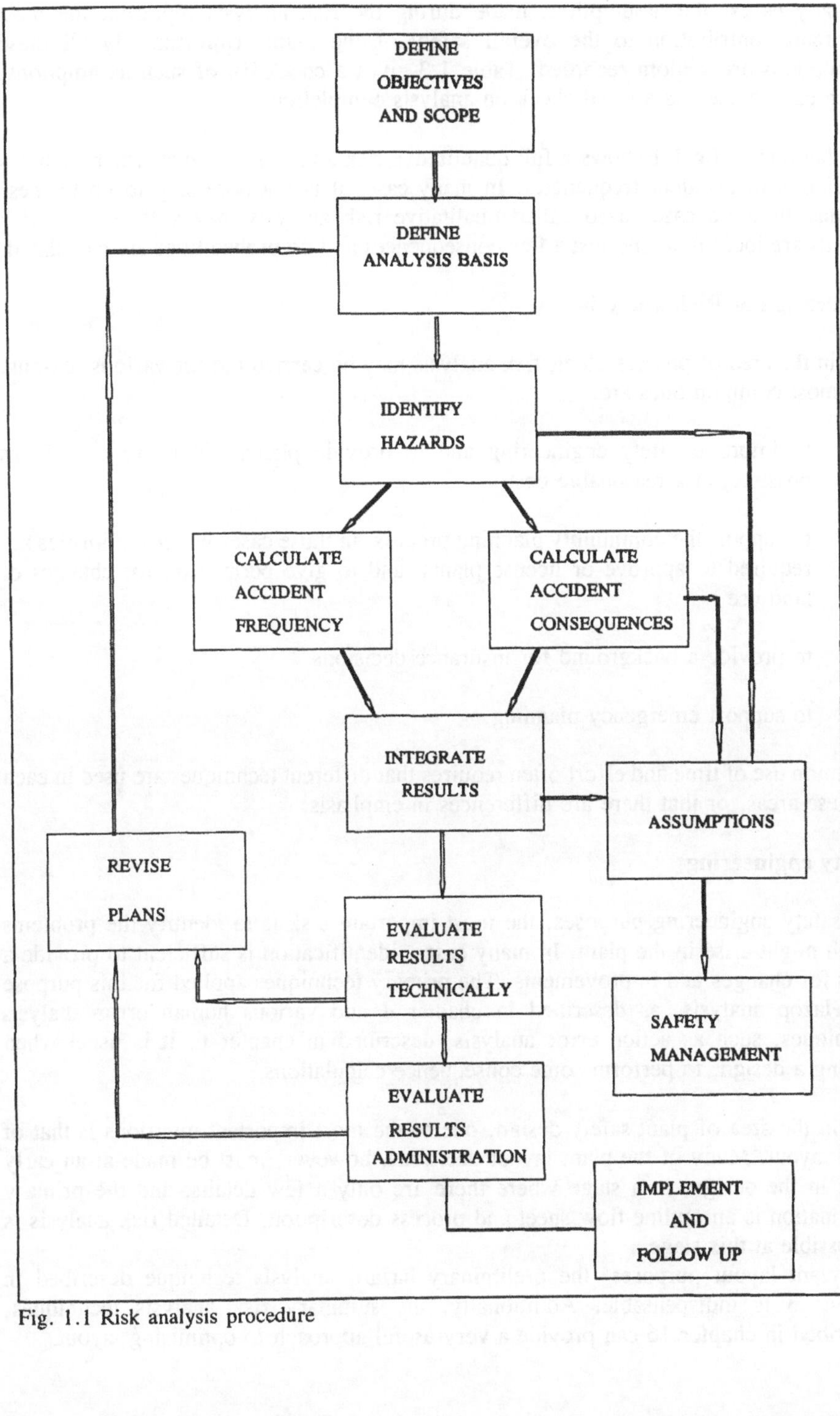

Fig. 1.1 Risk analysis procedure

In many cases, the assumptions made during the risk analysis represent the most important contribution to the overall safety of the plant. Unfortunately all these assumptions are seldom recorded. Table 1.3 gives a check list of such assumptions, which can be used as a rapid check on analysis completion.

The diagram in fig 1.1 shows a full quantitative risk analysis, i.e. one which includes calculation of accident frequencies. In many cases it is not necessary to go to these lengths. In these cases a so called qualitative risk analysis may suffice, in which hazards are identified, and just a few consequences and no probabilities are calculated.

Application of Risk analysis

Within the area of process plant, risk analysis may be carried out for various reasons. The most common ones are:

- to improve safety engineering and to provide plants which are as safe as possible, at a reasonable cost

- to support the community planning process, in those cases where authorities are required to approve or license plants, and to give permission for changes of land use

- to provide a background for insurance decisions

- to support emergency planning.

Optimum use of time and effort often requires that different techniques are used in each of these areas, or that there are differences in emphasis.

Safety engineering

For safety engineering purposes, the most important task is to identify the problems which might arise in the plant. In many cases, identification is sufficient to provide a basis for changes and improvements. The primary techniques applied for this purpose are Hazop analysis, as described in chapter 4 and various human error analysis techniques, such as action error analysis, described in chapter 6. It is usual when making a design, to perform some consequence calculations.

Within the area of plant safety design, one of the most important questions is that of plant layout. Many of the plant layout decisions, however, must be made at an early stage in the design, at a stage where there are only a few details, and the primary information is an outline flow sheet and process description. Detailed risk analysis is impossible at this stage.
For plant layout purposes, the preliminary hazard analysis technique described in chapter 3 is indispensable. Additionally, the summary risk analysis techniques, described in chapter 16 can provide a very useful approach to optimising layout.

Planning approvals

As a support for community planning, risk analysis is a useful tool, but the reader should be aware of its limitations. Some authorities are prepared to accept risk analyses as a legitimate basis for decision. Others become confused. In many cases, if community planning involves a phase of public review, the analyst, and the plant manager, need to be aware that the analysis will be only a small part of the information used for decision making. Questions of local employment, future land use, and general amenity will usually be involved. Risk analysis results are often used spuriously, in order to pursue other goals, which are not directly related to risk. For this reason, risk analyses intended for authority review and community planning need to be made with great care, and the presentation of the analysis needs to be made equally carefully. A company can scarcely afford to present to an authority anything less than the very best analysis of which it is capable.

The author's experiences in this area have so far been good ones. In the cases where public scrutiny has been important, the analyses have contributed to solving problems rather than exacerbating them. The following points may help to improve the presentation:

1. Always strive to get as near to the truth as possible in your analyses and presentations.

2. Don't try to explain away risk - make it clear that you take even the smallest risk seriously. Don't try to "defend" your plant, since this implies that the public is your enemy. In any case, what you regard as a small risk will be regarded differently by others.

3. Write in a language which can be understood, and keep technical descriptions short. Give reasons as well as facts. Use a natural, explanatory style, making it clear that your prime interest is in safety.

4. Tell what accidents have happened and what you have done to prevent them in the future. (If you don't, someone will probably find out, and accuse you of concealing).

5. Tell what problems you found during the analysis, and how the design has changed to avoid them.

6. Use the safety barrier diagram presentation given in chapter 3, to make clear what you have done to prevent accidents.

7. Make an uncertainty analysis, and discuss the strengths and weaknesses in your own results.

8. It is almost certain that the analysis report which is to be made openly available will be a reduced version of the internal, technical report. Effort should be given to making this well balanced and easily understood.

Insurance applications

Technical risk analysis for insurance purposes has generally proved to be a disappointing field. Insurance companies tend to rely on earlier experience and statistics, and on competitive pressure from other companies in setting insurance premiums. The idea of calculating your level risk, and determining premium levels on this basis, is an attractive one, but unfortunately, only rarely realistic. Performance of risk analysis, along with other loss prevention, safety training, and safety management techniques can be used in order to establish a good image, and hence a good insurance rating. Nevertheless, the best approach to low insurance premiums is still to have a good safety record.

Risk analysis has nevertheless become a very important tool for the company risk manager and insurance manager. Insurance premiums have risen astronomically during the 1980's in response to the increasing scale of process plant accidents. In some cases insurance has proved difficult or even impossible to obtain.

In these circumstances it is important to know

- if you are underinsured
- if a self insurance approach is worthwhile
- what levels of insurance and self insurance are reasonable.

Detailed technical analyses are rarely feasible for these purposes - the scope is generally too large. The preliminary hazards analysis procedure described in chapter 3 is a useful one, and can be applied with a minimum of effort. However, quantification is needed to make the analyses realistic.

The summary analyses, described in chapter 16, can be applied for a more detailed approach. The effort involved can be as little as a few days, and the understanding which results can be of value far exceeding the cost.

Emergency planning

For emergency and contingency planning purposes, the techniques required resemble those for plant design. However, detailed identification of potential accident causes is seldom necessary. A simple unit level accident assessment and a preliminary hazards analysis is generally sufficient for identification. Detailed calculation of consequence scenarios is necessary, however, and the effectiveness of the emergency measures needs to be evaluated. Chapter 3 is devoted to a description of these techniques.

Safety analysis legislation

Some kinds of hazard assessment are required by law. One of the major spurs to the development of risk analysis techniques has in fact been the major hazards directive of the European community. (Generally referred to as the "Seveso Directive", to the annoyance of many industry managers and risk analysis professionals) (EC 1981). This directive required all countries in the European community to implement legislation requiring, among other things, that companies with over a certain level of inventory of hazardous substances make a "safety report". This safety report is not necessarily a fully fledged risk analysis, but it at least requires a full list of the accidents which can occur, and the measures which are adopted to prevent them.

The legislation and administrative practice in different countries varies. The actual practice for some countries which have completed a large part of their safety reporting, and the general approaches used are given in table 1.2. Table 1.1 summarises the requirements of the directive, as far as safety reports are concerned.

A good example of the implementation of the EC directive is the British CIMAH regulations.("Control of Industrial Major Hazards Regulations", HSE 1984 and 1988). These give detailed information on reporting, and also on emergency planning". The German "Störfallsverordnung" provides another set of reporting regulations, with a difference of emphasis. Requirements for plant description in this regulation are similar to those in the other European countries, but there is a much greater emphasis in the German practice on norms, and norm requirements, and a correspondingly lesser emphasis on risk evaluation. The German regulations contain an impressive check list of activities and technical measures for ensuring plant safety.

In several European countries, reporting according to these regulations has been completed for all major hazards plants, or is approaching completion. Although at the time of writing it is too early to tell what the overall effect has been, some observations can be made, based on a sampling of results. Almost all plants analysed have been modified to reduce risk, with significant increases in safety as a result. There has been a very significant increase in safety awareness, even for those plants which had high standards of safety management beforehand.

In USA, legislative requirements for risk analysis of process plant, have followed much later than in Europe, a fact which is perhaps surprising, since USA pioneered the field of risk analysis for nuclear plant and for weapons systems. In USA emphasis has been much more strongly on "loss prevention" a subject which is related to, but wider in scope than risk analysis. Much background work has been done on chemical plant risk analysis methods, including the very early work on chemical plant fault tree analysis, work by the US Coast Guard and US Bureau of Mines, and work by Dow Chemical Company on the Dow Index, Monsanto on fault tree analysis. More recently, the Centre for Chemical Process Safety has produced a long series of method guides and research reports to support risk analysis.

In 1989 the State of California Governor's Office of Emergency Services published its Guidance for the Preparation of a Risk Management and Prevention Program The guideline declares itself to be just that. It is neither law nor a regulation. It does show a way in which some of the requirements of the state Health and Safety code can be met. It requires, as a minimum, the procedures set forth in the 1985 Guidelines for Chemical Hazards Evaluation Procedures. Requirements include:

- a description of accidents within the previous 3 years
- a description of the nature, age, and condition of equipment
- design, operating and maintenance controls to minimise accidents
- record keeping for audits and inspections.

In July 1990, OSHA announced hearings on proposed federal rulemaking for "Process Safety Management of Highly Hazardous Chemicals", 29 CFR part 1910. The proposed rules make requirements for "process safety information" which correspond to a very thorough qualitative hazard analysis, and also significant requirements for a safety management programme. The safety management requirements are described in chapter 16. At the time of writing, the rulemaking hearings are in progress.

In USA too, there is an act, the "Right to know Act" which requires declarations to be made to the public for hazardous plant, in particular concerning which accidents can occur, and giving the basis necessary for emergency planning.

The actual techniques for risk analysis for this kind of official analysis vary. The basis of the methods actually used for risk analysis for authority reporting is generally Hazard and Operability analysis (chapter 3), or for small plants, check list analyses (chapter 16). In some countries, quantitative analyses, in which accident probability calculations are performed, are encouraged, or even required. For this purpose a whole range of techniques is available, but generally require extensive use of computer programs (see e.g. Nussey 1988).

Recently, there have been a number of developments in the requirements for risk analysis. Requirements for offshore system have been an essential part of oil and gas platform engineering in Norway and Denmark since the beginning of the 1980's, but with the publication of the Cullen report in response to the Piper Alpha catastrophe, the requirement for risk analyses has been extended to the British north sea.

At the time of writing, prepatrations are being made to update the European major hazards directive. Some of the technical advances over the last five years will certainly be included. In particular, it can be expected that much more emphasis is placed on safety management (see Ch. 17)

Anumber of developing countries have passed legislation requiring safety cases or risk analysis for new projects. Some of these are quite demanding in nature, and represent a better practice than in Europe or USA, at least in the requirements. The actual practice of risk analysis often follows somewhat behind the requirement, but to see the highest quality of risk analysis work, one should at present look outside Europe or USA. Examples of countries with very advanced work are Brazil, Venezuela, and Indonesia.

1. Information on substances,
   chemical name
   CAS number
   IUPAC name
   Empirical formula
   Purity, and identity of important contaminants
   Detection and analysis methods used.

   b. Stages at which substances occur or can enter processes

   c. Amounts of substances

   d. The form in which the substances exist

   e. Other dangerous substances, which can affect the degree of hazard

2. Information concerning the plant

   a. Geographical placement, meteorological conditions, and special hazards arising from the placement.

   b. A general description of the technical processes

   c. A description of the parts of the plant which are particularly important from the point of view of accidents, of the sources of hazards, the conditions under which accidents can arise, and the steps taken to prevent accidents.

   d. The measures taken to ensure that technical equipment intended to prevent failure and ensure safe operations, is functioning properly.

3. Information concerning possible serious accidents:

   a. Emergency plans, and the safety equipment, alarm systems, and equipment which can be used inside the plant.

   b. The name of the person, and the deputy, or the body, responsible for safety and authorised to initiate emergency action, and to inform authorities and the police.

Table 1.1 Typical reporting requirements for regulations in accordance with EC directive 82/501 with amendments 19 March 1987 and 24 November 1988 (For illustration purposes only. Consult the regulations for each country for more specific information).

1. Introduction
   - Declaration of the subject of the report, and its purpose
   - Safety objectives
   - Methods and approach

2. Conclusion
   - Major sources of risk
   - Assessment
   - Recommendations
   - Sources of uncertainty
   - Future development

3. Description of the plant
   - Layout and location (a map with the plant and local community shown)
   - Diagram of the processes involved
   - Short functional description of the plant
   - Chemical reactions
   - Plant operating conditions
   - Internal transport
   - Listing of safety measures and safety equipment
   - Safety and construction standards used
   - Inspection and test methods and test intervals for safety equipment
   - Meteorological conditions (wind rose and stability data)
   - Manning of the plant, and location of persons

4. Description of substances
   - Names
   - Physical properties
   - Toxicological properties
   - Detection and analysis methods (possibly also antidotes and emergency treatment description, or a reference to these)
   - Inventory of substances and form in which it occurs
   - known byproducts, waste products, and accidental products

5. Description of hazards
   - Hazard identification for each plant section
   - Description of event sequences for an accident, and safety measures which can prevent or reduce consequences
   - Consequence calculations for typical accidents
   - Description of accidents which have occurred in the past

6. Description of Safety Management System
   - Organisation
   - Policy
   - Training
   - Quality assurance for equipment and operations

7. Description of emergency plan
   - Organisation
   - Alarm procedure
   - Equipment and methods
   - Training.

Table 1.2 Typical safety report contents

Preparing for risk analysis

A good risk analysis can take anything from one or two man months of effort for a small plant, to a man year, or even more for a large plant. It is worthwhile preparing carefully. Staff should be trained. Many short courses are available, which can provide an introduction. It helps if the staff have plant operations or commissioning experience. If you can, include an experienced operations foreman in the analysis team.

Determine your objectives clearly. It may be possible to avoid a lot of work and expense by knowing the purpose to which the analysis will be put.

Obtain good drawings, and good functional descriptions of the plant. This can be difficult, if the plant has been operating for several years, since drawings are sometimes out of date. For plant under design, drawings may not be available.

Obtain data for the substances involved. I generally try to have a copy of the Chemical Engineers Handbook (Perry and Green) available, but other data such as those on toxicity, are also necessary (see chapter 14). Be prepared also to invest in safety improvement. Risk analyses reveal problems, and many of them require solutions. Out of over 100 analyses in which I have been involved only one has resulted in no changes to the plant.

The analysis as a living document

Even when the immediate purpose of an analysis is to obtain authority approval or planning permission the ultimate objective is always to achieve a higher level of safety. Analyses should never be just completed and then put on a shelf. They can and should, be used as a basis for operator training, for emergency preparedness training, and for accident and incident follow up.

Whenever accidents, or near misses occur, a review should be made, and a comparison made with the accident report. This can be systematised (fig. 1.2). The report should be updated when incidents occur, if any new factors are found. Making reviews in this way increases safety awareness and reduces the problems of incompleteness in the analyses.

An approach used by one company represents something near to an ideal. Initially, a fairly thorough risk analysis was made using outside consultants. On the basis of this, a list of safety improvements was made, and this list was implemented over a period, as allowed by maintenance shutdowns. For each plant modification or new plant unit a new, supplementary analysis was made. After 5 years and 10 years new analyses were made, each improving on the earlier analysis. The later analysis being made by company staff. Throughout the period, a status list of problems and safety improvements was maintained. The results of the analysis, and the associated safety management actions have been apparent - a more than ten fold reduction in the number of incidents.

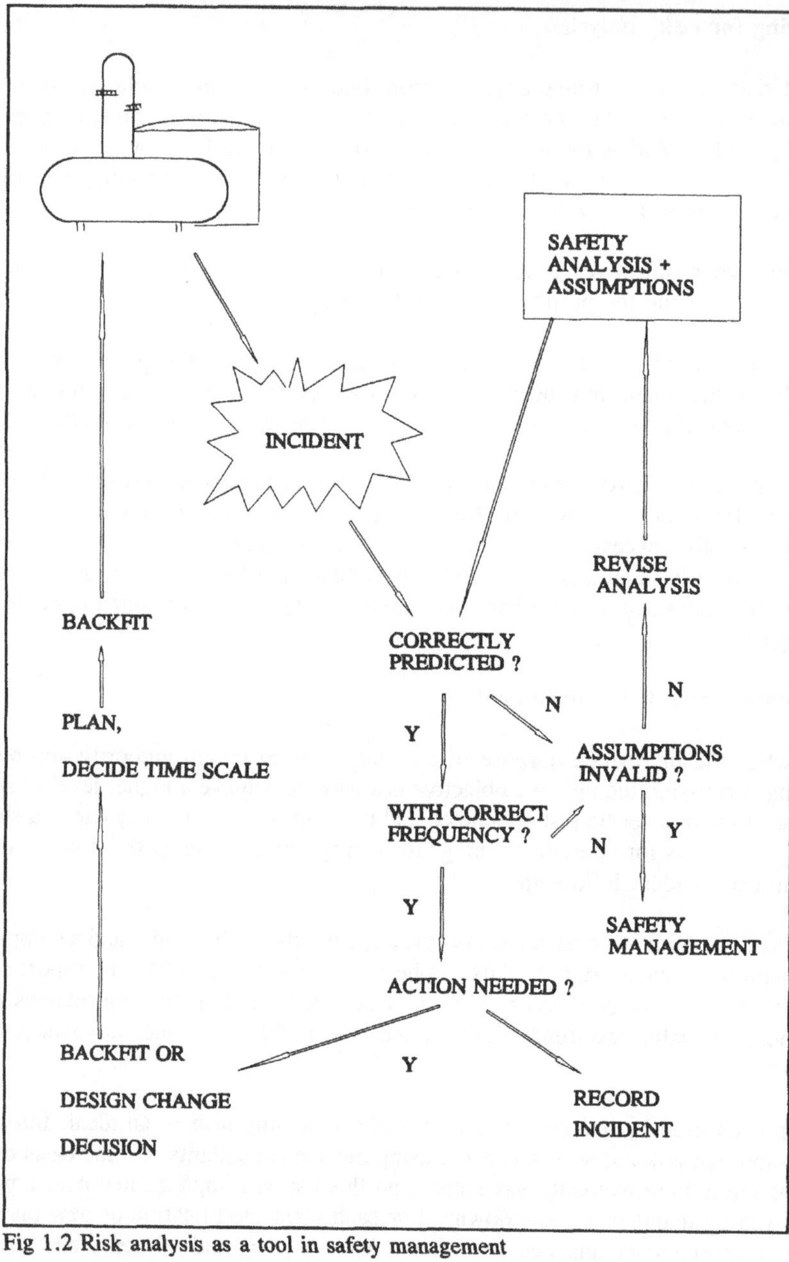

Fig 1.2 Risk analysis as a tool in safety management

Such an approach need not be expensive. Modern computer equipment allows data bases incorporating hazard analyses and allows calculations to be performed rapidly. Problem lists, and safety improvement plans can be maintained continuously, for less than the cost of lost time injury.

1. The maximum inventory of substance ____________________ will be__________ in vessel _________________

   etc.

2. The maximum operating temperature for vessel ___________ will be_____

3. The maximum operating pressure for vessel ______________ will be_____

4. The maximum heat flow rate for exchanger ______________ will be_____

5. The maximum production rate for unit _________________ will be_____

   The normal production rate for unit __________________ will be_____

6. The manning level for unit __________________ will be ____ persons, of which __________ will have the following experience:

7. The following safety and emergency equipment will be available: -----------

8. The plant unit will be shut down when the manning levels, and emergency reserves fall below the following levels:

9. The plant / unit will be shut down when the following contingencies arise

   External events:

   Safety equipment failures:

10. The following failure rate data are assumed to hold for the plant: (reference)

11. The following test and maintenance intervals will apply for the plant:

Table 1.3 Typical assumptions for a process plant risk analysis

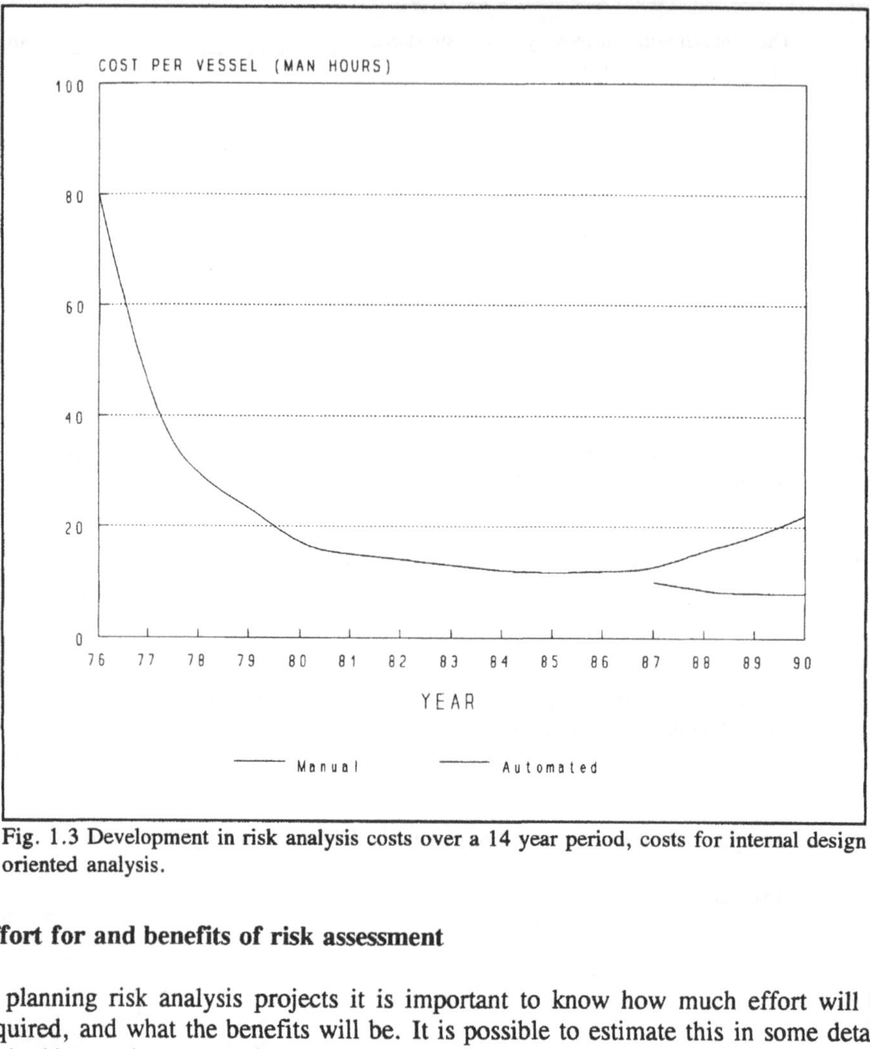

Fig. 1.3 Development in risk analysis costs over a 14 year period, costs for internal design oriented analysis.

Effort for and benefits of risk assessment

In planning risk analysis projects it is important to know how much effort will be required, and what the benefits will be. It is possible to estimate this in some detail, by looking at the records from earlier analyses.

The actual effort involved for an analysis varies of course according to the size of plant. Effort has been found to be very closely related to the number of vessels, that is, tanks, pressure vessels, and heat exchangers actually analysed. To make the assessment it is necessary to subtract from the total number of vessels those which are not analysed because they are not hazardous, and those which are similar to each other.

Effort can then be analysed in terms of man hours used per vessel. The amount of effort varies according to the style of analysis, and especially, according to the style of reporting. A limited analysis using the Hazop method (chapter 3) and reported summarily for internal use within a design group, can take as little as 8 man hours per vessel, even for a very thorough analysis. If some consequence calculations are required, and a full reporting for management, the effort grows, typically to 20 to 25 man hours per vessel for an experienced analysis team. For a full quantitative analysis, including probabilities, a complete range of consequence calculations, survey of population, and reporting, the additional effort is typically of the order 200 man hours per plant, assuming that good computer aids are used. (This kind of analysis is virtually impossible without computer aids). To all of this, meeting time and presentation time must be added, especially if authority or management decisions are involved.

Effort required depends also on experience. The most important factors here are the availability of background reference material, calculation methods, and computer aids, which cut down the effort required to start up a project. Figure 1.3 shows the development in costs of risk analysis from the mid 1970's. The big reductions in effort have been obtained by the availability of standardised methods and data.

The benefits of hazard analysis arise in two direct ways and a third indirect way. For analyses carried out during design, modifications found to be needed during the analysis allow changes to be made on paper, when they are cheap, rather than later when backfitting is required for implementation. The author, in a series of controlled studies, found a rate of about two design changes per man hour of effort in analysis, and a payback period for the costs of the analysis of one week, due to the savings in plant commissioning time (Taylor 1988, 1991). These studies used the most effective hazard and operability analysis approach available, and a minimum of reporting.

After the plant is built, benefits are largely from reductions in accidents. These are difficult to quantify, since often very rare accidents are involved. Calculated risk reductions of a factor of 10 are commonly obtained by concentrating on removing the worst hazards. This kind of reduction has been documented for one plant, over a period of 16 years, with a fall in accident rate for small and medium accident of a factor of more than 10 (Taylor 1991) (No major accidents were involved). Of course, the risk reduction depends on the initial level of risk existing, but in over 100 analyses, the author has only observed one plant which was judged by the plant management not to require improvements.

The actual benefits of risk analysis projects are extended indirectly, if plan management is seen to support them. The process of analysis increases hazard awareness and communication. Additionally it gives safety professionals in the plant a "hard science" basis for their work which can increase their effectiveness. This has been observed to give benefits outside the scope of the analysis itself.

There is one major potential disadvantage of risk analysis. It can be used, and in the past often has been used, to "prove" that a plant is safe, and to provide a cushion on which management can rest. It is important to remember that analysis involves a large number of assumptions, not least of which is that safety management standards remain high.

References

HSE 1984	Control of Industrial Major Hazards Regulations, Health and Safety Executive, HMSO,1984, and 1988
EC 1981	Directive on the Major Accident Hazards of Certain Industrial Activities (82/501/EEC), *Official Journal of the European Communities*, No. L230 5 august 1982, 1-18
Dept. Labor	Process safety management of highly hazardous chemicals, Notice of proposed rulemaking, 29CFR Part 1910, *Federal Register*, Tuesday July 17, 1990
F.A. Lercari, G.R. Schimke	Guide for the Preparation of a Risk Management and Prevention Program, California Office of Emergency Services, November 1989.
Perry, Green	Chemical Engineering Handbook, Perry,R.H, Green,D.W, Maloney,J.O., Sixth Edition, 1984
Pape,R.P, Nussey,C. 1985	A basic approach to the analysis of risks from major toxic hazards, in *The Assessment of Major Hazards*, Institution of Chemical Engineers Symposium Series No.93, 1985
Taylor, J.R. 1984	Costs, Quality, and Benefits for 6 Risk Analysis Procedures, in *Ermittlung und Bewertung industrieller Risiken*, ed.S.Lange, Springer Verlag, 1984 (in German)
Taylor, J.R. 1991	Risk Analysis Techniques in the Chemical Industry, Int. Conf. Probabilistic Safety Assessment and Management, Beverley Hills, 1991
Garrison,W.G.1989	Large Property Damage Losses in the Hydrocarbon - Chemical Industries, a Thirty Year Review 12th ed. Marsh & McLennan 1989
Pietersen,C.M. 1986	Analysis of the LPG Disaster in Mexico City, 5th Int. Symp. Loss Prevention and Safety Promotion in the Process industries, Cannes, 1986
Bateman, Small, and Snyder, 1973	Loss Prevention, 66 Annual Meeting AIChE, 1973
Ernst, 1989	Severe Explosion in a Paint Factory, 6th Int. Symp. Loss Prevention and Safety Promotion in the Process Industries. Oslo 1989

2 Risk and Risk Levels for Process Plant

Some definitions

Risk is defined as the possibility of loss. In process plant, the losses may be of many kinds, such as damage to equipment, loss of production, or environmental damage, as well as injury or death.

With the term risk we express the fact that we cannot predict the consequences of an event. A consequence may be good, and desirable, or bad and undesirable. It is the unwanted consequence which is usually associated with risk. The sequence which leads to the final consequence consists of random events. Some events are more probable than others. A unique consequence of the sequence of events cannot be predicted.

The term risk therefore involves two measurable parameters: consequence and probability.

The assessment of risk depends on the point of view, whether risk is considered from the point of view of the individual, a group, or society as a whole. Individual risk is defined as the risk level for a specific individual, usually the most exposed neighbour for the plant, or the most exposed member of the plant personnel. It can be measured in terms of the frequency of injury or fatality for that person.

Group risk is defined as the risk for a group of persons, for example the maintenance team on the plant, or the plant employees as a whole. It can, for example be measured in terms of the frequency of accidents of a certain size affecting the group. Note that group risk is averaged over the group, so that one individual with a very high individual risk, will tend to "disappear in the crowd". The measure of group risk for single person accidents will be the product of the number of persons in the group, and the average individual risk. Group risk for multi person accidents is also a relevant measure.

Societal risk is the risk for society as a whole. It may be measured in terms of the frequency of accidents above a certain size. Societal risk for single person accidents will be the product of the number of persons exposed and the average individual risk. Societal risk may be high when there are many persons exposed, for example in cases of transport of dangerous goods, even though the individual risk is low. Societal risk is usually considered in the case where accidents can involve many persons.

Other major distinctions are between voluntary and involuntary risk. Voluntary risk is that undertaken willingly by a person, such as risks in motor sport, risks of smoking, etc. Involuntary risk is the risk imposed on us by activities beyond our control.

Major hazards are often defined as any hazard which can result in accidents affecting persons outside the factory fence. For practical purpose, a major hazard is defined here as a potential cause of accidents which can affect 10 persons or more, which causes material damage or production loss of over US$ 10 million, over 100 serious injuries, or environmental damage over a large area which takes more than 5 years for recovery.

Risk analysis is the systematic investigation of all the undesired consequences of events. In the analysis, the various event sequences which can lead to an undesired consequence are described. At the same time, an impression or measure is obtained of the probability of the undesired event.

Measurement of risk

It is rare that probability and consequence can be directly measured in a technical installation. The two risk parameters must generally be calculated on the basis of theory concerning accidental events and consequences, and statistics of failures and accidents. Risk analysis requires a detailed knowledge of the plant under study.

If a risk analysis shows that there is a potential for accidents, it becomes necessary to judge whether the plant is tolerable as it is, or whether the plant should be modified. If the risk analysis is to be used as a direct basis for risk acceptance, it becomes necessary to measure the risk in a well defined way. In order to evaluate acceptability, the risk level must then be compared with a general level of acceptability, which corresponds to the method of measurement. This kind of comparison is called risk evaluation.

The immediate consequences of an accident in a plant can for example be dispersal of poisons, heat radiation, pressure waves (explosions) and explosions fragments. The evaluation of consequences usually involves calculation of these effects. Beyond this, an evaluation is needed of the extent to which persons and the environment can be damaged. Consequence is therefore not a simple quantity which can be measured with a single number.

Just which measurement is used in evaluating risk is in principle irrelevant, provided that it in fact reflects all the risks involved in an activity. In most risk analyses, the probability of immediate fatality has been used as a measure of risk. Fatality risk is convenient to use, because it is easy to define, and is generally representative of other risks such as injury. This measure is not usable if the activity also involves a disproportional possibility for accidents with delayed or environmental consequences. Several attempts have been made to measure risk in terms of a single quantity, namely money. A measurement of this kind is extremely difficult, since the principles for valuation of human life, sickness, and environmental damage, are unclear (see e.g. conference papers by Jones-Lee 1981).

The sequences of events which result in unwanted consequences generally involve both mechanical failures and human error. Human error may involve not only traditional operation errors, but also faults in safety management - inadequate maintenance, inadequate training, poor housekeeping etc. For the most part there are good statistics concerning mechanical failure, but only poor data concerning human error, especially when the error does not involve simple operation error.

For each type of error and failure, a probability value is required. The probability of failure is usually measured in terms of frequency. The measure may for example be: the average number of accidents over a period of 1 million years.

Assessment of risk

The two (calculated) values, consequence and probability are regarded in risk assessment as known values, with an associated level of uncertainty. In the assessment of risk, the two values are combined, and possibly supplemented with other factors. The values are then compared with the established accept criterion.

Attitudes to risk are affected by a large range of factors. Individual persons, interest groups and administrators can place different weight on these factors. This is reflected in the way the measurable factors are combined, or in the choice of acceptance criterion.

A number of factors can be mentioned which affect the way in which risk is assessed. Public and political reaction to large accidents is normally much more violent than the reaction to a series of small accidents, event when the total number of injured is the same. This means that the requirements placed on safety for the largest accidents are disproportionately strict. An average here, in which risk is simply regarded as the product of consequences and probability, does not reflect public attitudes.

An example of this is the contrast between the reactions to fires in the Hafnia hotel, Stengade, Copenhagen in 1975. This cost 41 lives. The accident resulted in a fire protection programme which cost several thousand millions of kroner. This does not correspond to the amounts spent on traffic safety (roughly 2 to 3 million kroner per saved life). More is generally invested in preventing a large accident than in preventing a corresponding series of small accidents.

Some activities are regarded by decision makers as necessary for society. Examples are military installations, power plants, storage for gas and plant for chemical waste destruction. Other activities may be more questionable in some minds, for example plant for production of insecticides and PVC (There is for example a campaign for voluntary reduction of the use of PVC in Denmark). If an activity is regarded as necessary it may correspondingly be necessary to accept an increased level of risk.

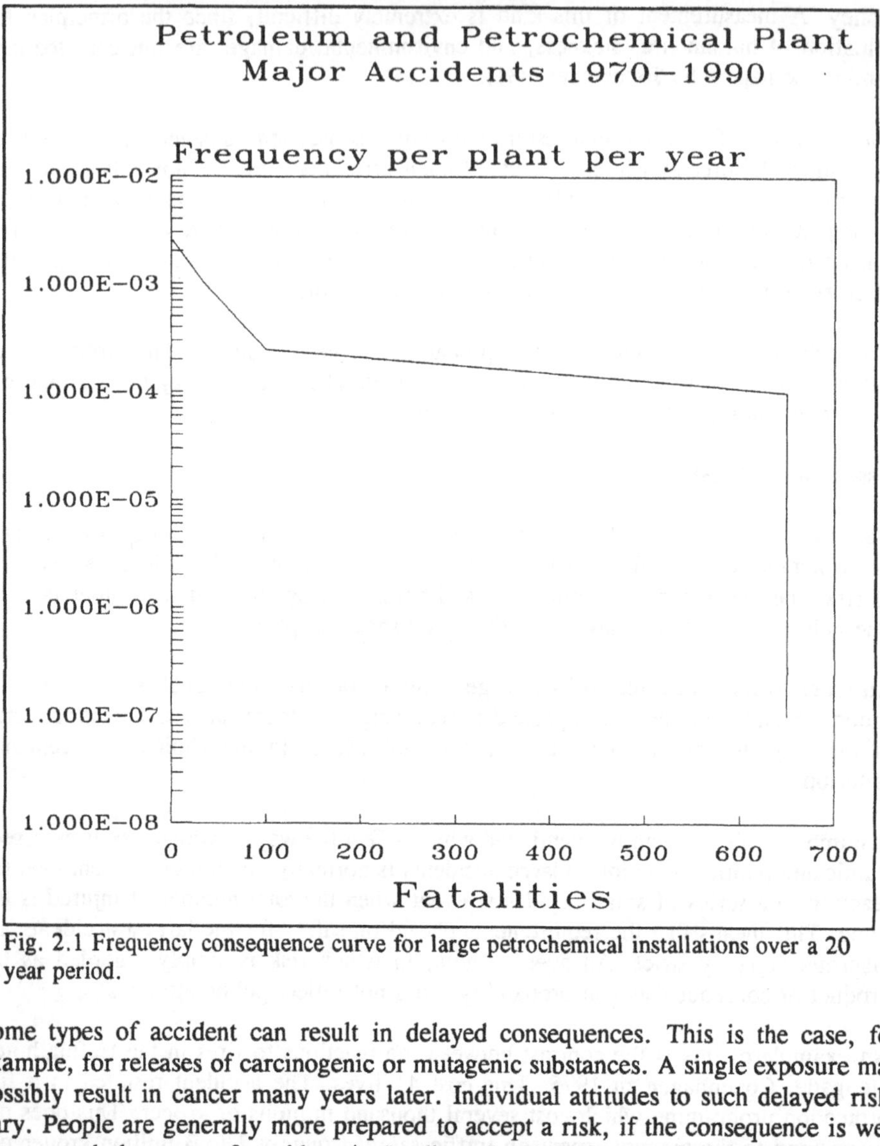

Fig. 2.1 Frequency consequence curve for large petrochemical installations over a 20 year period.

Some types of accident can result in delayed consequences. This is the case, for example, for releases of carcinogenic or mutagenic substances. A single exposure may possibly result in cancer many years later. Individual attitudes to such delayed risks vary. People are generally more prepared to accept a risk, if the consequence is well known and the type of accident is "ordinary". New forms of risk, such as new diseases, accidents with new technology, or new factories, will often provoke a strong response. This means that it must be expected that there will be different criteria for well known and for special or novel forms of risk.

A person may be prepared to accept a risk, if that person has recognized the risk, and has decided to allow himself/herself to be exposed to it. A similar risk, to which the person is exposed with no chance of avoidance, will be less acceptable.

A person will be more ready to accept a risk, if the result of the activity produces benefits for him/her.

Experience gained in the course of time concerning a company's way of dealing with risk weighs more heavily than the results of a risk analysis in the public mind.

In some cases, both decision makers and public will accept a higher level of risk if the activity creates valuable employment. Questions of national balance of payments can have a certain influence too.

Engineers traditionally assess plants on the basis of information concerning the plant design and operation, that is to say "local information". The objective of this technical evaluation is to determine whether the level of safety on the plant bears a reasonable relation to the consequences of accidents. The criteria used will often be comparison with similar plant which has been accepted earlier, sometimes by comparison with risk levels which have been accepted earlier. The decision makers' basis for evaluation is generally wider than the engineers; local population and public risk, and economic and political factors, as well as the engineers' and scientists' own technical evaluation.

Criteria for risk acceptance are formed in the light of these aspects. Unequivocal acceptance criteria are (as a result) desirable, but very difficult to achieve.

Existing levels of risk

The most important method that is used in evaluation of risk is comparison. In order to undertake comparison, it is necessary to know what the actual levels of risk are. Luckily, several studies have been made of this subject over the last 20 years. The most useful comparisons are for fatality risks, since other kinds of risk cannot be easily compared on a uniform basis.

Data concerning individual risk have been collected by Pochin, Sowby, Gibson, Crouch and Wilson, Dohan and Lee, Fryer and Griffiths, and Kletz. Results from their work, and of the author, are shown in table 2.1. The ultimate sources of these data are national statistics and reports from the respective ministries in each country, together with reports covering injuries affecting several people. These data are primarily from UK and USA.

In this context, the frequencies associated with natural catastrophes present one of the most relevant comparisons (because they are related to very large accidents). These levels are used in various recommendations for accept criteria. This is especially so for the criteria which state that the risk for neighbors to an industrial plant should be less than the risk arising from natural sources. The fatality risk level is about $5*10^{-6}$ pr. year for USA (table 2.1) and about $2*10^{-6}$ pr. year for Norway.

Societal risk for major accidents has been studied by Fryer and Griffiths, on the basis of data for USA and Great Britain. They give frequency consequence curves for major accidents for complete countries. It is more convenient however to plot frequency consequence curves for individual plants since these can then be used for comparison with calculations. This has been done for worldwide experience with petroleum and petrochemical plants for a 20 year period, in fig 2.1. Note that the very large numbers of fatalities arise from just three accidents.

From these statistics, a clear pattern can be seen.

- a frequency which falls rapidly at first, then levels out, finally falling very rapidly, as the maximum accident size is reached.

- a maximum of ca. 600-900 fatalities, as a limit for the largest accident from petrochemical systems. For chemical plants, the maximum is larger.

It is worth noting that the last figure as a limit for the maximum accident size, is dependent on place and circumstances. In cooler countries, persons are usually protected by being in their homes or workplaces, with limited ventilation. This provides protection which can limit the maximum number of fatalities.

By comparison, the largest natural catastrophes, such as floods and earthquakes, have killed over 50.000 persons in single events.

One of the main problems with statistical studies of this kind is that they cannot be related to specific process plants. If risk is to be related to a single plant, there is generally insufficient statistical material to form a basis. On the other hand, there are several public available predictive risk analyses which can be used. A cross section of these is shown in fig. 2.4 (page 40). The curves are based on risk analyses carried out by the author.

The risk acceptance process.

Different types of risk are accepted every day. Buildings are built, roads and factories are planned. Technical developments are made. Every one of these activities results in new risks, and each involves some form of acceptance. The process by which risk is accepted is normally complex.

The basis for everyday acceptance of risk, is that the new construction satisfies existing norms, standards, and laws. It is seldom that there is a direct political influence in this. The administrative apparatus, that is to say the community and county technical administrations, the labour inspectorates, and the fire inspectorates, check that requirements are fulfilled. This check is in some cases very thorough, and in others is based on a system of trust in the responsible engineer (with corresponding periodic checks). The system of trust is in some cases formalised in the form of an authorisation of particular persons or organisations (for example for pressure testing), or in a properly audited "self control" program.

Type of risk	Reference	Fatality risk pr.m.years	Assumed exposure hour pr yr.	Fatal Accident Accident Rate pr.100 m.hours
UK				
All accidents	g	345		
Road transport	g	145		
Fires	g	18		
Falls	g	111		
Choking, suffocation	g	16		
Poisoning	g	16		
Drowning	g	11		
Electrocution	g	2.4		
Medical misadventure	k	1.6		
Natural and environmental	k	3.6		
Lightning	k	.17		
Falling aircraft	k	.02		
Meteorite	k	.00006		
Supernova explosion	k	.0001		
Staying at home	k			3
Air transport	g	3.3		240
Rail Transport	g	3		5
Road transport	k	145		57
Bus transport	k			3
Water Transport	k	2.6		
Motor cycling	p	20000		660
Run over by vehicle	k	60		
Transport petroleum and chem.	k	.02		
Industry	gb	73.6	1840	4
Fishing	gb	644	1840	35
Steel industry	gb	147.2	1840	8
Construction industry	gb	1232.8	1840	67
Coal mining	gb	736	1840	40
Chemical industry	gb	64.4	1840	3.5
Flat racing	s			10000
Denmark				
Iron, steel, transport equip.	t	80.96	1840	4.4
Building and construction	t	66.24	1840	3.6
Chemical industry (1975)	t	58.88	1840	3.2
All industry	t	53.36	1840	2.9
US				
Fire fighting	cw	800	1920	42
Coal mining	cw	630	1920	33
Farming	cw	240	1920	13
Rodeo	cw	30	200	15
Police work	cw	50	2100	2.4
Flood	st	2.2		
Earthquake	st	1.7		
Tornado, mid west	st	2.2		
Storms	st	8		
Falling aircraft	b	.1		
Transport petroleum and chem.	k	.05		
Run over by vehicle	k	50		

Table 2.1 Individual risk levels for a range of occupations and involuntary risks.

Key

k	Kletz, T 1978
s	Sowby, 1975
a	Danish Labour inspectorate, 1985
cw	Crouch and Wilson
p	Pochin, 1975
b	Bulloch, 1975
gb	Gibson, 1976

Political influence is for the most part exercised indirectly. In cases where a specific source of risk is noticed, and arouses public interest, the authorities will be influenced. Ministers and governments can then initiate inquiries, and can pass new laws. There is usually considerable technical work required, before such a law can be interpreted and expressed in the form of regulations.

Research, which brings new knowledge concerning risks to light, has an influence on changes in norms and standards. Authorities themselves start research and development in order to gather a basis for new requirements in the case of new technical developments, or in areas where risks are higher than average (for example risks from crane operation, protection of ground water resources).

Engineering organisations have a continuing standards making activity, and produce new norms with associated safety requirements. Some of these become legal requirements, by being mentioned in laws (For example electrical standards).

Public and political influence on the acceptance or approval process often arises as a direct result of accidents. An accident which leads to a public or a political discussion, can result in a change in administrative practice, and in some cases, to new lawmaking. An example of this process is the response to the Hafnia hotel fire. Another is the reaction to cases of poisoning.

A public or political reaction to a specific case may be a sign that there has been a change in attitude to risk acceptance. There is hardly any reason to doubt that such changes have taken place over the last 20 years, especially in the areas of environmental and working environment protection. A public reaction can, though, be a sign that the risk management process does not live up to expectations.

The process described above is complicated, and risk acceptance criteria are often implicit and clouded. It is necessary to study the individual norms, regulations, and requirements in detail in order to extract the safety principles on which they are based. Where safety principles are explicit, interpretation is often required.The norm and law system functions well, in that it allows people and organisations to carry out their work and build new plant, while holding the everyday risk at a normal level. There is only limited communication between various parts of the system, however, and therefore very little uniformity in the safety levels achieved, and little overall optimisation.

Potential major accidents have a special position in this context. There is normally at best poor, most often no, experience in the country concerned with the different kinds of accidents which may be of concern (in the particular situation). Existing norms are generally based on "normal" production, and "normal" safety, rather than on major accident potential. There are only a few norms which cover the field of major hazards. This results in a need for new acceptance criteria, or at least, a need to broaden the traditional norms. It also results in a desire for clear acceptance criteria.

Principle for risk acceptance

There are a number of philosophical/ethical principles, which can be used as a basis for risk acceptance. Some of these were described in "the green paper" (AT 1985) (This is a basic source document for the Danish approach to major hazards regulation. It was the result of a 5 year study be a committee appointed by the minister of labour, and chaired by professor K. Østergaard).

1. It is good practice, for the activities that involve a risk of major accidents, that a safety level is aimed at which is no worse than that in comparable countries.

2. The risk to which one is exposed in daily life should not be changed significantly by participation in industrial activity.

3. The risk to which one is exposed by natural phenomena should not be increased significantly by activities, such as industrial production, which are created by others without our acceptance.

4. In planning the establishment of new plant, there should be an investigation of whether there are alternative processes which can provide similar products, with less risk.

5. The resources which are available for safety purposes should primarily be used where they give the best overall result.

Other principles exist in various laws, for example the principle of compensation for any injury caused or for losses such as reduction of property value. There is an old principle of jurisprudence, "De minimis" or "The court is not concerned with trivialities", which can perhaps be used to eliminate the smallest risks from consideration. A common feature of these ethical principles is, that they cannot be used directly in a technical context. They require interpretation.

There are a number of practical principles which can be as a basis for approval and which provide an interpretation of the basic principles:

1. A plant may be approved on the basis of comparison with similar plant.

2. A plant may be approved if it satisfies all relevant standards requirements.

3. A plant may be approved if it has a high level of safety, expressed in terms of the number of safety measures in relation to the possible accidents sequences.

4. A plant may be approved, if a thorough and well based risk analysis shows the risk level to be acceptably or tolerably low.

5. A plant may be approved if the benefits from its use are greater than the risks.

6. A plant may be approved, if it uses the best possible practice.

Principle 1, "experience", is used quite generally, in many types of plant. The principle can hardly be used in plant with a potential for major accidents, since the principle requires that experience of major accidents is available.

Principle 2, "standards", is a basic and necessary part of every approval procedure. One can, under everyday engineering circumstances, neither have trust in safety devices, nor calculate their reliability, unless they are built in accordance with some standard. Existing national and international norms (API, BPI ASME, etc) are recognised as a good basis (for major hazards plant design), but are seldom sufficiently broad. Most large oil and chemical companies have their own internal standards. These, often encompass many thousands of pages which nevertheless are sometimes recognised as inadequate, and supplemented with risk analyses.

Principle 3, can be referred to as "defence in depth". The principle recognises that no safety measure is perfect, and therefore requires that there are several safety measures for every accident.

The principle is simple and transparent (easy to understand), but fell into miscredit in the 1970's as a result of several accidents in nuclear power plant, in which several safety devices failed at the same time. These so called "common cause" failures resulted in much rethinking of the design of safety systems. The principle has more recently been used in a modernised form, in which earlier weaknesses are reduced or removed (Taylor et al. 1989).

Principle 4, quantitative or probabilistic risk analysis, is an operational approach to risk assessment. As described above, however, it cannot be used alone.

Principle 5, "cost /benefit analysis", has been used especially in USA, as a step in the evaluation of transport risk, risk in the use of medicines, in work environment and environmental protection. The major disadvantage is that "cost/benefit" analysis is extremely difficult, and time consuming. The principle can at present presumably only be used for analysis of the most important and difficult decision. The principle requires also that a position is taken in the question of the value of human life. It is therefore difficult to use in a technical context. (Attempts have been made to define this factor for use in connection with traffic safety, and in connection with medicinal and hospital treatment, see e.g. Jones-Lee 1981,1985).

Principle 6, "as safe as practicable", is used today as in some facets of nuclear safety, and is regularly used in environmental and work environment questions. The principle is described below.

All of these principles have relevance in some area or other. An operational use of the principles requires, though, that the principles are supported both by methods and by specific criteria, and these must be usable, operational, and reasonably easy to understand.

Risk reduction

No risk should be accepted, if it is easy to remove it. Risk reduction is one of the most important safety principles.

The requirement for risk reduction is reflected in the criterion "As low as Practically Achievable". The principle is used especially in environmental protection.

In practice, a goal is set for risk or pollution reduction, and invitations are invited concerning the way in which these improvements can be achieved. There follows a period of practical trials and hearings. Hereafter, the new methods (provided they work) are declared to be "practical" and are built into new legal or administrative requirements.

The principle is also known from the British Health and Safety Executive for example, which produces notes on the best practicable way of fulfilling environmental requirements.

The "as low as practicable" principle is useful in cases where risk is accepted as being too high, but where a reduction is difficult, and requires technical development. The procedures in USA ensure a steady and balanced improvement of safety and environmental protection methods. There are many areas where the principle is impractical and even unusable. In chemical process plant it is almost always possible to reduce risk, irrespective of how low the existing level is. One can for example move the plant further from dwellings, build higher and thicker walls, install more sprinklers etc. Using all possible safety measures uncritically, even in less hazardous plants, would result in an economic overload, completely out of relation to requirements in other areas. The problem lies in the fact that "what is practical" is often much more extensive than "what is reasonable".

A second principle is used in these situations, ALARA, "as low as reasonably achievable". The idea is, that risk is reduced as far as reasonable, rather than as far as possible. What is "reasonable" is interpreted in different ways. In a supreme court judgement (see Lave and Romer 1981) it was interpreted as "investment in safety equipment to such an extent that it did not threaten the existence of the company, but required its entire profit margin". Whether such a principle can survive in the long run is doubtful. There are many companies which do not have profits, but where safety equipment is nevertheless necessary. It is also impossible to operate companies in a free market economy, unless there is some profit.

A second interpretation of the ALARA technique is that the costs of safety equipment are balanced against the "values" of the increased safety. Here, engineers become involved in a most direct way in political decisions in that it is necessary to give a value for an increase in safety, and therefore, for the value of a life saved, or an accident avoided.

Within public decision making, it is possible to investigate the "value" of increased safety and to use this to set a list of priorities for different safety measures. Those which prevent the most accidents for the same investment, are set highest on the priority list. At this stage, investment can be sought, and investigate which measures can be afforded. The measures which fall outside the approval budget are thereby "unreasonable", at least at the time when public money is allocated.

This approach is used especially in traffic investment. It ensures a maximum of political influence on technical decisions, but is at the same time consistent and technically straight forward. Traditional separation of decision processes have until now prevented the use of this approach across industry and across authority divisions.

It is doubtful whether one could use the same approach directly in private industry, in that there is no fixed limit for what could (in principle) be invested in safety. Public investment in safety should give an indication, though of what is reasonable. Additionally, it is possible to calculate the effect and cost of standard safety measures, and make a decision on this basis. What can be done is to compare plants and to find a "generally" acceptable" level by comparison with what has been accepted earlier.

Inherent Safety

There is a design principle for technical systems called "inherent safety", which has much to do with risk reduction (see Kletz 1982).

The principle means that during design of a plant, risk is reduced as much as possible by removing potential causes or contributors to risk. Different methods can be used. Some of the more important are:

- replacement of hazardous substances by less hazardous ones. An example is the replacement of processes where organic solvents are used by others where water is used

- use of reactors and concentrating units (stills, evaporators etc.) which have a high conversion rate and therefore small amounts of substances in a critical state

- use of integrated plant in which the substances are used as rapidly as they are produced, and which therefore do not require storage. An example is the use of phosgene in isocyanate production in which modern design eliminates the buffer tank with the large phosgene inventory.

- use of reactions which are stable or self limiting, so that runaway reactions are impossible. An example is the use of a low boiling point solvent, so that boiling limits the maximum temperature in a reaction to below the critical value.

Inherent safety is an old principle, which is experiencing a renaissance especially in a nuclear context, where systematic exploitation of the concept has been started, especially in Sweden. It is desirable that the principle be used more widely in chemical plant than is the case at present.

Quantitative Criteria for Risk Acceptance

Quantitative criteria for risk acceptance involve calculation of probabilities and consequences for accidents, and comparison of the results with one or more numbers, which describe what level of risk is acceptable.

When once it is recognised that no activity is risk free, the questions arise, how large is the risk, is it too high, and can and should it be reduced? Quantitative risk calculations can give the specific values of risk level for a plant. The next question is then - How large may the value be?

In design of aircraft, a risk level of 10^{-7} per hour for the frequency of major accident due to technical causes (FAA). This corresponds to a fatality risk of about 10^{-4} per year for those for whom flying is a profession. The criterion is applied to the part of the risk stemming from technical failure, thereby avoiding the difficult question of quantifying human error probabilities.

For offshore oil and gas platforms, the Norwegian Petroleum Directorate has established a set of guidelines which divide accidents into six (or nine) categories. Each category can at most contribute to the risk of major accidents, with a total of one large accident per 10,000 years. The accidents of concern here are those which can prevent evacuation or cause fatalities beyond the immediate neighbourhood of the initial release.

In USA quantitative criteria may be used in the approval of nuclear plant. If quantitative criteria are used, then a frequency of 10^{-4} per year is set as an upper limit for the frequency of accidents leading to damage to the core and partial meltdown.

Experience with use of criteria of this kind is in general good. They are in general useable and result in design improvements beyond those achieved with the use of standards. Results within the aerospace and military area are especially good, and the results are only slightly dependent on the individual experience of the analyst (see for example FAA 1985). The results show a good correlation between experience and calculation. The reason for these good results is that many resources are used in collection of data and standardisation of methods, and that maintenance and operating standards are high in these areas.

Analyses in the offshore areas are often less encompassing, and therefore give less uniform results. The main reason for this is that the analyses are for the most part made at the conceptual design phase when many design and operational details remain unclear. The offshore example illustrates the conflict between the desire to use uniform and evenly operating acceptance criteria, and the need to use risk analysis at an early stage in design, in order to achieve the highest degree of safety. Norwegian practice has therefore been to perform several analyses at different stages of plant design and construction.

Generally, the analyses show a need to balance between a standardised calculation technique, so that different analyses are comparable, and the need for freedom in the calculation structure, so that new developments and new knowledge within the discipline of risk analysis can be exploited.

Several public authorities have evaluated actual chemical plants on the basis of risk analyses.These evaluations can be used to show which levels of risk have been regarded as acceptable under various conditions. This is especially true in the cases where a risk analysis is undertaken for a plant both before and after recommended improvements have been carried out so that it is known that the plant represents a limiting case. The calculations which have been made for this kind of approval are of varying kinds. Some calculations result in accident frequencies, without necessarily involving fatalities. Others result in a value for individual risk, and still others, a value for societal risk.

One of the precedents is especially relevant, for Canvey Island, in the mouth of the Thames. Here lies a medium sized town and five petrochemical plants. In 1976, a commission of enquiry was appointed, to investigate the reasonableness of building a new refinery in the area. The commission considered the risk to be unacceptable and recommended several improvements in the conditions existing at the time. A new analysis was undertaken after modifications had been made, and the result was judged to be acceptable for an already existing plant, but that the risk level would probably not be acceptable for a new plant.

The evaluation was justified on the basis of individual risk, and the commission did not place any special weight on societal risk. The commission took into account the fact that the calculations were "conservative", that is to say, that the overestimated the risk. Conservatism was incidentally a requirement placed by the commission on the analysis process.

Recommendations from the literature

Various persons have in the course of time recommended acceptance or tolerance criteria, and justified them. One of the first was Gibson (1974) who recommended a value for human life of £ 200 000, with the justification that investments in protection of human life corresponding to more than this amount could normally be better used elsewhere (implicitly to improve safety elsewhere). This is an argument based on existing normal good practice in a large company with a thorough awareness of safety problems. (The corresponding investment level would be larger today because of inflation, if for no other reason).

Kletz (1982) recommended a maximum level of fatality risk for employees of $4*10^{-8}$ per hour of exposure, and pointed out that no part of a plant should contribute more than a fraction of this amount. He also recommended a value of at most 10^{-6} per year for fatalities for the nearest neighbour. The justification for these levels was comparison with existing levels of risk in well designed plant. The levels described by Kletz have been used routinely for many years in internal risk analyses in several companies in England, USA, and Denmark.

In a presentation of risk analysis for a large pipeline, Ramsey recommended a maximum level of $4*10^{-6}$ for fatality risk for the nearest neighbour. His justification was that this level would not change the overall risk level for the neighbours by any significant amount, compared with their existing background risk.

Authorities in Groningen, Holland, formulated a criterion for societal risk in 1979. The starting point for the criterion is a risk curve, with an initial level of 10^{-6} per year for a single fatality, falling in proportion to the inverse square of the number of fatalities, and with an absolute maximum of 10 fatalities for clear acceptability. An important point was the introduction of a "grey area", within which improvement in safety should be sought. The absolute upper level for this grey area, was 1000 fatalities.Risk curves of the type used by the Groningen authorities give the frequency of accidents for at least N fatalities. Curves of this kind are called cumulative risk curves.

The Dutch criteria were studied by Styhr Petersen, and presented in a revised form, with justification based in part on evaluations made by the Royal Society Group on risk assessment (ref. Styhr Petersen 1984, Royal Society 1983), in part on work of Lans and Bjordal (1984), and in part on study groups results carried out by a group organised by the European Federation of Chemical Engineers.

The Dutch criteria have been revised in the course of time. The Dutch authorities mention the difficulties of fulfilling the criterion for existing plant, and higher acceptance levels have been established for existing plant. The criteria have been used in evaluating many plants in Holland (van Kuijen 1988).

The British criteria for land use planning (HSE 1989) give explicit advisory limits on risk, for use in new planning applications. Planning authorities are advised to give more weight to safety if:

- housing developments provide for more than 25 people with a calculated individual risk greater than 10 in a million per year.
- housing developments provide for more than 75 people, with a calculated individual risk greater than 1 in a million per year.
- for retail community and leisure areas, no hard and fast rules are proposed. For moderately sized developments the 10 in a million per year limit is suggested, for large development 1 in a million.
- limits are revised downwards for areas where institutions are involved.
- advice against planning permission is more likely to be given if the planned development sets a precedent for further growth or encroaches within any pre-existing cordon sanitaire.

Risk index

The information in the above curves can be expressed as a risk index as follows

$$R = F N^2$$

Where

R is the risk index
F is the accident frequency
N is the expected number of fatalities or permanently seriously injured.

The index gives a possibility for a summary of the information contained in the risk curves in a single number. The index can be used to provide a priority listing of hazards.

Plant size

A remarkable aspect of the risk criteria is that they are seldom if ever related to plant size.

It is implicit in the acceptance criteria described above that they apply for all sizes of plant. This leads to a curious situation, in which small plants are allowed to result in just as large risk levels as very large plant. This situation is not likely to prove reasonable in the long run, and results in immediate problems in the evaluation of transport, and small storage units. Acceptance should presumably be related either to the size of the production (with cost benefit implications) or to the size of the community benefit (with absolute or humanitarian implications). Unless principles of this kind are used, criteria could be met by dividing a large unit into several small, dispersed units, with little overall effect on risk. The idea of "spreading the risk" by dividing into smaller units, has been used on some plants.

A possibility for overcoming this problem is to relate societal risk to societal value. Possible measures of societal value might be salaries or taxes paid or number of persons employed. For some types of plant, such at those producing medical products, or destruction of chemical waste, this approach is unreasonable, but there are many types of plant where this approach gives a direct relation between risk and societal benefit, assessed here as tax and salary payments, and product value added.

In order to create a relationship between risk and societal benefit, the risk levels for a petroleum refinery unit, a chlorine production unit, an LPG storage unit, and gasoline transport were investigated. These units and activities were chosen because actual values were available for risk, taxes, and salary payments.

In order to be able to make simple comparisons, risk was expressed as a risk index, as described above. Table 2.2 shows the results.

Table 2.2 Risk index per million dollars societal benefit for five specific chemical industry activities.

Plant or route	Risk index per million dollars societal benefit
Alkylation plant	$6*10^{-7}$
Chloro soda plant	$7*10^{-7}$
LPG storage	$35*10^{-7}$
Gasoline transport	$3*10^{-7}$

The values are surprisingly close to each other, and even though there are many assumptions involved in the calculations, the relative values should be comparable. The closeness of the results presumably reflects a degree of uniformity in attitudes to safe design, of which we at present have little systematic knowledge.

Examples of quantitative criteria

In 1988 a team of analysts was asked by the Danish Environmental Agency to propose suitable acceptance criteria for risk "acceptance" (Taylor et al 1989). As a result of an extensive study, the proposed criteria were:

- an individual risk level for fatality of at most 10^{-6} per year for the most exposed neighbour.
- a societal risk level for fatality of 10^{-4} per year for accidents with at least 1 fatality.
- the societal risk level with a frequency falling in proportion to the square of the number of fatalities (see fig. 2.2).
- a grey area lying above the "acceptability" or tolerance curve where the risk should be "As low as Reasonably Achievable".
- risks which can reasonably be reduced or eliminated,should be reduced or eliminated, irrespective of the risk level.
- the possibility for serious or permanent injury should be taken into account together with the possibility of delayed injury. The most straight forward approach would be to regard serious permanent injury either on the same basis as fatality or with a percentage evaluation of the kind used in insurance or compensation cases.

- no clear recommendation for accidents occurring with a frequency less than 10^{-8} pr. year.

Experience shows that even though plants are built and operated in a way corresponding to good engineering practice, these will often not be able to achieve frequencies less than 10^{-8} per year and coming near this level will in many case require everything which is practicable is done to reduce risk). Moreover documenting accident frequencies at the level concerned in the region 10^{-8} per year is beyond the present state of the art. Other criteria, especially safety management criteria, will normally be the most important aspect of the decision basis in cases where risks are calculated to be below these levels.

The following assumptions are required in order that the criteria can be applied and standard industry data can be used in the analysis:

- The plant to be analysed is built to a standard which is normal in the industry.

- The equipment is tested and maintained to a normal industry standard.

- There are no unusual possibilities for human error in operating the plant.

These assumptions are important in that different levels of safety management can result in differences in accident frequency with a factor of up to 100. This can be seen by comparing accident statistics from different companies within the same branch (see eg. Joschek 1983).

It is possible that plants which fulfill the above criteria will prove unacceptable in any case, for example because of their effect on other planning, because of difficulties of contingency plan implementation, or lack of benefit from the plant etc. The decision will depend on who makes the decision, and which responsibilities and prerogatives the decision maker has.

It is also possible that a plant which does not fulfill the above criteria will in any case be approved. A typical reason could for example be that the plant fulfilled a valuable or necessary function and that additional risk reduction would be impossible.

Despite these difficulties, an analysis which shows that the above criteria are fulfilled will be a good starting point for risk assessment.

The criteria have so far been applied to some 60 plants in many countries and to 10 "reference plants", that is, fictional plants which have been designed to illustrate various analysis and design principles and which are typical of actual plant practice (Taylor 1991). Some results from the reference studies are shown in fig. 2.3. The frequency consequence curves shown here are for a refinery with LPG storage spheres 1 km form the nearest town, for a plant storing cryogenic ammonia 1.5 km from a town, for an alkylation plant with a 2 km safety zone, and for a chlor - alkali plant with a 1 km. safety zone. The plant standard for the analyses represents an ordinary level, before any special risk reduction effort is applied, beyond the normal fulfillment of standards requirements.

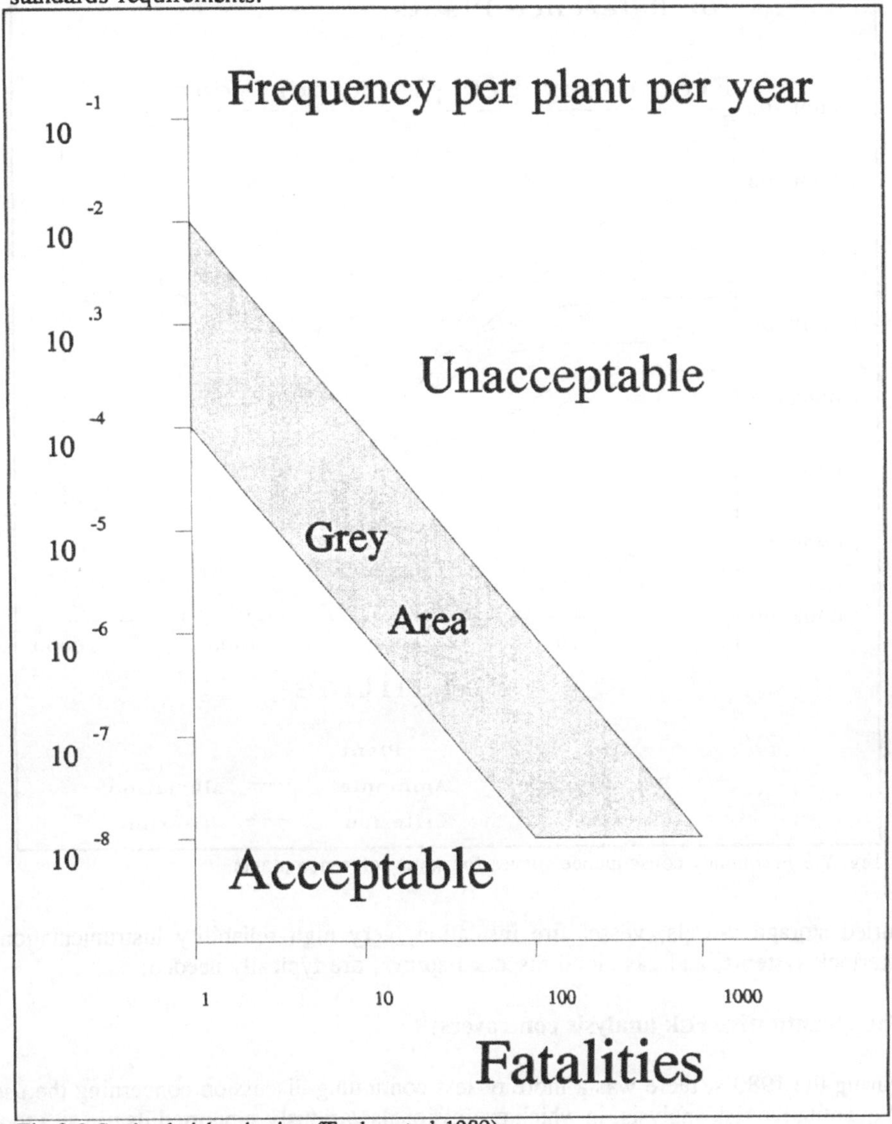

Fig.2.2 Societal risk criterion (Taylor et al.1989)

The experience in applying the criteria has been that they are difficult, but not impossible to meet. For well sited plant (e.g. a refinery with a 500 m. safety zone) the criteria can be met by providing the best conventional safety technology, and a good level of safety management. Lower standards bring the plants into the "grey area". For plants which have difficult siting, such as liquefied gas plants with dwellings at the plant boundary, and little or no safety zone, it is extremely difficult even to bring risk levels down to the "grey area". Unconventional safety techniques such as mounded or

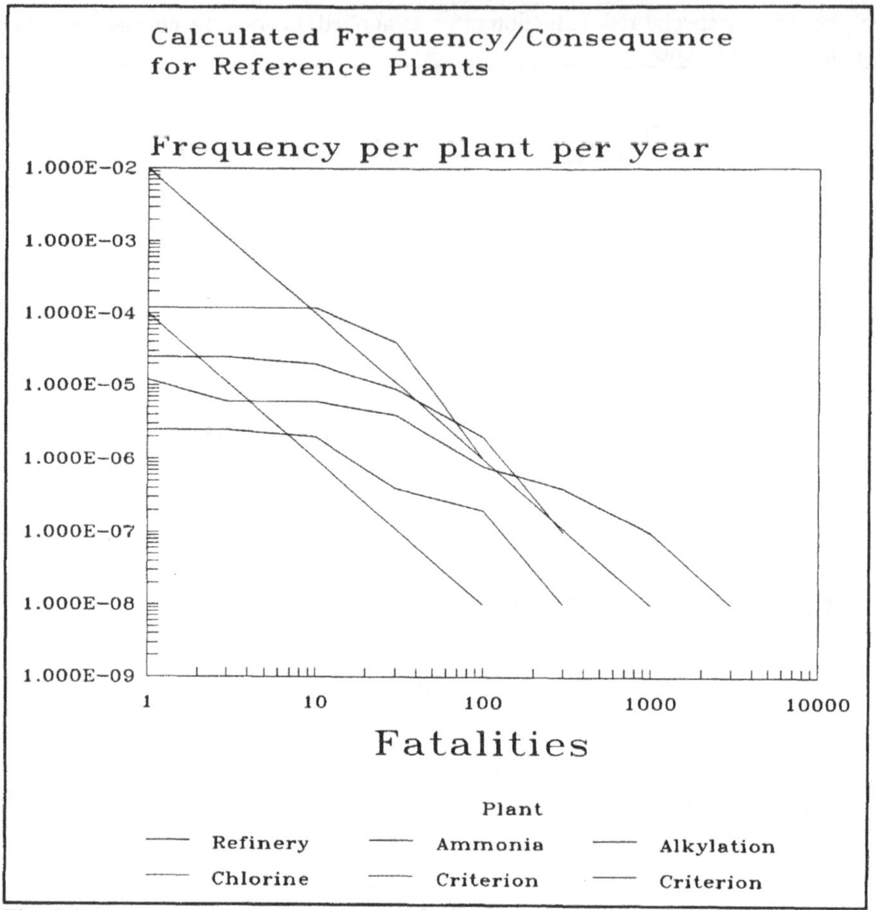

Fig. 2.3 Frequency consequence curves for some reference plants

buried storage vessels, vessel fire insulation, very high reliability instrumentation, interlock systems, and gas cloud dispersal sprays, are typically needed.

The Quantitative risk analysis controversy

During the 1980's, there was a more or less continuing discussion concerning the use of quantitative risk analysis, in which professionals variously supported its use as a tool for plant assessment, denied that could be useful, or recommended restricted use within design groups (see e.g. Lans and Bjørdal, 1983).

The arguments against the use of quantitative risk analysis are,in summary

- it is costly (This argument is not so strong as it once was, since improvement in techniques have reduced quantification costs to a small fraction of safety assessment costs)

- the methods used involve a great deal of uncertainty

- the methods encourage a rather mechanical approach to plant assessment

- the results can be understood by a layman in principle, but are extremely easy to misinterpret.

There is little doubt too, that many professionals have been afraid that risk analysis results would be used in local political opposition to plants. Risk analyses which are published can undoubtedly be used for such purposes, since they generally require quite detailed descriptions of the accidents which could happen. Quantitative risk analyses, moreover, require that the full scope of potential accidents, be described, including the largest.

The arguments for the use of quantitative analysis have been:

- the quality of analysis, and of resulting design is improved by use of quantitative analysis

- the process of assessment is made much easier when quantitative analysis is used.

- assessment of different plants can be made on a uniform basis. Authority approvals can be checked to avoid setting precedents which will be unworkable elsewhere.

The approach presented in this book is based on experience from Denmark, Eastern Asia, south America, and USA, where quantitative analysis has been found to be extremely useful in many practical projects. The conclusions drawn are:

- For design purposes the main effort should always be placed on qualitative analysis. If engineering effort or time is limited, it should be used primarily in hazard identification and risk reduction

- For design purposes quantitative analysis is very useful for comparing design alternatives, but for this purpose it can generally take a quite limited form. For difficult and expensive design decisions, quantitative risk assessment is an extremely useful tool.

- For authority approval purposes qualitative risk analysis is necessary, and should ideally be performed by plant employees

- For authority approval purposes quantitative analysis is extremely useful, provided that results are not interpreted too rigidly. Quantitative analyses allow effort to be focussed on areas of real significance, rather than on minor problems. They also help to avoid precedents being set which will cause problems later. For example a refusal of planning permission at one site may be used as a basis for out of hand rejection of a proposal for another site unless there is a method for differentiating the two cases.

- Probability calculation for emergency planning is useful, but not essential. Consequence calculation is generally essential.

- Process plant risk analysis for insurance purposes is generally meaningless, unless it is performed quantitatively.

- In all cases, the primary effort should be on good hazard identification. and risk reduction.

References

Ramsay,C.G. (1980) Assessment of Hazard and Risk, WHO Europe, Seminar report, Aberdeen 1980

Kletz, T.A. (1978) The application of hazard analysis to risks to the public at large. SRS Quarterly Bulletin, October 1978

van Kuijen, C.J. Risk management in the Netherlands, a quantitative approach. Ministry of Housing, Physical Planning and Environment, 1987

Styhr Petersen, H.J.(1984) Acceptable Risiko, Dansk Kemi 1, 1984

Royal Society Study Group (1983) Risk Assessment, The Royal Society, London 1983

Lees, F.P. (1982) Loss Prevention and Safety Promotion in the Process Industries Butterworths, 1982

Ministry of Housing, Physical Planning, and the Environment Environmental Program of the Netherlands 1986 - 1990

AT (1985) "Den grønne betænkning", Betænkning om betydelig risici, Betænkning Nr. 1045, Arbejdsministeriet, 1985

Gibson, S.B.(1985) Risk Criteria in Hazards Analysis, Chemical Engineering Progress, Feb. 1985

Lave, Romer,(1981) A Survey of Safety Levels in Federal Regulations, NUREG/CR-2226, US Nuclear Regulatory Commission, 1981

Fryer, Griffiths (1978)	Incidence of Multiple Fatality Accidents in the UK, SRD 1100 UKAEA, 1978
Cohen, Lee (1979)	A Catalogue of Risks, Health Physics, Vol. 36, pp 707-722, 1979
Bowen, J.H. (1976)	Individual Risk and Public Risk Criteria, Chemical Engineering Progress, 1976
HSE (1978 and 1980)	Canvey - an investigation into the potential hazards from operations in the Canvey Island, Thurrock area, HMSO 1978 Canvey - a second report, HMSO, 1981
EC (1982)	European Community directive of 24 June 1982 on risk for major accidents in industrial activities.
Garcia Tavel, N., Taylor, J.R.(1989)	Major Hazards Assessment for Contingency Planning, 6th Int. Symp. on Loss Prevention and Safety Promotion in the Process Industries, Oslo, June 1989
Joschek, H J,(1985)	Welcome Address, 5th Int Symp. on Loss Prevention and Safety Promotion in the Process Industries Cannes, 1985
FAA (1985)	Committee on FAA Airworthiness, 1980 Improving Aircraft Safety, National Academy of Sciences, USA
H.Lans, E.Bjørdal,	Quantified risk analysis in the process industries. EFCE working paper, ISGRA working group, Presented at Int. Loss Prevention and Safety Promotion Conf., Harrogate, 1983
Pochin, E.E, 1975	Why be Quantitative about Risk?, BM Bulletin, 1975, pp. 184ff.
Kletz, T.A.	SRS Quarterly Digest, October 1978
Sowby (1975)	Health Physics, 1975, p 869 ff.
Bulloch (1975)	I Chem E Symposium Series No 47, 1975
HSE 1987	The Tolerability of Risk from Nuclear Power Stations, HMSO 1987
HSE 1989	Risk Criteria for Land Use Planning in the Vicinity of major Industrial Hazards. Health and Safety Executive, HMSO 1989

Taylor, J.R., Becher, P., Pedersen, K.E, Pedersen, C.G., Kampmann, J.,Schepper, L., Kragh, E., Selig, R. Quantitative and Qualitative Criteria for Risk Analysis, Miljøprojekt 112, Danish Environmental Agency, 1989

Linnerooth, J. The Evaluation of life Saving: a Survey IIASA Research Report PP-75-21, 1975

Kletz, T Cheaper, Safer Plants Notes on Inherently safety and Simpler Plants. Inst Chem Engineers, 1984

Joscheck, H-J. Safety and Loss Prevention in the Process Industries, Opening Address, 5th Int Symp. Loss Prevention and Safety Promotion in the Process Industries, Cannes, 1986

3 Hazard Identification

The objective of hazard identification is to provide a list of failures and failure combinations which can lead to accidents. There are very many different techniques which can be used. They vary in:

- the basis for search (piping diagrams, flow sheets, electrical wiring diagrams, the plant itself)
- the starting point
- the direction of search (forwards or backwards in time)
- the way combinations of failures are taken into account

Hazop Analysis

One of the simplest approaches to hazard identification is hazard and operability analysis or Hazop analysis (HG Lawley, Operability Studies and Hazards Analysis, Loss Prevention Vol.6 1972, and CIA). This involves a vessel by vessel or pipe by pipe review of a plant. For each vessel, a series of disturbances is suggested, for each disturbance, potential consequences are described, and written out in a table. If the consequences are serious, potential causes are also tabulated. If the consequences are serious, possible cures are also tabulated.

This method is usually applied by a group of 3 or more people, such as

- the design engineer

- an operations foreman

- an instrumentation engineer

The idea of using such a group is to get a range of expertise applied to a single problem. One person is selected as "secretary" in such analyses. His job is to pose "leading questions" to the analysis team, and to record the results.

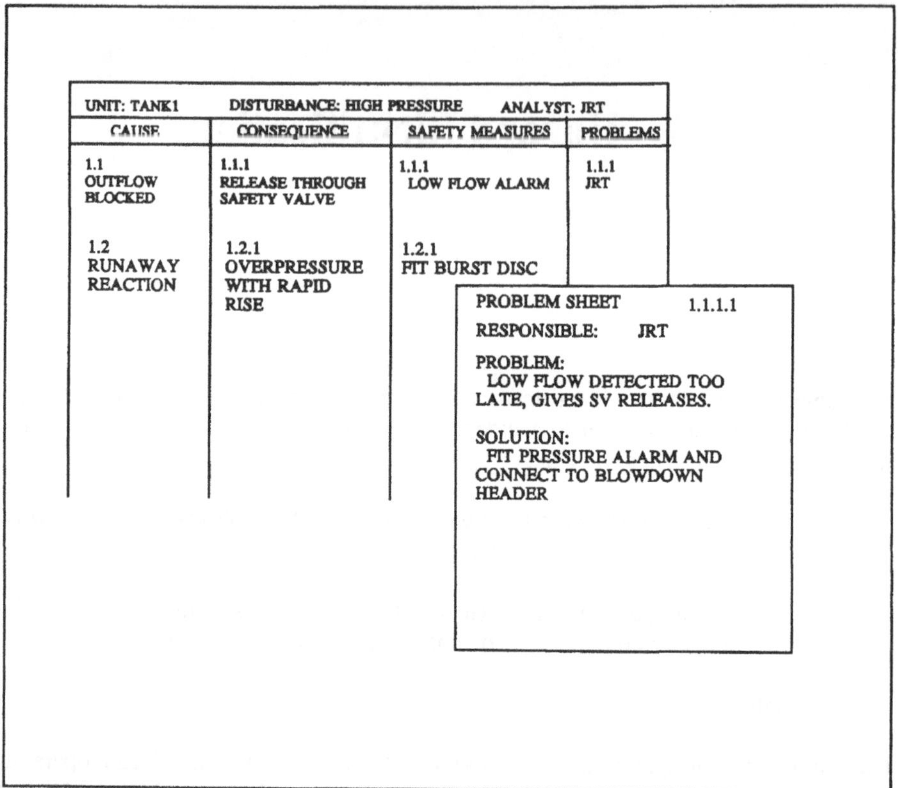

Fig 3.1 General principle for recording Hazop results

The disturbances suggested, and the questions posed during Hazop analysis greatly affect the quality of the result. The original list of questions,suggested in the chemical industries association guidelines is given in table 3.1.

The intention of this list is to stimulate the imagination, and relies very much on the expertise of the persons performing the analysis.

Another set of questions is given in narrative form later in this chapter, (p 53 - 6 . These check lists were made by taking the results from 10 Hazop analysis studies and combining them. The idea behind the questions is that any disturbance in a chemical plant can be described in terms of physical state variables. The state variables for a vessel, such as pressure or liquid level can be "too high" or "too low". These variables can in turn be affected by flow to and from the vessel, as shown in fig.3.1

The vessel-by-vessel analysis can be made quite rapidly. but if the piping system is complex it is advisable either to use a line-by-line analysis, or to use a fault tree analysis to determine possible disturbance causes.

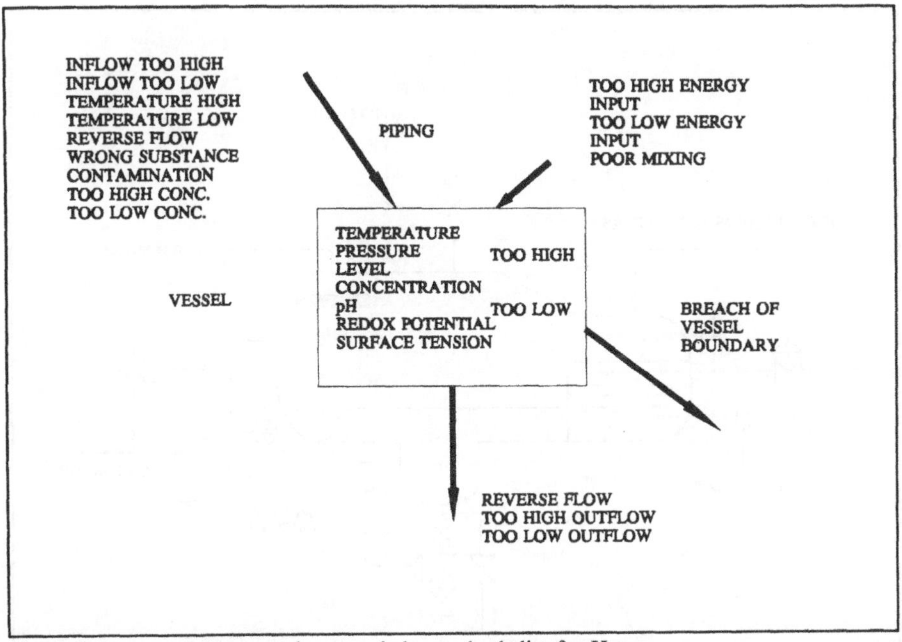

Fig. 3.2 Structure of mass and energy balance check list for Hazop

Table 3.1 Hazop guide words

Guide word	Meaning	Comment
NO, NOT	Complete negation of intention	No part of the intention is achieved.
MORE,LESS	Quantitative increase	These refer to quantities and properties such as flow rates and temperatures as well as activities like "HEAT,"REACT".
AS WELL AS	Qualitative increase	All the design intentions are achieved, together with some additional activity.
PART OF	Qualitative decrease	Only some of the intentions are achieved.
REVERSE	Logical opposite of	Mostly applicable to activities, for example reverse flow or chemical reaction. Can be applied to substances e.g. POISON instead of ANTIDOTE.
OTHER THAN	Complete substitution	No part of the original intention is achieved. Something different happens.
SOONER, LATER		
OTHER PLACE		

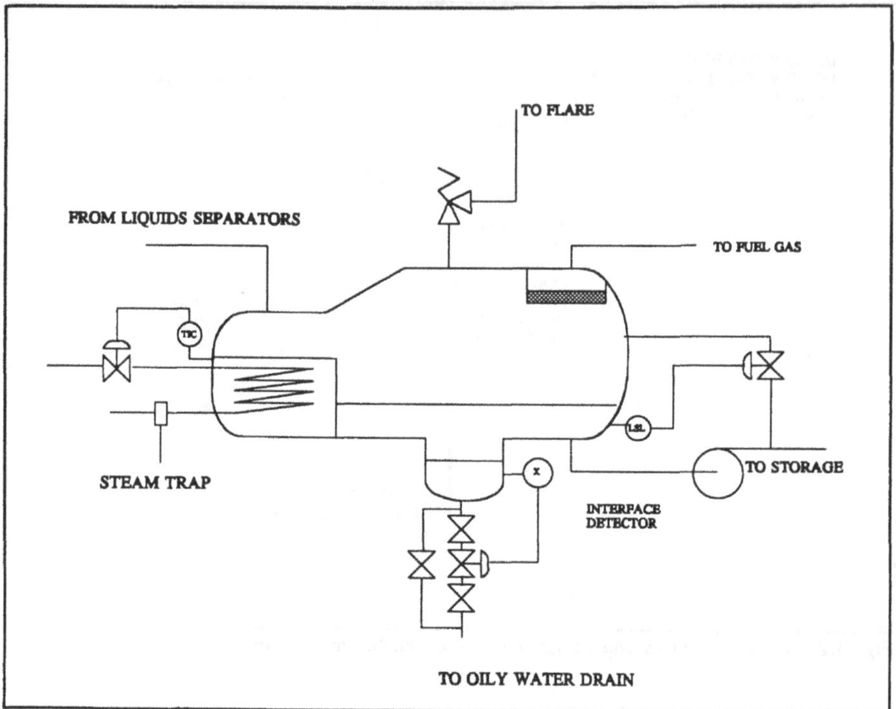

Fig 3.3 Separator liquids flash vessel. NOTE: INCOMPLETE DESIGN

Example of Hazop Analysis.

Figure 3.3 shows a separator and flash vessel for natural gas liquids. Liquids reach this vessel at a pressure of about 16 bar, after flashing through a pressure reduction valve from 80 bar. In the flash vessel, light vapour (up to butane) is further removed by heating the liquids, in a special section of the flash vessel. The vessel also serves as a separator/decanter for water.

In operation,the liquids tend to come in pulses, over a period of minutes, since the source of the natural gas is a fairly long pipeline with varying flow rates. Table 3.2 shows just one of the hazop analysis sheets for this example.

NOTE THAT THE DESIGN SHOWN IS NOT COMPLETE. IT CONTAINS SEVERAL WEAKNESSES.

<u>Lessons to be Learned from the Example</u>

1. A single vessel can involve more than one "volume", in the example there were four different liquid levels to take into account.

2. Once disturbances in a vessel have been found, it is important to identify the places where dangerous substances may cross the vessel boundary.

Project:Liquids stabilisation		Date:10/10/90	Sheet:
Vessel:Flash vessel		Analyst:JRT	1 of 10
Disturbance, Causes	**Consequences**	**Safety Measure**	**Problem**
1. High temperature			
1.1 High temperature upstream	1.1 High pressure in vessel, SV opens	1.1 Safety valve and blowdown	
1.2 Fire	1.2 High pressure, safety valve opens, vessel weakens, possible BLEVE		1.2.1 Sprinkler needed JRT
1.3 Temperature regulator failure on steam supply	1.3 As for 1.1		
2. High pressure			
2.1 High temperature, see above			
2.2 Failure of pressure reduction upstream	2.2.1 Safety valve opens		
	2.2.2 If upstream failure is massive, SV unable to cope, rupture		2.2.2 Increase pressure rating of vessel ? JRT
2.3 Vessel overfilled with liquid, and shut in	2.3.1 Safety valve opens		
	2.3.2 Light ends and water released, may freeze in safety valve	2.3.2 Alarm and shut down if fuel gas take off is closed	

Table 3.2 Example of filled out hazop table

Guidelines for Hazop Secretaries

The success of hazard and operability studies depends on the meeting secretary. It is his job to lead the meeting, to ensure that questions are asked in a creative way, and to ensure that results are recorded. There is a number of points it is useful to remember.

1. The meeting should be well prepared. Before the meeting information should be obtained, as follows:

    A complete set of drawings
    Sufficient analysis sheets, properly sorted
    A list of design pressures for vessels
    Vessel material and corrosion properties
    for each substance
    - boiling point
    - freezing point
    - toxicity
    - flash point
    - autoignition temperature
    - vapour pressure at and around operating temperature
    - density.

2. The vessels and pipes should be numbered, and should be crossed off the drawing (for example with yellow pencil or highlighter) as the analysis is completed.

3. Answers to questions should be recorded carefully, even if this takes a little longer, as they are often difficult to interpret later.

4. The discussion should not spend too long on each point. Five minutes is probably sufficient for discussing a disturbance. If a problem arises which cannot be solved quickly, it is better to

    - NOTE THE PROBLEM ON AN ACTION SHEET

    - APPOINT ONE PERSON SPECIFICALLY TO SOLVE THE PROBLEM

5. Hazop meetings are very tiring and should preferably not last longer than four hours, though this may be difficult if persons must travel to the meeting.

6. It is difficult to take good notes during a hazop meeting. The secretary should write the record of the meeting in fair form as soon as possible after the meeting.

7. At the next meeting, design modifications and actions should be reviewed as the first stage of the meeting.

Filling out the hazop tables

When actually undertaking a hazop analysis, the sequence of problems raised is controlled by the team leader. He or she will select the series of disturbances or deviations to be studied. The secretary notes the deviations, then possible causes of these are considered. If there are possible causes, the possible consequences are considered. Then the existing or designed safety measures are listed. (These should be specified as precisely as possible for example noting valve and alarm numbers, if there is any chance of confusion).

Once the accident possibilities have been described, the question may be asked, are the safety measures adequate? Generally, in a hazop analysis, this must be decided on the basis of general experience.

If it is decided that further measures are needed, these must be recorded. The actions column is used to record problems which cannot be solved completely during the meeting, either because they are too complex or because the discussion takes too long.

The hazop tables generally provide space for recording the initials of the person who is responsible. (It is important to allocate responsibility to a particular person within the team, otherwise some required actions will certainly be forgotten.)

Note that the consequences and safety measures are in general specific for each disturbance cause. For this reason the consequence description and safety measure columns must be filled out according to the cause. However, repetition can be avoided by using "ditto" or "identical to # 3 above" for example. Numbering the disturbances is a necessary aid to cross referencing. Each deviation cause in the entire analysis receives a unique number.

Which disturbances?

One of the basic ideas underlying hazop analysis is to allow flexibility and freedom for creativity. For this reason, no fixed check list is provided for each team member to follow. However, a check list has been provided for the team leader. Experience shows that the hazop analyses are more complete and systematic when the check list is used. Additionally, when the check list record is filled out, it is possible to see at a glance how far the analysis has progressed.

The check list provides for high and low values for the major process variables in a vessel or pipeline. In addition, there is space for "other" disturbances. The team leader should remember always to ask team members whether there are other disturbances in addition to the basic ones, which could lead to accidents.

The team leader should always ensure that service power failures are discussed. However, it is generally possible to treat an entire unit at one time rather than just one vessel at a time with regard to power and services

Vessel by Vessel or line by line analyses ?

Analyses may be made vessel by vessel, or line by line, or both. There is a good deal of overlap between analysis types, if both approaches are used, for example a high rate of flow through a pipeline will generally result in high level or pressure in a vessel.

For this reason, it is reasonable to choose a mixed approach. If just a line by line approach is used many accident types will be overlooked. Examples are reactions, overheating, lack of mixing, scumming, left over products, etc. in the vessel.
A vessel by vessel analysis is usually much faster to perform, too. If a line by line analysis is not made however, some accident types may also be overlooked. A list of problems which are specifically related to pipes is as follows.

- Liquid hammer:
- Two phase flow, and resulting vibration (particularly vertical two phase flow).
- Pipe and flange leaks
- Air or gas sucked into lines
- Overpressuring due to liquid expansion with shut in lines.
- Siphoning
- Reverse flow
- Freezing
- Air locks
- Pressure reduction failure, especially due to low flow or zero flow.

For these reasons, line by line analysis should be carried out at least to identify line specific problems. Also, when there are branching lines, a line by line analysis of the network is almost essential.

Following up the analysis

Once an analysis has been completed, it is the actions which have been noted, which ensure that decisions have been taken, and safety improvements made.

Modifications may in some cases be marked in red on drawings during hazop meeting, if someone with sufficient design authority is present. This may be necessary if changes are major, in order to allow hazop study to be completed. The hazop drawing in this way becomes an important design document, and should be dated and initialled. Copying of modifications to master drawings is essential.

Following up the hazop analysis is fairly straight forward. The solution to all action requirements must be known, and can be checked on the plant drawings. Additionally it should be possible to check the actions recommended on the plant itself.

Plant disturbance phenomena

Many of the disturbances which can affect a plant are fairly obvious, but some are very obscure. Many of the following accident phenomena have been found by studying accident records, particularly the Manufacturing Chemical Association records, and Loss Prevention Bulletin.

High Pressure

High pressure in a vessel can be caused by high inflow to the vessel. Typical causes of high inflow are

- Continued pumping, after the vessel is full
- Too high a pumping rate
- Control failure on a pump
- Loss of pressure reduction.

Typical safety measures here are:

- Use of centrifugal pumps with a maximum pressure less than the design pressure of the vessel.
- Use of a safety valve to "spill" the pump flow back to the inlet of the pump.
- Use of additional pressure sensors as a pump trip

Pressure reduction is often performed by throttling through a pressure control valve. If either the valve or the controller fails, then the downstream pressure may be dangerously high.

A similar, but less obvious, cause of high pressure, arises from liquid level controls in gas liquid separators. If the level controller, or the valve, fails then liquid level may fall to zero. and gas then flows through the level controller. Even though the mass rate of flow of gas will be lower than for liquid, the volume may be too high for the outflow from the downstream vessel, allowing pressure to rise.

Typical safety measures for these kinds of problems are:

- providing an orifice plate which limits maximum flow.

- providing shut off valve(s) (so called slam shut valves, which close rapidly) and additional pressure trip(s) to drive it (them).

- providing a safety valve on the downstream vessel which can take the maximum flow.

- making the downstream vessel so robust, that it can take the full upstream pressure.

Other causes of high pressure are:

- blockage of vents, particularly if there are condensing vapours,or liquids which can solidify after an overflow. Animals and birds sometimes block vents. Painters and welders sometimes deliberately block vents to prevent entry of dirt into a tank, then forget to remove the blockage.

- high temperature (see below).

- low outflow.
  In some processes, pressure control depends on a balance between inflow and outflow. If the outflow is blocked (see below) then flow into the vessel may be restricted. This may result in loss of pressure reduction by throttling.

- scumming

  If scum forms in a gas or vapour flow, resistance to flow rises considerably and can cause overpressuring.

- gas generation

  Some liquids decompose, generating gas which can overpressure a vessel which is closed in. Acid in tanks, in particular, can corrode tank walks (generally very slowly if the design is well done) and overpressure a tank with hydrogen.

- liquid expansion

  If a tank is filled completely with liquid, and then closed in, a small temperature rise can cause overpressuring.

- runaway reactions (see below)

- reverse flow

Overpressuring in a downstream vessel can cause back flow and overpressuring. Typical causes are blockage in a common closed drain, or overpressuring in a blow down line or flare line. Less common is reverse flow from a common receiving vessel or reaction vessel.

Overpressurisation can be caused by pumping the wrong liquid into a vessel.

- pumping hot water or hot oil into a cold vessel containing light oil.

- leak of water from a cooling or heating oil into hot oil or fat.

- pumping light oil into a vessel containing hot water or hot oil.

A special mechanism, roll over, is associated with liquefied gases, and oil, of different densities and temperatures, in the same vessel. A hot, heavy fraction at a low level is accompanied by a cool, light fraction at a higher level. If these layers are disturbed,or the temperatures change, then the layers may tend to change place. When the layers mix, boiling takes place, and both the rate of mixing and boiling becomes violent. Just a few degrees temperature difference can in some cases destroy a vessel.

The most obvious consequences of high pressure are bursting of vessels, or leakage. Less obvious consequences are: - formation of condensate compounds. - upset of vapour concentrations (eg. light fractions in heavier fractions, LPG in gasoline, gasoline in fuel oil etc.) and dissolving of gases in liquid. These might later release toxic or flammable compounds downstream in the liquid flow,or cause overpressuring.

Overpressuring sometimes distorts vessels, without causing rupturing directly. One case is known of a butane tank "ballooning" slightly (rising a few inches from its foundation). The resulting distortion caused cracking of nozzle entries to the tank.

Runaway reactions

A runaway reaction occurs typically when reaction rate increases with increasing temperature, or increasing concentration of one reactant. Typically causes are then:

- poor temperature control or temperature control failure.

- loss of cooling water. This can be caused by water supply or power failure blockage etc.

- loss of cooling surface due to encrustation, crud etc.

- poor mixing or mixing failure, resulting in loss of cooling, hot spot formation, or errors in temperature measurement.

Temperature measurement is critical for such reactions, and typical failures are:

- simple failure of the instrument

- coking or deposits on a thermowell so that the sensor becomes insulated.

- placement of the sensor in an outflow line or recirculation line. Measurement becomes incorrect when the flow stops or is reduced.

Some reactions are autocatalytic, and accelerate with time. Holding a batch reaction for a period can then be hazardous e.g. nitration reactions, or the TCDD reaction which was put on hold over the weekend at Seveso.

Another typical cause on runaway reactions is disturbance of ratio control or mixing control in a continuous reaction feed. If one reaction component is interrupted in some way, its concentration will fall. When the supply is restored, most controllers will try to compensate for the low concentration, and supply an excess of the reactant, which may then cause a runaway reaction.

An unusual but possible cause of a runaway reaction is over addition of catalyst, or presence of a stray catalyst. This may, in particular, enhance an unwanted, unstable reaction, such as decomposition of a product.

Typical safety measures for runaway reactions are:

- emergency cooling (independent of normal cooling).
- burst discs to relieve pressure
- burst discs to provide cooling through boiling.
- addition of quenching water or solvent
- dumping of the reaction to a dump tank filled with water or solvent.
- venting of the reaction to a scrubber.
- depressurisation, to provide cooling by boiling.

Blockage

Pipelines can be blocked in many ways, most of them obvious, some obscure:

- valve closure, due to human error, control failure, or valve failure
- forgotten blanking plates, spades etc.
- blockage with foreign material impurities in the flow
- sand
- broken parts from vessels internals
- column packing
- catalyst

- welding rods, welding slag
- paper from packing
- stray objects such as beer bottles, footballs, containers, packaging
- dead animals or insects
- tree roots (in drains)
- freezing/solidification
- emulsion formation
- crystallisation
- solid forming reactions (eg. isocyanate/water)

Blockage often occurs when pipe flow is shut down, and is very difficult to detect at this time. (Restart is then hindered).

High temperature

The most frequent causes of high temperature are:

- failure of temperature control, due to instrument failure (steam, heating oil fuel supply).
- formation of hot spots due to flow blockage, or deposit formation.
- reduced flow in heat exchangers.
- too high flow rates for heating flows, or, correspondingly, too low flow rates of reactants to he heated.
- reduced cooling flows (blockage, control failure, pumping failure).
- too high temperature in cooling flows. (eg. in unusually warm weather, due to excess hot water flow to cooling ponds,failure of brine chillers or cooling towers).

More obscure causes of high temperature are:

- heating by the sun, or by radiation from furnaces or flares.

- heat tracing, when flows in pipelines are reduced. Electrical heat tracing, in particular, can suffer from control failure.

- continued agitation of shut in liquids

- pumping against a closed valve (dead head). Many pumps have a bypass to allow continued circulation in case of closed outlet valves or flow blockage. It is important, in such cases, to ensure either that pumping power cannot lead to overheating, or that the flow returns to a supply tank with sufficient heat capacity, and not to the pump inlet.

- stray electrical currents.

- blanketing with dust, product, rags, tarpaulins, coats, so that convective cooling is blocked. Applies particularly to electric motors.

High temperatures makes expansion of vessels and pipes. If the piping and supports are not designed for this, the expansion can cause overstressing, cracking, or distortion of supports, valving and pumps. Distortion of pump piping may in turn cause misalignment of motors, vibration, and pump breakdown or rupture.

- Friction
  particularly in belt drivers, gears,transport belts, and mixers

- Fire
  this is the most common cause of overheating

The main consequences from overheating are:

- overpressuring (see above)

- runaway reactions, (see above)

- fire

- release of vapour

Other problems are:

- break down of equipment

- coking of liquids, or evaporation, leading to blockage, hot spots

- vessel weakening
- expansion, leading to overstressing.

High level

The main causes of high level in a vessel are fairly obvious - too high inflow, or too low outflow.

Typical causes are:

- control failure, particularly level sensors
- errors in depth measurements
- errors in valving, so that liquid is pumped to the wrong vessel
- back flow from a common delivery pipeline which has high pressure, or is partially blocked.

More subtle causes of overflow are:

- liquid expansion due to temperature rise
- scumming and boiling

Level sensing is often made by measuring pressure differential between the top and bottom of a tank. Pressure sensors, such as diaphragm type sensors, are generally much more reliable than level sensors, which tend to rely on floats of some kind, and tend to stick. However, it should be recognised that such measurements are not a direct indication of level. A change in density due to a change in liquid or in mixture, can lead to errors in level measurement, and to overflow.

Crushing or distortion of vessels can cause overflow directly. This can occur as a result of a vessel crash, or as a result of pipe expansion. An insidious cause of overflow is reduction in vessel volume due to leakage of liquid (eg. water) through the tank base, or cooling or heating coils, followed by filling with a constant volume of liquid. Another is reduction in vessel volume due to sucking in (vacuum) so that the tank is distorted, followed by filling with a constant volume of liquid.

Transport vessels for liquefied gas are generally filled up to around 80% of volume - the actual degree of filling depends on the substance, and on temperature. Filling is generally checked by weighing. Errors in weighing lead to overfilling. Also, operators may try to overfill, through ignorance, or in order to minimise the number of vessels needed for a particular delivery. (Tank truck drivers have been known to overfill with LPG and to sell the excess later).

A special phenomenon causing overflow is "boilover". This occurs when a tank (generally a floating roof tank or cone roof tank which has opened in an explosion) is exposed to heat, typically from a fire. At some stage, the oil in the tank begins to boil. If there are lighter fractions in a layer low down in the tank, then these may begin to boil first, and throw oil out.

This kind of boilover can be very violent, throwing oil a long distance. Less violent boilovers can throw oil out repeatedly.

Another possibility is that boiling foam can leave the tank. This kind of boiling foam can flow very rapidly, and poses a threat to firefighters.

Low temperature

Low temperature in a vessel may be caused by:

- low temperature in a heating flow
- low temperature in a cooling flow (either of these may be caused by control failures).
- loss of heating (blockage, power failure, loss of heating fluid, valve closure).
- evaporation, especially of a liquified gas or volatile liquid.
- cold weather
- dissolving salts

Cooling may also cause changes in concentration, for example for volatile gases such as ethane and propane in gasoline. Low temperatures may cause condensation of LPG in pipelines, something which may be dangerous in LPG burners.

A rather unusual problem of this type is condensation of oxygen from the air into liquid nitrogen exposed to air.

Low temperatures, below - 20^0 C may cause changes in ordinary carbon steels, so that these become brittle. Piping which has become brittle in this way has been known to shatter as a result of vibration.

Low pressure

Low pressure may be caused by:

- pumping liquid out of an unventilated vessel.
- blockage of vents or vacuum pressure valves.
- connection of a vessel to another in which there is a vacuum.
- cooling

Operators of large closed tanks are for example obliged to take account of cooling during rainstorm.

Long drain lines can cause considerable underpressure (corresponding to the height of the drain line) if draining takes place from an unventilated tank.

An unusual cause of low pressure is gas absorbing reactions. Rusting on the inner surface of a closed vessel can cause underpressure sufficient to cause tank collapse. Ammonia gas, dissolving in water can cause very low pressures.

The most common consequences of low pressure are:

- collapse of vessels
- sucking back liquids

A case is known of a steam line which was cooled by water. The condensing steam created a vacuum and sucked back condensate. The resulting water hammer resulted in pipe rupture. Several similar cases are known for ammonia gas pipelines taking return gas from ships.

Low pressure can cause air to be sucked into vessels, and, for example to be mixed with flammable vapours, with explosions as a possible result. A typical place for this is pump intakes, especially if intake filters become blocked.

Low level

Low levels in vessels may be caused by:

- loss of inflow or low inflow due to blockage, closed valves, or control errors
- loss of supply for an inflow
- leakage or pipe bursts

- too high outflow
  (control failure, erroneous opening of valves)

- evaporation, or boiling dry

Consequences of low level may be:

- boiling dry, in boilers, evaporators etc.

- disturbance of deposits, or water, at a tank bottom, so that these pollute the flow.

- break through of gas into a liquid drain or separation line. This can cause a high volume flow of gas into a low pressure line resulting in overpressuring or to two phase flow and resulting vibration.

Gas can be drawn into liquid drain or separation lines, even when the liquid level is above zero, through formation of a vortex. This can often be prevented by fitting a "vortex breaker", that is, one or two vertical plates, into the drain line.

Too high or too low concentration

High or low concentration of a reactant can cause:

- high vapour pressure and overpressure

- too rapid reaction or a runaway reaction

- a wrong reaction giving a toxic or flammable product

- a toxic, acid, alkaline, oxidizing or flammable end product

- a toxic, acid, alkaline, oxidizing or flammable residue

- an excessively viscous product which causes pumping difficulties, or blockage.

High or low concentration of a substance can be caused by:

- control failures

- blockage in addition lines

- delay in adding a component

- settling out of a solid component of a mixture

- insufficient mixing

- human error in addition of a component
- labelling error leading to addition of an inert component
- a residue of one component from a previous batch, diluting the new addition.

Wrong substance

Stray or erroneously added substances can cause a wide range of disturbances or accidents:

- unwanted or unexpected reactions giving toxic, flammable, acid, alkaline, oxidising or high vapour pressure substances
- catalysing effects, leading to runaway reactions
- solidification, and overpressurisation
- impurities in a product, causing production waste, product discoloration
- catalysing effects, leading to spoilt product, decomposition etc.

Impurities can come from many sources:

- erroneously added raw materials
- wrongly labelled materials
- back flow
- leakage from supply tanks
- corrosion products
- products or residues left from previous production batches
- inflow (sucking in) of air, due to vacuum
- leakage from heat exchangers, cooling coils etc.
- flooding
- absorption
- unwanted or unexpected reaction products.

Breach of vessel boundaries

Many accidents are associated with the vessel itself in the initial stages. Breach of vessel boundaries can be caused by:

- weakening of the vessel. This can be due to high temperature, vibration fatigue, temperature or pressure cycling fatigue, vitrification (plastics), or creep.

- loss of toughness, due to aging (plastics) low temperature, or hardening (stress hardening, temperature, curing, crystal growth).

- loss of strength due to change in crystal structure

- growth of cracks, arising from lamination or impurities in original material or welds.

- corrosion. This can be superficial, crevice, electrocouple corrosion, stress corrosion, or pitting. Unusual or unexpected levels of corrosion can arise from unplanned operating conditions, impurities, or design using the wrong materials, or installation of the wrong materials.

- high temperature due to fire can give overpressure and vessel weakening at the same time, for example giving a BLEVE.

- Overstressing

  This can arise from overpressuring. It can also arise from thermal expansion or contraction. Supports should be designed to prevent significant overstressing eg. spring loaded pipe supports, free sliding shoes. Overstressing then arises if temperature changes are larger than expected, or if supports jam or sieze. In one case observed by the author, an asphalt transport pipe was repeatedly subject to thermal expansion, and overstressing on cooling. The 30 m pipe had, over a period of years, extended by 1/2 m.

- collapse of foundations can cause overstressing or collapse of a vessel

  Foundation collapse can occur as a result of poorly designed foundations, flooding or washout of foundations, corrosion of supports, or impact from a vehicle crash.

  An unusual cause occurred in the case of a chlorine "bullet" vessel. A sudden gas flow caused a wave of liquid to form in the vessel, and the swash of liquid against the end of the tank knocked the vessel off of its foundations.

A recent major foundation collapse of an oil tank resulted from a leak from a water pipe under the tank.

- unusual weather conditions.

  Tornadoes, and hurricanes can cause vessels or columns to collapse, and can create missiles which penetrate tanks. Flooding can wash away foundations, and can cause partially filled tanks to float away. Hurricanes can cause excessive pipe vibration and rupture. Lightning can cause holes in vessels, and can ignite fires.

- Earth quakes, tsunamis, and land slides can cause vessel collapse.

- water hammer/steam hammer can cause rupture in pipelines, bellows etc. Two phase flow,particularly in vertical pipelines, can cause so much vibration that cracking is inevitable.

- explosions, and explosion fragments, can cause vessel rupture

- mishandling of vessels can cause rupture. Examples are use of nozzles and pipelines as support or as scaffolding attachments. Welding on operating equipment, and ill prepared repairs. More common is attempted repair of the wrong vessel.

- crashes of cranes, trucks, trains, and aircraft can damage vessels. A particularly common cause is a crane which drives through a pipe bridge with a raised boom.

Columns

Columns are subject to some special accident types.

- flooding of the column, when the reflux rate is too high. This can lead to overpressuring, but more often to disturbance of, or damage to column packing or trays.

- steam explosions due to disturbance of water in traps in column trays, and pump around take offs.

- cooking or build up of tar on the packing, leading to overpressure.

Heat exchangers

Heat exchangers are subject to some special disturbance types.

- leakages across tube plates and tubes due to corrosion

- fretting (wear) between tubes due to vibration against foreign objects or loose baffle plates.

- fatigue cracking due to flow induced vibrations

- tube ruptures due to jamming of floating heads.

Valves

Most valve failures can be described in terms of failure modes rather than plant disturbances. These are listed here.

- valve body leakage due to cracking or corrosion

- valve seal packing leaks

- body seal leaks in split body valves
- broken valve items, leading to valve closure, or valve opening, depending on valve seat position.

- valve jamming in open or closed position (due to crud, overtightening, dried out packing, foreign bodies in the valve seat.

- valve spring breakage, leading to failure in open or closed position, or simply to failure to close on loss of activation.

- activator diaphragm leakage leading to fail open or fail closed

- activator piston jamming

Traps and pitfalls in Hazop analysis

The gap between a good hazop analysis and a poor one is about 50%. That is, by taking care over the performance of the hazop analysis, an extra 50% of hazards can be found. This statement is based on a reading and assessment of 35 safety reports submitted to authorities, and a parallel (independent) cross check analysis of each. Of the 15 analyses studied, several contained omissions which added further hazards of this order. When the risk level associated with the omissions was calculated, it proved that some of the Hazops had overlooked the most significant risks!

It might be thought that this just results from inexperience on the part of the analysts, but many of the problems arose in work carried out by experienced teams of plant engineers and safety analysis consultants. The problems arose from various traps and pitfalls into which the analysts fell. Anyone can fall into these traps. I have observed all of them in my own work.

Neglect of banalities

When you are trying hard to study a complex chemical plant, there is a tendency to focus on the technical and the complex, and forget the simple things. For example that a pipe may leak due to inadequate flanges, or simple corrosion, as well as due to overpressuring. Or to forget that a crane crash can be a significant cause of accidents. A check list of such banal points may be of some help. I use such a list to follow up and check the quality of my own work.

Know your phenomena

Sometimes very curious phenomena are the cause of accidents. Examples that I have met are:

- blockage of cooling intakes by a cloud of ladybirds
- failure of a pipe which had extended and contracted so often due to heat cycling that it had "grown" by over 1 m.
- "ballooning" of a tank due to overpressuring, resulting in damage to foundations.

There is really no guarantee that you can ever know everything which might happen in your plant, and even if you did, some of the possibilities are so remote that it is unlikely you would guard against them. It is important to know as many as possible, however. Reading publications, such as "Loss Prevention Bulletin" is a very good help. Some of the odd phenomena which <u>must</u> be remembered are:

- cavitation
- vacuum due to tank draining
- scumming
- cold embrittlement
- vibration fatigue
- zinc embrittlement of steel in fires
- aluminium sparking
- aluminium fire reaction with chloroflourocarbons

Completely unexpected phenomena

Irrespective of the quality of an analysis, there is always the possibility that something new will show up, which has never been seen before by anyone. This is especially true of chemical and plant material properties, and human behaviour. Example: an operator was found "smoking" paper soaked in solvent because he "liked the feeling".

This kind of thing cannot be taken directly into account in an analysis, but it is wise not to make too strong statements about safety in your report conclusions.

There is always a limit

In any hazard analysis, there will be a limit to how far you can go in looking for causes. An example which occurred for me was a possibility of a leak of water into hot oil from a glass cold trap. This was missed in the hazop analysis, but found using computer assisted analysis. The reason for omitting the cause was to find the problem, a search had to be made through a piping network with a total of 30 valves and pipe sections. To find this in a uniformly detailed analysis would have required several years of effort.

The sneak path analysis procedure can help in reducing the effort required to find some of these problems, and is a useful and cheap follow up procedure. (see page 204).

Exhaustion (of patience and concentration)

When you are carrying out a hazop analysis you will often find your attention wandering, or that you have missed a part of the discussion. This is a sign of exhaustion. It often occurs after a particularly strenuous piece of logic, or just after lunch.

If this does happen, relax take a cup of coffee, and trust in the procedure. The procedure is a great help because to a large extent, it does not let you go too far wrong. Be very suspicious however, if you find an area where you have recorded that there are "no problems".

Secretarial and recording oversights

It is all too easy for the secretary to get left behind in the recording. He or she may then miss a point. If the secretary falls behind, he or she should interrupt the proceedings.

A similar problem arises when the secretary cannot read, or cannot understand his/her own notes after the meeting.

Assumption that the design is correct

In a hazop analysis it is generally necessary to assume that components are correctly designed e.g that pressures are correct, that vessels, piping, and supports are sufficiently strong, that pumps are sufficiently powerful. This may in fact not be the case. (Example: a tank had to be raised on an earth mound to get sufficient suction head, because an engineer worked with absolute and not relative pressure)

Design reviews should check many of these points. I don't think the hazop procedure can be expected to take everything into account.

Too much experience

It is impossible to have too much experience, but it is sometimes tempting to put too much trust in it. Experienced engineers have often seen plants work and assume therefore that the next one will, also. Or, even more typically " I saw that happen once, and nothing serious followed". Experience is a great help in showing what can happen, but be careful when it is used uncritically to reject a problem.

Overshadowing or masking

Sometimes a serious problem is found, which requires a lot of analysis effort. It is very easy to then proceed quickly to the next disturbance, or the next vessel, ignoring other causes or consequences in the same area.

Repetition and analogy

One of the most effective techniques for getting through an analysis is to use analogy, or repetition, in order to say "this item is just like the last one". This may lead to traps. For example, a new unit may be just like the previous one except that is more dangerously sited. Or that the combination of the two units may cause an excessive load.

When using analogy or repetition, a good practice is to ask "in what way do the two cases differ", and "is the combination more dangerous than the separate units alone".

The diagram is not the plant

There can be an enormous difference between the diagrams which are used as a basis for analysis, and the plant itself. Some of the most obvious ones are that the height of units may be different from what appears on the diagram. This may cause flow, drainage, and pressure problems. Other problems are the actual construction quality, use of wrong materials, and construction short cuts. An inspection is a necessary follow up to a hazop analysis, if the best quality of work is to be achieved.

Achieving high quality analyses

The quaility of hazop analyses, as described above, and in chapter 13, varies greatly. It seems doubtful whether results which are more than 95% complete can be achieved using a single hazop process alone, even for the kinds of hazards which are succeptibl to hazop analysis. Techniques which have been found to improve quality are the use of check lists, the use of computer aids, and careful post analysis documentation and follow up. The most effective methods, however involve the use of other procedures. Of these the hazard tree analysis and preliminary hazards analysis procedures are very effective.

Preliminary hazards analysis

The objective of preliminary hazards analysis is to provide a very rapid summary of plant risks, as a starting point for more detailed analysis. The procedure can also be applied at a very early stage in plant design, when only very broad layout and flowsheet information is available.

The basic idea of preliminary hazards analysis is to select an object for study, and to identify which problems might arise. It can be used as a team method, with a very free structure allowing a wide range of problems to be raised. It is easiest to apply to plant types for which experience already exists.

The basic material required is minimal, e.g. a flow sheet or a sketch of the plot plan, plus information on the substances to be processed. The analysis can obviously be more detailed if the basic material is more extensive e.g. initial piping and instrumentation diagrams.

Given this background material, the analyst, or the team, considers the possible hazards, first freely, then with a check list of hazard types (table 3.3 gives a suggestion). For each hazard, the relative frequency and consequences are considered, and potential problems are identified.

A semiquantitative approach to analysis is very beneficial in setting problems in perspective. The selected problems are assessed in terms of expected frequency and degree of damage. The assessment is intended to be a very rough one. For this reason frequencies and consequences are classified only into ranges,for example using the table given below. Note that the scales are essentially logarithmic, so that high numbers represent serious hazards.

Frequency	**F**
Not possible	0
Rarer than 1 per 1000 years	1
Between 1 per 100 and 1 per 1000 years	2
Between 1 per 10 and 1 per 100 year	3
Between 1 per year and 1 per 10 years	4
More frequent than 1 per year	5

Harm to people	**H**
No injuries	0
Serious injuries	2
Fatality	3
Several fatalities	5

A risk index may then be calculated as follows:

$$R = F + H$$

The risk index indicates seriousness, and is quite rough, since the various consequence classes are only roughly comparable. The risk index is used to establish which hazards require special effort, and to set the scope for detailed analyses. It is useful too, at this point, to consider risk reduction measures and inherent safety (se Ch. 2)

Table 3.3 Hazard check list for preliminary hazards analysis

Situations	**Places**
Normal	Plant unit
Holiday	Pipe bridges and trenches
Strikes	Storage
Shift change	Water supply
Sickness	Drains
	Control room
Operations	
	Hazards
Startup	
Shutdown	Release
Normal operation	Fire
Unusual operation	Explosion
Maintenance	Missiles
Overhaul	Toxics
Transport	Product contamination
New construction	Plant damage
	Shutdown
External causes	
	Causes
Storm	
Flood	Technical failure
Snow	Operator error
Freezing	Maintenance error
Earthquake	Power supply
Landslide	Services
Vehicle crash	Sabotage
Crane crash	Raw materials
Fire	**Other Causes**

Failure Mode and Effects Analysis

Failure mode and effects analysis was one of the first systematic techniques for failure analysis. It was introduced as an aid in designing and documenting military systems in the late 1950's.

The principle of the method is to examine every component in a system and ask the questions:

- How can this component fail?
- What will happen if this component fails?

As an aid in completing the analysis, results are recorded in a tabular format. Several different forms are currently in use. The one shown in Fig 3.1 is a good general-purpose format, based on the one given in IEEE std, 352, "General Principles for Reliability of Nuclear Power Generating Station Protection Systems". The format given here has been extended to cover various phases of plant operation, under different modes of subsystem operation. This extension is frequently necessary in process plant analysis.

A study of failure criticality is a useful final step in completing FMEA forms, picking out those failure modes which are most serious.Rogers gives the list of criticality categories in table 3.4. (Adapted by the author).

The success of failure mode and effects analysis depends on the analysis having a good conception of what failure modes can occur within a process plant, and what modes it is necessary to distinguish to ensure a reasonably complete analysis.

Tables 3.6 and 3.7 give some examples of failure mode lists.

Fig. 3.4 shows part of a circuit for a fail-safe alarm system. The output circuit of the alarm should open, and the output signal disappear if either one of the measurement circuit signals disappears, or if one of the alarm circuit components fails. Fig 3.1 shows part of the failure mode and effects analysis for this circuit. The circuit has the property that all simple failure modes should lead to activation of the alarm, by de-energising the relay.

Failure mode and effects analysis suffers from a number of drawbacks, from the point of view of plant component at a time, and does not treat multiple failures, or the effect of latent failures. If there are many phases of plant operation which are significantly different, as can happen for plant startup, shut-down, or batch processes, then the list of "effects" can grow very long.

Perhaps the worst drawback is that <u>all</u> component failures are examined and documented, <u>including those which do not have any significant consequences</u>.

For smaller systems, this kind of analysis can yield quite interesting facts. For large systems, especially those with a fair degree of redundancy, the amount of unnecessary

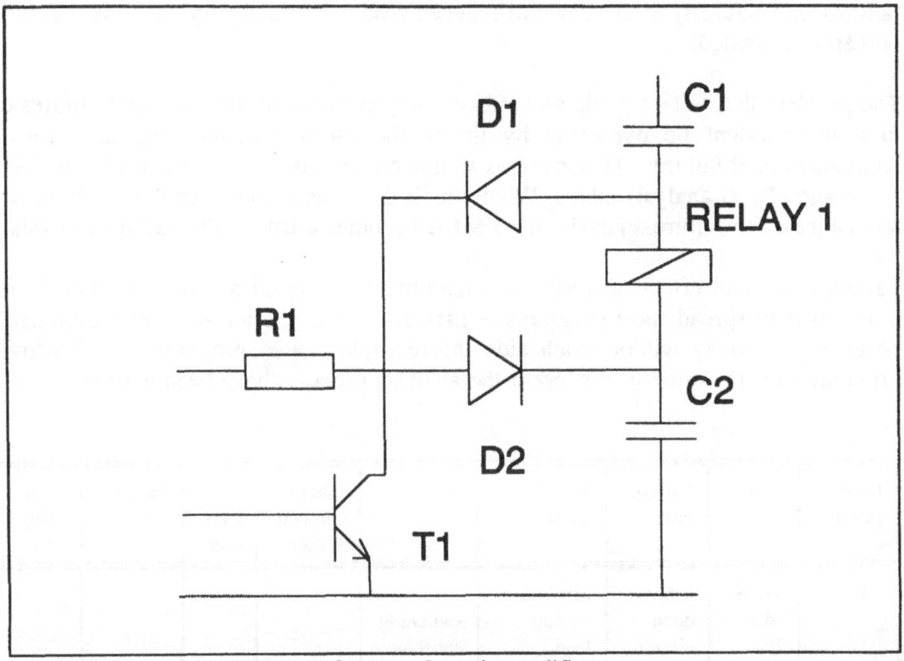

Fig 3.4 Diode pump output stage from a safety trip amplifier

Table 3.4 Criticality categories.

Criticality Category	Effect
Catastrophic	System loss multiple injuries, death
Critical	Injury, equipment damage, hazard requiring immediate corrective action
Marginal	No major system degradation, no functional damage to system, no personnel injury
Negligible	No major system degradation, no functional damage to system or personnel injury, no immediate need for repair

documentation work is a major disadvantage, as far as safety is concerned. For reliability analysis, the treatment of all components is not such a drawback, since most failures lead naturally to shutdown or reduced production, and very little analysis effort is therefore wasted.

The problem that failure mode and effects analysis does not treat multiple failures can to a large extent be overcome by listing the safety measures provided for the consequences of failures. This involves adding one or more extra columns to the failure mode and effects analysis tables. When this is done, and when also the safety devices are subjected to failure analysis, the FMEA becomes a true multi- failure analysis.

Failure mode and effects analysis can readily be supported by computer based aids, in the form of spread sheet programs or data bases. Large scale analyses would hardly be attempted today without such aids. More sophisticated programs which provide prompting for the analyst to support the analysis process, have been applied.

Com-ponent	Type	Failure mode	Possible causes	Effects	Inherent compen -sation	De-tect-tion	Failure rate	Criti-cality
T1	Trans-istor	Emitter open circuit	.broken chip lead .broken connector	No switching constant voltage at output Relay opens				
		Emitter short circuit	burnt out transistor	No voltage at output Relay opens				
D2	Diode	short circuit	Transistor short Overload	Conducts both ways No voltage across relay relay opens				
RL1	Relay	fails closed	Over magne tisation	Fails to open on de-energis-ation				System fails Hazard Catas-trophic

Table 3.5 FMECA table

Mode	Typical Causes	Effect
contacts stuck cannot open when called on to do so	contacts welded by arcing, dirt, corrosion in armature overmagnetization (lack of use) dirt, oil in armature, dirt, foreign bodies in contacts, mechanical damage	
delay in opening contacts delay in closing contacts	oil, paint, dirt in armature overmagnetization, excessive coil current	does not switch at right time, can switch after circuit has become unsafe
chattering (rapid opening and closing)	leaking current when relay should be open, low current when it should be closed	many different effects, in both contact and coil circuit
short circuit between contacts	mechanical deformation short pieces of wire or solder	depends on circuit
short circuit to ground	mechanical deformation short pieces of wire or solder	depends on circuit, but can overload other components
short circuit to signal line	mechanical deformation short pieces of wire or solder	depends on circuit
short circuit to power line	possibility of arcing	
contacts stuck open	dirt, oil, in armature, dirt, foreign bodies in contacts, mechanical damage	cannot close when called on to do so

Table 3.6 Failure Modes of an Electromagnetic Relay (to be continued)

Mode	Typical Causes	Effects
coil open circuit	mechanical damage, faulty soldering, vibration earlier short circuit	same as "contacts stuck open" but can also affect circuits in series with coil
coil low resistance	short circuit in coil	similar to short circuit, but less predictable
coil short circuit to ground	mechanical deformation, small pieces of wire or solder	depends on circuit
coil short circuit to power line	mechanical deformation, small pieces of wire or solder	depends on circuit, arcing and overload of power supply
coil short circuit to signal line	mechanical deformation, small pieces of wire or solder	depends on circuit
coil internal short circuit	water, corrosion, overheating	same as "contacts stuck closed, but can also affect circuits in parallel and can cause overheating

Table 3.6 Failure Modes for Electromagnetic Relay (continued)

Mode	Typical Causes	Effect
stuck closed	overtight packing lubrication dried out, overtight closure, foreign body in valve, leak in or breakage of positioner (valve with close on failure mechanism)	cannot open, no flow through valve, possible breakage or over load of positioner
stuck open	overtight packing, lubrication dried out, overtight closure, foreign body in valve, leak in positioner (pneumatic valve), breakage of positioner or valve piston (valve with open on failure mechanism)	cannot close cannot regulate possible breakage or overload
stuck partly open	overtight packing, lubrication dried out overtight closure, foreign body in valve or leak in positioner (pneumatic valve) breakage of positioner or valve piston (valve with open on failure mechanism)	cannot regulate possible break age or overload of positioner
leak across valve	corrosion of valve seat, foreign body in valve, errors in positioner adjustment	flow while valve is nominally shut off
valve body breakage	physical damage, fire operator stands on valve, design error (valve underdimensioned or overdimensioned, positioner adjustment error). Corrosion cracking of valve body	release of con trolled fluid, loss of pressure
chattering	high-pressure drop across valve, flashing of vapour, resonance in pipework, loose parts in valve	noise excessive load on downstream and upstream equipment

Table 3.7 Failure Modes for a Control Valve

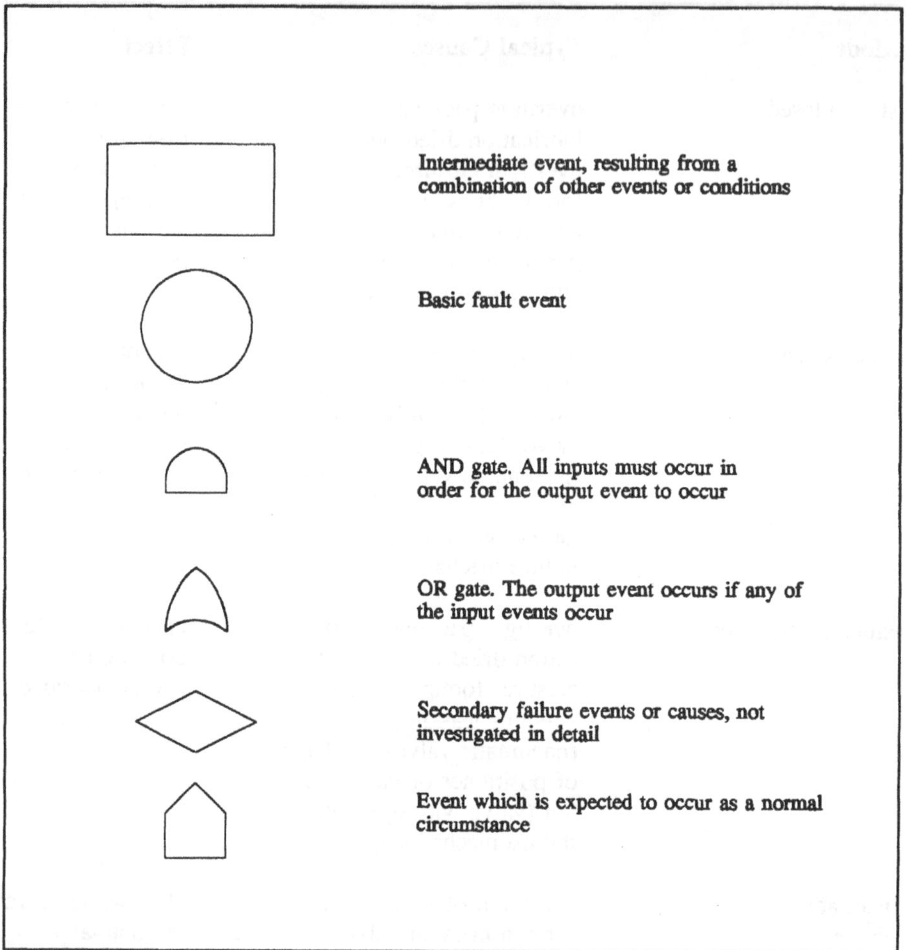

Fig 3.5 Fault tree symbols.

Fault Tree Construction

Fault tree analysis is one of the classic methods for identification of risk sources. It is especially useful in identifying failure combinations which can lead to accidents.

There are many variants of the method. Common to all of them is a notation for describing accident causes.

A fault tree has a so-called TOP event which is the main unwanted event. Connected to this event may be an OR GATE with several inputs. The OR GATE specifies that the output event of the gate can occur if any of the input event occur. It may also be an AND gate. This expresses that the output only event occurs if <u>all</u> of the input events occur.Other symbols are used such as triangles to express a complex and uninvestigated complex of events, a circle to express basic events - primary or spontaneous failures, and a house to represent normal or expected events.

These then are the symbols which can be used to describe the alternative combinations of failures and errors which can cause a particular top event. Figure 3.5 shows the symbols used.

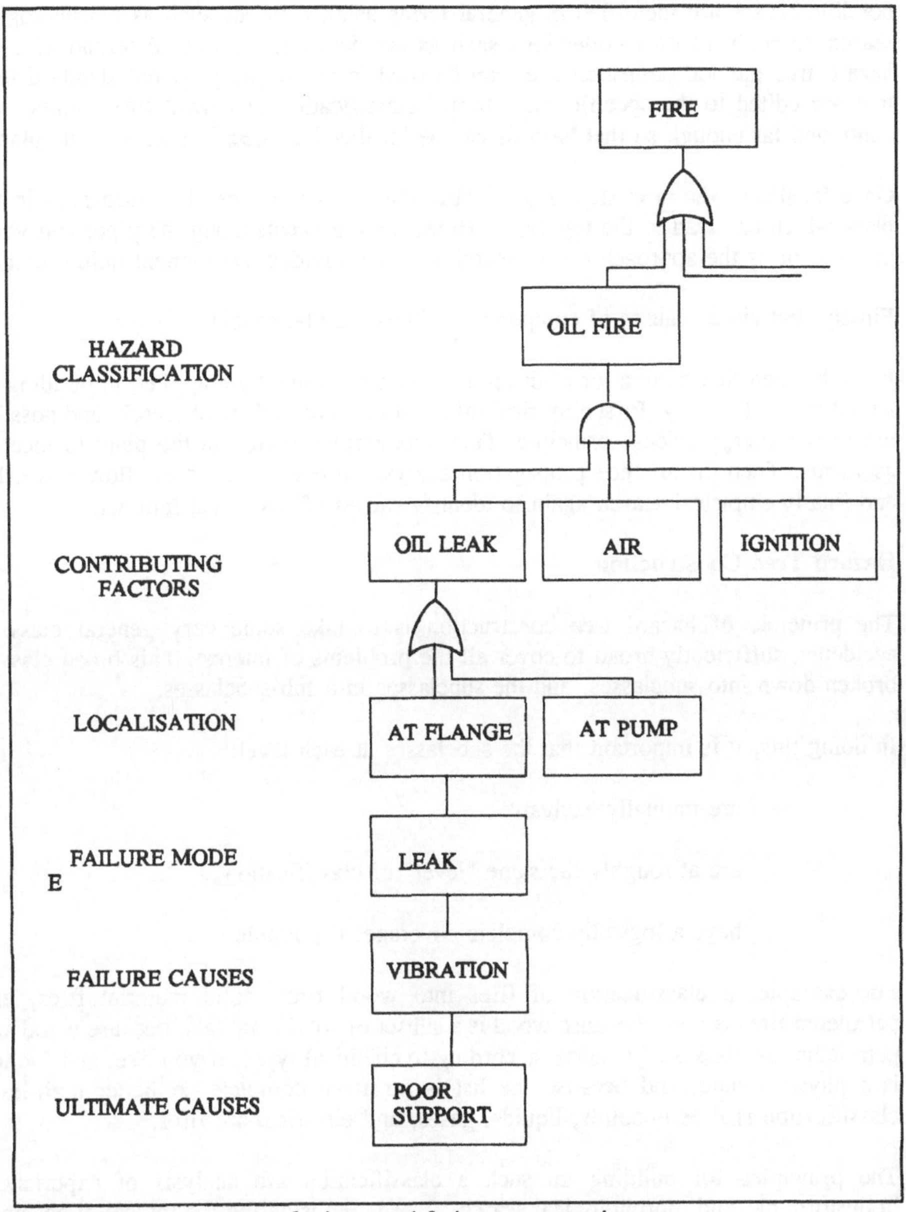

Fig 3.6 Choosing top events for integrated fault tree construction

A major question is how to choose the top event, and how actually to build up the tree itself. There is very little guidance to this in the literature. Perhaps the best description is that by Lambert (1975). The method described here has been chosen to provide as systematic an approach as possible.

The approach is in four or five stages. In the first stage a top event is chosen, and its possible causes are identified in general terms using methods such as morphological search or study of data collections such as accident case stories. Alternatively, the hazard tree method described later can be used in which pre-prepared standard fault tree are edited to the specific use. Hazard classification or hazard tree methods are continued far enough so that hazards can be localized at specific places in the plant.

Once localized, the next step is to identify the various chains of disturbances in the plant which can lead to the top event. Here, tracing events along the pipes and wires in the plant is the approach taken, searching for individual component failure modes.

Finally, individual causes of component failures may be sought.

It can be seen that quite a lot of different modes of thought are applied in building up a fault tree in this way. First empirical information, morphological search, and possible use of the energy release principle. Then topographic search in the plant to localize accidents. Then disturbance propagation analysis in energy and mass flow networks, and finally empirical search again to identify causes of individual failures.

Hazard Tree Construction

The principle of hazard tree construction is to take some very general class of accidents, sufficiently broad to cover all the problems of interest. This broad class is broken down into subclasses, and the subclasses into sub-subclasses.

In doing this, it is important that the subclasses at each level:

- are mutually exclusive
- are at roughly the same "level" of classification
- have a logically complete coverage, if possible.

For example, a classification of fires into wood fires, solid material fires, and petroleum fires is poor, because wood is a subset of solid materials, because wood and petroleum are classes of material according to chemical types, if you like, and "solid" is a physical state, and because the list is far from complete. A better high-level classification is fires in solids, liquids, gases, and electrical arc fires.

The principles for building up such a classification are analysis of experience, brainstorming, and morphological search, (that is, considering the logical dimensions of the problem and investigating all variations within each dimension). It is hard to convince oneself that such a construction procedure can ever be thorough.

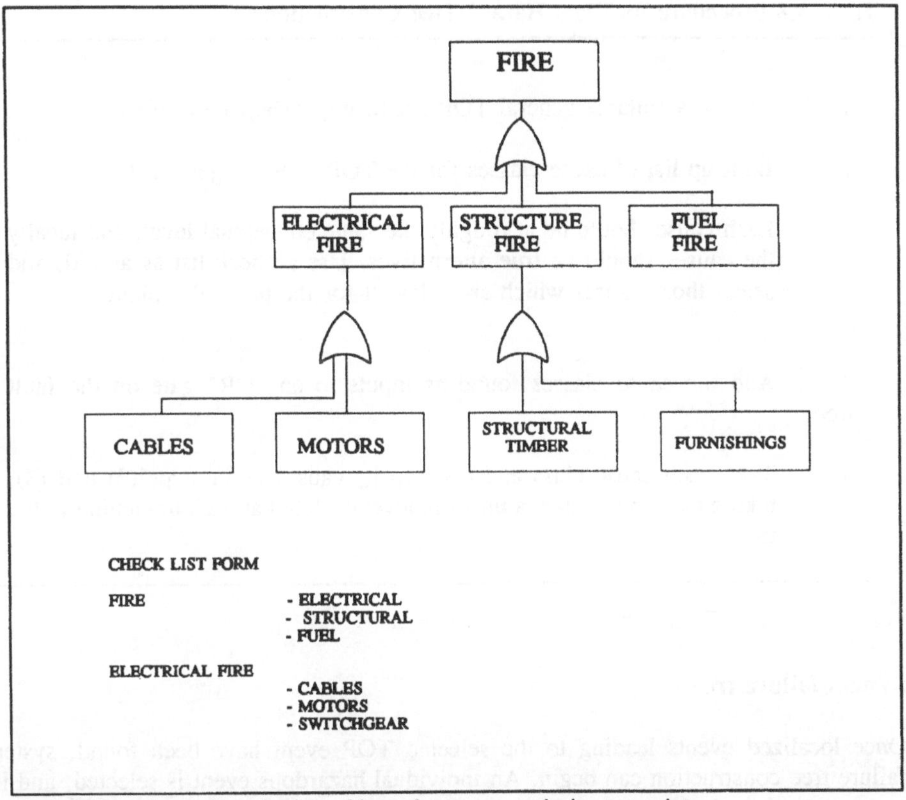

Fig 3.7 Hazard tree for selection of hazard types as analysis start points.

One approach which can be used is to build up standardized trees very carefully over a period of years, covering a broad spectrum of plant hazards. The tree can be supplemented from experience. In use, such standardized trees can be used as a check list. One takes such a tree, and a red pencil, and underlines the relevant parts.

A tree of this kind can be constructed for generic use, first using imagination and experience alone. Then each accident case story from a suitable source (e.g."Loss Prevention Bulletin") in turn is marked on the tree. If the accident cannot be identified from the tree, the tree is extended.

Such a general purpose tree was constructed by the author and used for computer aided hazard identification. In use, the tree has provided a useful supplement to other methods. It is better than a simple check list since it is structured. In any particular analysis, some branches of the tree can be abandoned at the outset, without detailed inspection.

Table 3.8 Procedure for Plant Hazard Tree Construction.

1. Choose a suitable general TOP event, e.g. "Damage to plant"

2. Built up list of cause classes for the TOP event in general terms.

   Each cause should be at roughly the same conceptual level, and ideally the causes should be true alternatives. Use a check list as an aid, and select those causes which are relevant for the particular plant.

3. Add the cause classes found as inputs to an "OR" gate on the fault tree.

4. Take each cause class and develop its causes as in steps (2) and (3), taking care to maintain a uniform level of detail at each branching in the tree.

System failure tree

Once localized events leading to the selected TOP event have been found, system failure tree construction can begin. An individual hazardous event is selected, and its causes traced.

An event in a component such as "Loss of product flow" in a pipe will be termed an output event, because this event represents a change which affects, or is felt by the rest of the system. The output event may be caused by a failure event or some other change within the component. The cause may be a spontaneous internal event, that is without any explicit causal explanation. An example might be "valve steam breaks, valve becomes blocked". If the spontaneous event is a failure, it may be termed a spontaneous failure.

An output event may alternatively be caused by a component failure or event which is in turn a result of a secondary cause.

Examples of secondary causes are:

- design errors, including incorrect application of components
- fabrication errors
- installation errors

- maintenance errors
- overload
- environmental damage (vibration, overheating, stress deformation etc.)
- operation error.
- secondary effects from other components (contamination, overload, fire flooding, physical damage, vibration or oscillation, overstressing)

It should be clear that the distinction between a spontaneous failure and a failure resulting from secondary failure cause will often be one of convenience or judgement depending on how much detail is required in the analysis. Secondary events are particularly important in the search for common cause failures described later.(Ch 4.)

A third possible way which an output event may be caused is by an <u>input event</u>, or <u>command fault</u>. The term command fault has generally been used in the literature (see e.g. Lambert) and seems appropriate for control systems. It seems less appropriate for chemical process plant though, where events such as loss of cooling are not easily envisaged as "command faults". An input event is any event at the inputs to the component which may cause the output event to occur.

In searching for causes of output events in a component, all the alternative types of cause given above - spontaneous internal events, secondary events and input events are considered as cause of output events.

All the different possible input events and spontaneous events are added to the failure tree as inputs to an OR gate, as causes of the output event. In some causes (as for example, the disappearance of flow from the outlet of a tree junction) the output event requires a combination of input event in the same example, (failure of two input flows). In such cases, the combination of events is added to the fault tree as inputs to the AND gate. Once a new set of cause events has been added to the fault tree, the causes of the input and secondary events must be found. Each of these in turn is regarded as a corresponding output event, and the procedure repeated. The search for causes of input events proceeds along the physical connections in the plant flow sheet, piping and instrumentation diagram, wiring diagram, flowing pipes, wires etc. seeking the component within which the input may be caused. A more detailed "recipe" for fault tree construction is given in fig.3.8.

AND gates are used in the fault tree wherever there are two alternative supplies. This includes both liquid supplies, power, and cooling supplies. AND gates are also used when there is a safety device which prevents the propagation of an event chain causing an accident. The propagation requires both the initiating event AND failure of the safety device.

It should be noted that in fault tree construction normal events and plant states are often preconditions for a hazardous event to occur.For example, it is often the case that particular hazard can only arise while the plant is running, never while the plant is shut down. It is generally the custom to suppress such preconditions, avoiding expressing them on the fault tree, unless their suppression affects the logic of the fault tree appreciably.

The procedure described here allows a fault tree to be constructed in a simple and comprehensive manner. It requires a degree of judgement at each stage, however, to decide just what causes of a hazardous event are possible. The procedure does not take account of timing and sequence effects, nor does it allow for feed forward or feed back loops in the plant. For further description of these problems, the reader should refer to later sections of this chapter.

A major advantage of fault tree analysis, compared with failure mode and effects analysis or hazop analysis, is that only event sequences leading directly to the TOP event are investigated. Event sequences without serious consequences are never inspected, reducing the amount of work to be carried out. Another advantage is that combinations of failures, rather than just single failures are investigated.

If two engineers set out to construct fault trees for the same plant by hand, they will almost never reach the same result, even if they use the same procedure, as in Fig. 3.6. This is because their way of describing individual components will differ. A method of construction standardized fault trees is to regard steps 3, 4 and 5 of the procedure as "looking up" event causes in a library of standard component models. By defining standard models for components, standard phrases for event description, and automating the piecing together of parts of the fault tree, it has proved possible to make the fault tree construction process repeatable, giving identical results from different analysts (Taylor and Olsen 1983). However there are, even with this approach two kinds of judgements which lead to differences in results. Firstly, two analysts may not agree on whether a particular event can occur at all, or whether it can produce a particular effect. Here, the fact that the fault tree is only an approximation to real life is important. Loss of cooling may for example cause a temperature rise, but will this rise be significant ? The answer may be yes, no, or sometimes. Answers to such questions could be given by detailed calculation, but are more often based on judgements, which can vary.

Secondly, analysts vary in the degree of detail which they put into trees. If repeatability is to be achieved in constructing fault trees, it is necessary to define the point at which the analysis should terminate. For example the analysis can be taken to the failure mode level in individual components, or to the failure mechanism level (see tables 3.6 and 3.7) or even to the level of secondary causes (design, maintenance, unavoidable physical effect, misoperation).

1. Start with a flow sheet of the plant (sufficient detail so that failure rates are known or can be determined)

2. Construct a plant hazard tree, with localized events at the lowest level

3. Take a localized event and investigate its causes as follows:

4. Take event in the particular component and investigate if failures or spontaneous events in the component can cause the chosen event

5. Investigate events at the input to the component which could cause the chosen event. Trace the path along which these events could be caused on the flow sheet

6. Take the input events found in step (5) and investigate their causes in turn, as in steps (4) and (5)

7. Repeat steps (4) ,(5) and (6) until there are no more unexplained events on the tree

8. Some components require two events at their inputs for the output event to occur. The event is added to the fault tree via an "AND" gate

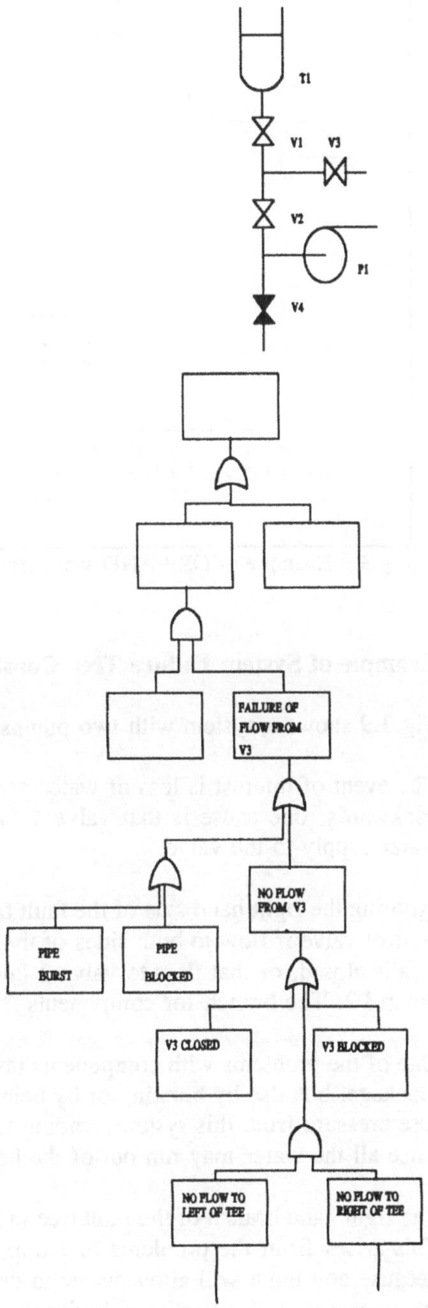

Figure 3.8 Procedure for fault tree construction

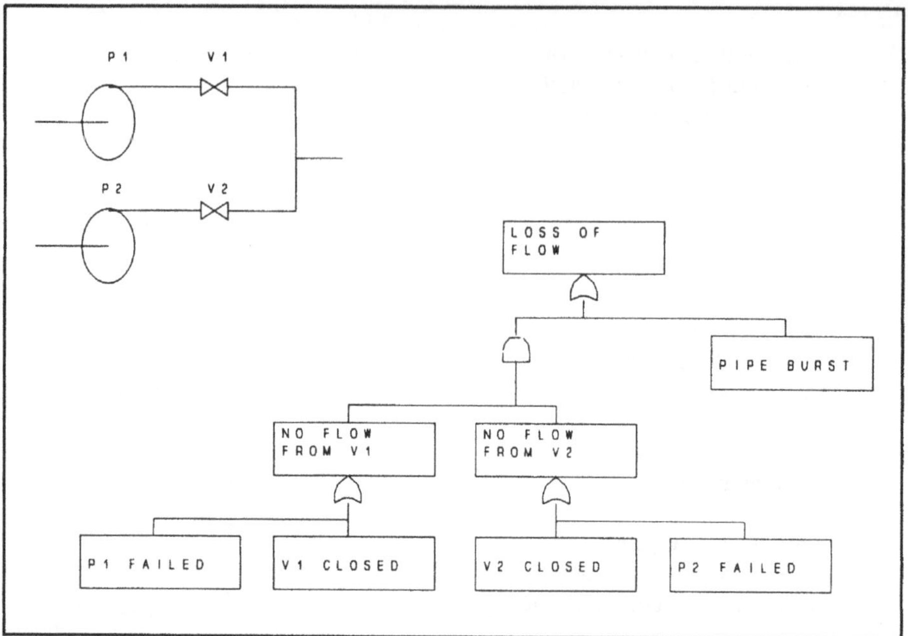

Fig 3.9 Example of OR+AND structure for a flow junction

Example of System Failure Tree Construction

Fig.3.9 shows a system with two pumps supplying water through a control valve 1 .

The event of interest is loss of water pressure at the outlet of V1. Tracing the causes backwards, one cause is that valve 1 fails closed. Another cause is that there is no water supply to the valve.

Ignoring the right hand side of the fault tree for a moment, there will be no flow to the control valve if flow to both sides of the tee junction ceases. This might be that valve 2 fails closed, or that flow to valve 2 fails. This in turn might be caused by failure of pump P2. The branch for components P1 and V1 is developed similarly.

One of the problems with components involving fluid flow is that they fail not only by blockage, but also by bursting or by being erroneously opened to atmosphere. We can lose pressure from this system, leading to total loss of flow, if valves V1 or V2 burst, since all the water may run out of the hole.

The right hand branch of the fault tree in Fig. 3.9 shows this starting with the OR gate. This arises from the problems of pump, valve or pipe bursting. The problem arises because any burst will allow water to drain from the system. The redundancy which the system provides applies only for the pump failure and blockage failure modes.

This kind of problem arises everywhere in piping systems, and a corresponding effect, shorts to earth, occur in electrical and electronic systems. The problems are in practice reduced by fitting check valves (or diodes for electronic systems), but this does not simplify the fault tree, rather the reverse, it means that more complex subtrees are required below the OR gate.

In general, it is best to treat the problems of bursting and loss of pressure separately from the problems of blockage. The logic of the fault tree for the two cases for tee junction is completely different. The same is true for electrical circuits subject to short circuit and open circuit.

Fault trees for large systems

The techniques described above for construction of fault trees are systematic. For large systems they are also quite tedious. The results for large systems can be enormous fault trees, which are very difficult to understand.

One way of simplifying the construction process is to treat whole groups of components as an integrated unit. For example, if there is a whole series of valves in a pipeline, a simple fault tree might include the subtree "Failure of Valve 1 OR Failure of Valve 2 OR.......". This results perhaps in five or six gates and events in the fault tree. Simply changing to a single event "Failure of any of the valves in pipe A" will then reduce the size of the tree six fold, and remove a lot of unnecessary detail.

This idea can be used systematically. What it means is that the whole fault tree is built up on the basis of subunits, rather than individual components. The subunits can then be analysed separately, in more detail. The analysis is thus carried out in several levels. It is generally possible to keep the size of individual fault trees down to a single A4 or A3 page.

<u>Example</u>

Fig.3.10 shows a fire water system for a petrochemicals plant. Fire water can be pumped from the firepond using either an electric pump or a diesel driven pump. Both pumps supply a ring main with hydrants placed around the petrochemical plant. (The fire water piping is buried, at a depth of about 1/2 m.)

The ring main is kept filled with water, and pressurized by two "jockey pumps". These pumps operate continuously, and there is a continuous bleed of water back to the fire pond, via two pressure control valves.

The electric pump is started on loss of pressure in the ring main. Loss of pressure can for example be caused by opening one of the hydrants. The diesel pump is also started in this way, but with a lower set point pressure for the switch, so that the diesel only starts if the electric pump fails.

Fig.3.11 shows a fault tree for the fire water system, with "no pressure at hydrant 8" as the top event. The first part of the fault tree shows that loss of pressure requires blockage at both sides of the ring main, loss of supply to the ring main, or a burst in the system without closing off the burst. The two halves of the ring main, the jockey pump system, the diesel pump and the electric pump are each regarded as a subunit. The problems of blockage and of bursting are treated separately.

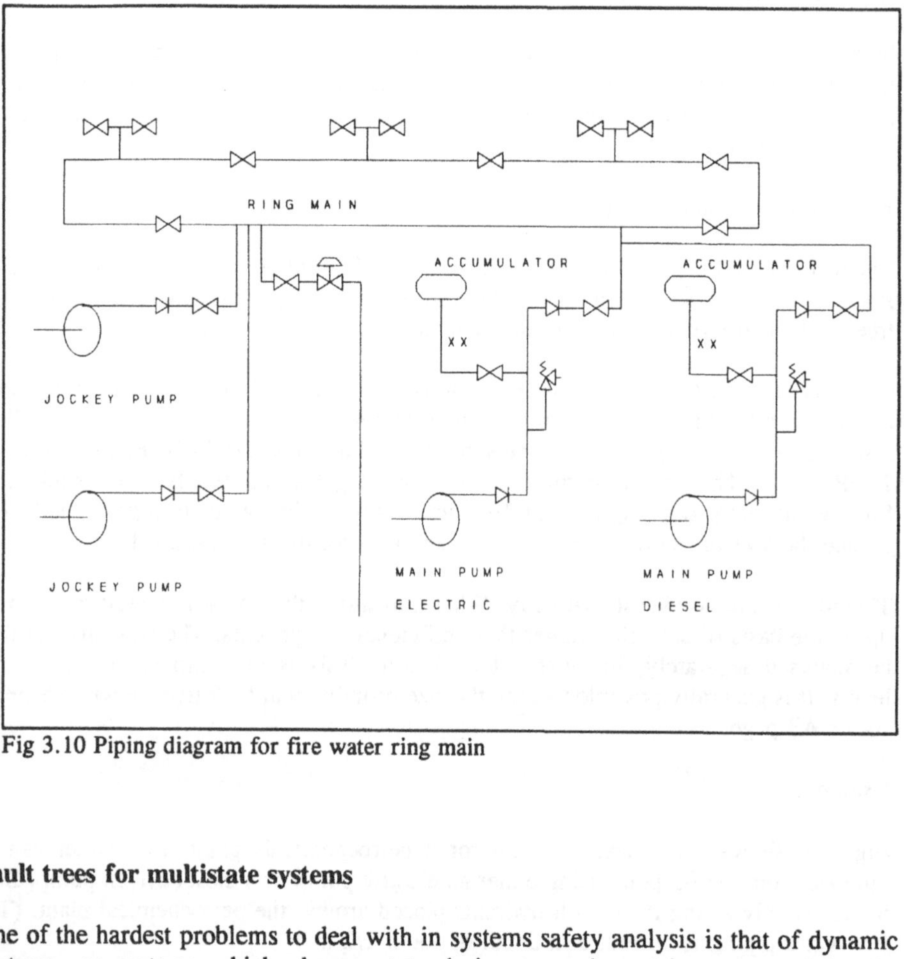

Fig 3.10 Piping diagram for fire water ring main

Fault trees for multistate systems

One of the hardest problems to deal with in systems safety analysis is that of dynamic systems, or systems which change state during operation. After many trials, the following approach has been fond to be reasonably effective:

1. Build up the fault tree using the system failure tree procedure, as described on page 85.

2. When failure effects are dependent on component include the effect or the failure into the tree, ignoring whether there may be incompatible states in other parts of the tree.

3. Record any component states assumed into the fault tree.

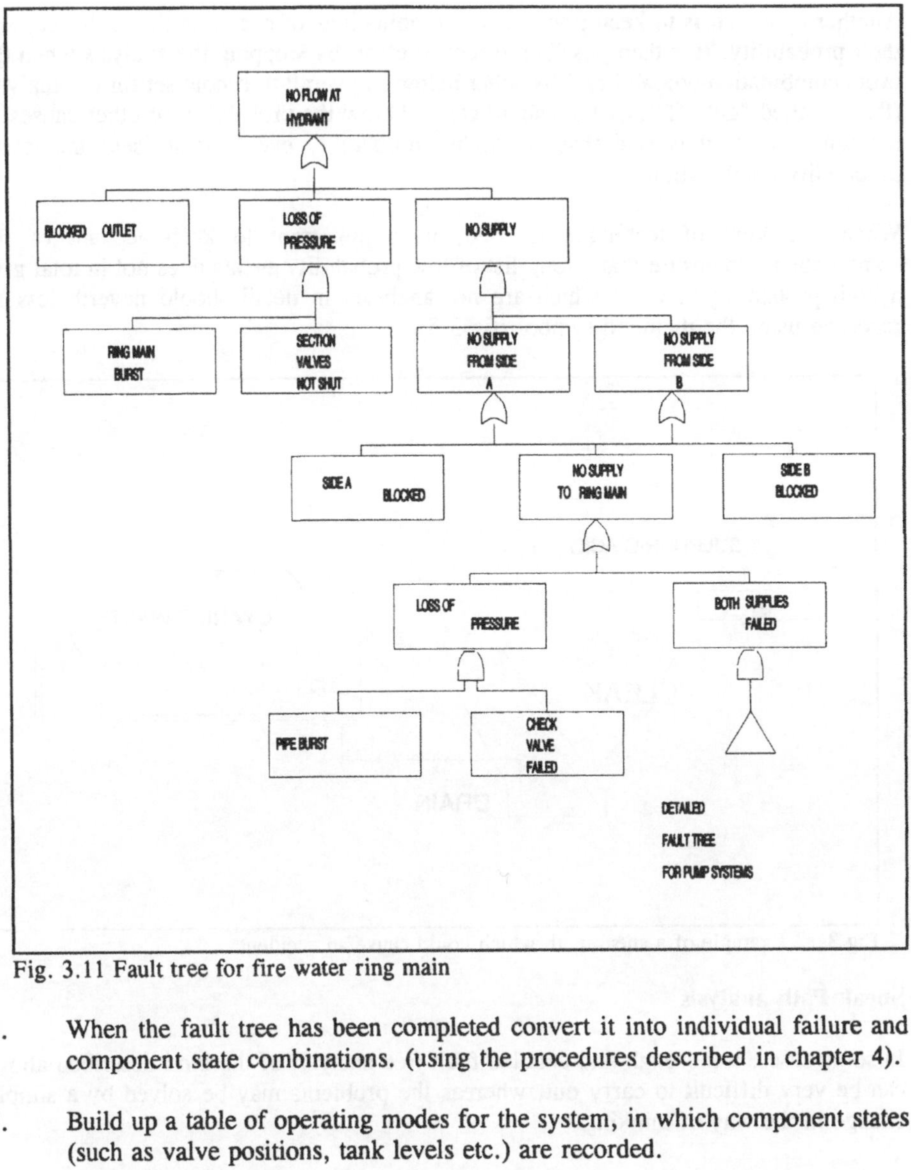

Fig. 3.11 Fault tree for fire water ring main

4. When the fault tree has been completed convert it into individual failure and component state combinations. (using the procedures described in chapter 4).

5. Build up a table of operating modes for the system, in which component states (such as valve positions, tank levels etc.) are recorded.

6. Cross check the failure/component state combinations, to see whether there is any corresponding system operation mode.

Cut Off

For large analyses of complex systems, with large degrees of redundancy, fault trees can become very complex. For nuclear plant and military systems, for example,fault trees often involve thousands of events. Although the structuring of fault trees described above is s great help here, it is rarely sufficient for such large analyses.

Another approach is to keep track of the combinations of events in the fault tree, and their probability. It is then possible to reduce effort by stopping the analysis when the event combination probability falls either below an overall threshold set for the analysis (this is called "cut off"), or falls significantly below the probability of other causes of the same event (it is said that the higher probability event "majorises" the lower probability combination.

When this kind of technique is used, it is important to keep account of the combinations, to ensure that a long list of low probability events does not in total give a high probability. Events which are not analysed in detail should nevertheless be recorded using the diamond symbol (fig 3.5)

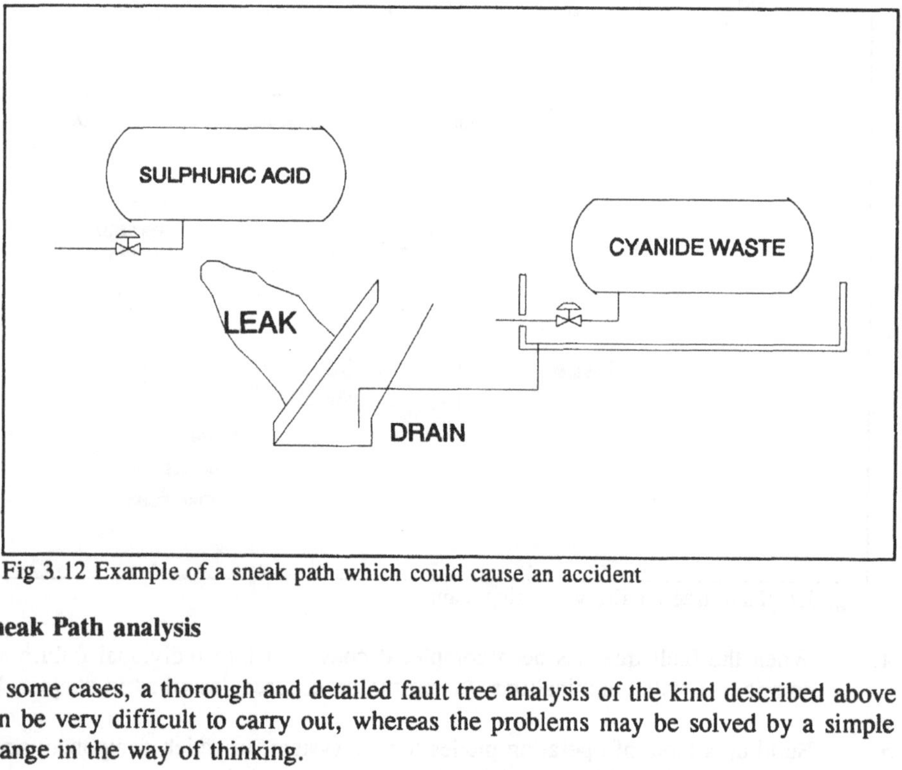

Fig 3.12 Example of a sneak path which could cause an accident

Sneak Path analysis

In some cases, a thorough and detailed fault tree analysis of the kind described above can be very difficult to carry out, whereas the problems may be solved by a simple change in the way of thinking.

An example of this occurred for the author in the analysis of a distillation plant. The plant was analysed in detail using the hazop method, and the analysis turned out to be very successful. However, a year later, while conducting a fault tree analysis using computer aids, a new fault type was found, involving flow of water from an environmental protection unit through many valves, tanks and heat exchangers, to a reactor filled with oil at 150°C. A failure would have caused an explosion, and the probability turned out to be high.

The failure had occurred in fact only a few days earlier, but had been saved from causing an accident by an alert operator.

The problem had been overlooked during the hazop analysis because of the complexity of the situation. The fault tree which demonstrated the problem was 34 layers deep, and involved over 1000 events.

The problem could have been foreseen quite easily, however, by noting that there was hot oil and water present in the plant. An immediate question is then "can these two substances come together". A quick analysis would easily reveal a flow route.

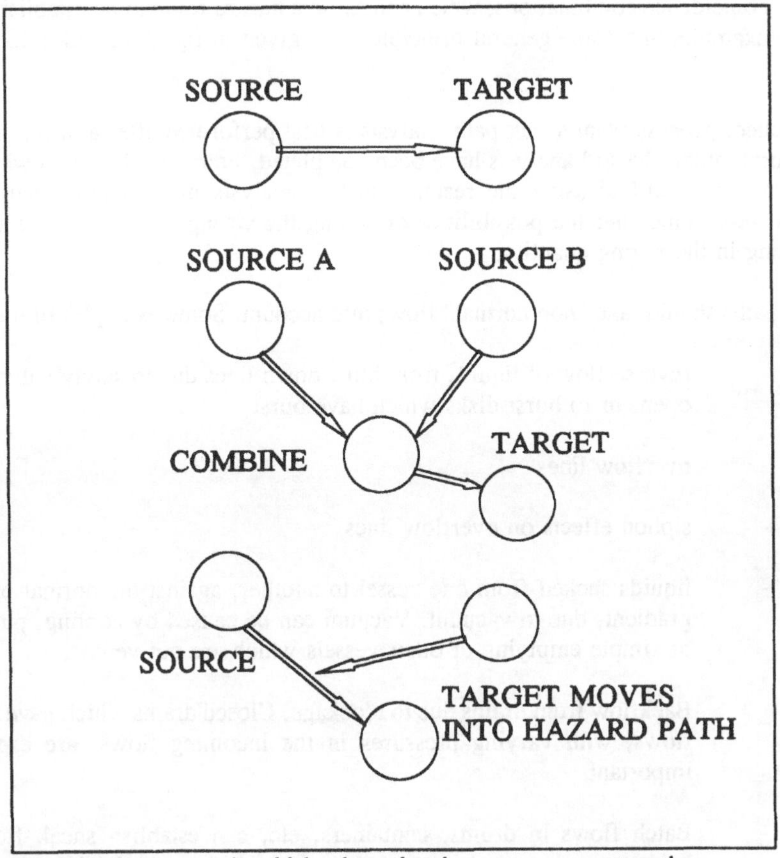

Fig. 3.13 Some ways in which a hazard and target can come together

Sneak path analysis is a technique based on this concept. The technique is described in detail in the chapter on human error. The principle involved is:

- Find whether the necessary elements of a certain type of accident are present.

- Find whether all of the necessary elements can come together, using fault tree techniques, for example.

- Find whether the path can be accidentally established, taking normal and reverse flow directions, and the patterns of fig. 3.13 into account.The technique is also known as route and barrier analysis. The idea here is to identify hazardous substances and energy, and to identify "targets" for these. Routes along which the hazard can flow are then traced, and the any safety barriers to flow are recorded. The analyst can then concentrate on finding failure and coincidence combinations, which can cause a hazard route to be established. An example, and some general principles, are given in fig. 3.12 and 3.13

In practice, process plant sneak path analysis is best performed after a hazop analysis and a preliminary hazard analysis have been completed. Pairs of substances which can react can be identified using the reaction matrix analysis process described below. Beyond this, remember the possibilities of adding the wrong substances to a mixture or adding in the wrong quantity.

Sneak paths should take "non normal" flows into account. Some examples of these are:

- reverse flow of liquids from blow down lines due to safety valves stuck open, or to burst disks which have burst

- overflow lines

- siphon effects on overflow lines

- liquids sucked from one vessel to another, against the normal pressure gradient, due to vacuum. Vacuum can be caused by cooling, pumping, or simple emptying of other vessels which are not vented.

- Backflow from drains due to blockage. Closed drains which have several flows, with varying pressures in the incoming flows, are especially important.

- Batch flows in drums, containers, etc, can establish sneak flows by being routed wrongly, or by retaining residues, or by being wrongly labelled.

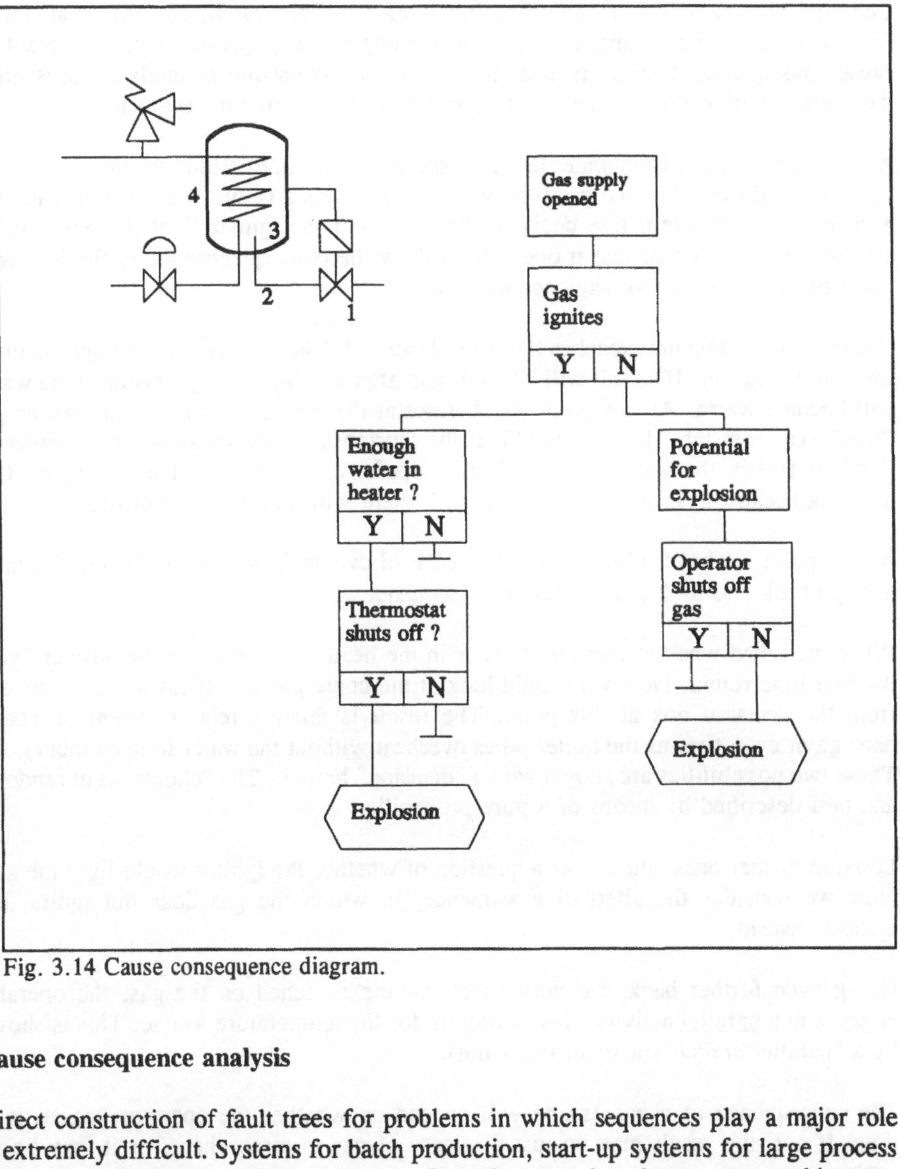

Fig. 3.14 Cause consequence diagram.

Cause consequence analysis

Direct construction of fault trees for problems in which sequences play a major role is extremely difficult. Systems for batch production, start-up systems for large process plants, and multi-stage safety systems such as those used on large steam turbines, or in nuclear power plants, provide many such problems. For these systems, there are many "phases" of operation, and several alternative sequences of event, depending on what kinds of failure have occurred previous to an initiating or trigger event.

In such cases it is preferable to work forward in time from some initial event, tracing the alternative sequences of events, parallel chains of events, and the different consequences of event sequences.

Consequence analysis is a technique involving forward search, starting at an initial "critical event", and tracing event sequence along causal paths in the process plant. A cause consequence diagram records the results of a consequence analysis, presenting the causal relationship between events, and indicating their time sequence.

As an example a gas-fired water heater is shown in Fig. 3.14. Here we choose as point 1, the critical event that the gas supply valve opens. As a result, the gas will reach the burner. The first alternative decision then is - will the igniter light the gas. Let us assume for the moment that it does, and follow the consequences along the left hand chain of events in the consequence diagram.

The chain of events now reaches the water heater, 4. There is a question now - is there water in the heater. If so, all will be well, and after a delay (triangle symbol) the water will become warm. At this point the thermostat (5) should switch off the gas supply ("yes" exit from the "decision box"). If the thermostat does not switch off, however, ("no" exit) then the water will overheat. Hopefully, the safety valve will open. If it does not because it is stuck or gagged, then there will be a boiler explosion.

At this point we have followed one sequence of events from start to finish. We must now go back and look at the alternative sequences.

When we asked whether there was water in the heater, we assumed the answer "yes" the first time round. Now we should look at the consequences of taking the "no" exit from the decision box at this point. The result is fairly direct. It might be heater damage or even fire, as the boiler pipes overheat without the water to keep them cool. These two possibilities are shown with a "decision" branch. The "choice" is at random, and best described by means of a pure probability.

Looking further back, there was a question of whether the igniter would light the gas. Now we consider the alternative sequence, in which the gas does not ignite, but escapes instead.

Going even further back, we notice that, having switched on the gas, the operator engages in a parallel activity, that is waiting for the temperature to rise. This is shown by a "parallel branch" (a small black dot).

The consequence of a gas escape will depend on whether the operator notices it in time. If not, the result may be just a release of gas to atmosphere, or it may be an explosion.

The example is now finished. The analysis could be made much more complex by taking further details into account, but few of these would affect the risk analysis significantly.

The example has shown us several things. First, it has shown us how we can start from an initial event and trace event chains, following the pipes, wires, control connections, and in general, all the causal paths through the plant. Secondly, it has shown us how we can deal with alternative and parallel branches in event chains. Thirdly, it has shown how we can fit small fault trees onto the diagram, to show the cause of abnormal conditions. The thing it has not done, is to show how to choose the starting point for the analysis.

Cause consequence analysis was developed be D. S. Nielsen (1971, 1975) based on the concept of a "critical event". A critical event is defined as

- an event which can in certain circumstances produce serious consequences and, moreover
- is a convenient starting point for analysis.

Critical events should preferably be chosen so that there is a complete coverage of plant hazards. At the same time, it is necessary to keep the amount of overlapping between the analysis of the events to a minimum. A further objective is to choose them in such a way that most of the sequential effects in a failure incident come after the critical events since it seems to be easier to treat sequential effects in time sequence, that is, in a forward going consequence analysis.

There are two main strategies in consequence analysis. One is to trace the sequence of events in normal (designed) operation of a plant, or of a safety system, and then to trace departures from this sequence. The departures are evaluated, by postulating that the normal events do not occur, because of failure or error, and considering what else occurs instead. This event which occurs instead of normal or designed operation are described and its consequences traced. This may be called "normal sequence strategy", (strategy B), and is very useful for batch or start up analyses.

The second strategy is simply to trace the spread of events through the process plant, starting at an initial event, and following the resulting changes and events along pipes, power supply cables, signal wires, ventilation ducts, and physical connections such as frameworks and supports. This second strategy may be called the "causal path strategy", strategy A). It is most useful for analysis of disturbances for continuous operation.

Often the normal sequence strategy will give a set of possible failure events along the designed sequence. The causal path strategy will be required to trace the consequence of these events further.

The way of choosing critical events are designed to make the best use of these strategies. One direct way of choosing critical events is the hazards analysis procedure - inspection each vessel or pipe in the process plant in turn, considering the effects of major variations in variables such as pressure, temperature or flow. Additionally, the effects or rupture should be investigated for each vessel or pipe. The causal path strategy is appropriate in tracing the consequences of this kind of events.

This method of choosing critical events is appropriate for safety analyses. For production reliability analyses or availability analyses, vessels and pipes may be considered for just the main production flows, rather than in the entire plant. Stoppage of motors and machines in these flows is an important kind of critical event.

In some cases an event such as closure of a valve can affect several variables within a vessel simultaneously, for example affecting both temperature and pressure by shutting off the inflow of a hot fluid. In such cases the further course of events may depend on the combination of variable changes. There may be several instruments measuring temperature, pressure, etc. and the changed measurements made by these instruments may be combined in a control or safety system, determining the further course of events. In such cases the event occurring just before a change in process variables will be good choice as a critical event.

An extreme method of choosing critical events is to start with the plant in its normal working state, and to regard each of the failure modes of the active components as a critical event. In this case, consequence analysis can be regarded as an extension of FMEA. (An extension, because combinations with several latent failures as well as the initial failure may be taken into account). This kind of analysis is most appropriate when each of the failure modes identified leads to a new kind of event sequence.

Both the "hazard analysis" and the "FMEA" methods of choosing critical events, if they are used, should be applied for all the plant states which extend over a period of time, such as shut-down state, normal production, reduced production, and unusual operation states which occur when part of the plant is broken down.

Events which lead to activation of safety systems involving switching and start up of safety equipment provide another straight-forward kind of critical event. For reliability analyses of batch processes, and plant start up and shut down, the initial event in the procedures serve as a convenient kind of critical event. For these critical events, the normal sequence strategy is most appropriate.

Cause consequence analysis may be used systematically throughout an plant to describe all disturbances. This is sometimes done for nuclear plant, and especially for operating procedures analysis (see Ch 6.) More often, though, cause consequence analysis is used to supplement hazop and fmea analyses, when consequence descriptions become too complex; and to supplement fault tree analyses when sequential effects are involved.

Cause consequence analysis is often confused with event tree analysis described below. Cause consequence analysis is a much more powerful technique than event tree analysis. Both parallel and alternative branching is supported, and loops are allowed. As a result it is possible to build up the cause consequence diagram at the same time as event chains are traced in the plant diagram, something which is impossible if only alternative branching is allowed in the diagrams.

PROCEDURE FOR CAUSE CONSEQUENCE DIAGRAM CONSTRUCTION

A. For plant disturbances

1. Start with a flow sheet, piping and instrumentation diagram, or wiring diagram for the plant.

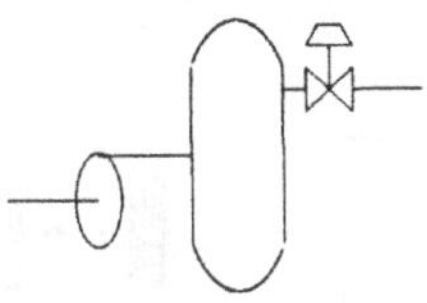

2. Select a critical event (see list).

3. Add changes caused in physically connected components to the diagram.

4. Add further events, tracing along physical connections in the plant.

5. Where there are two output events from a component, let the event chains branch.

6. Where a component may be in a failed or unusual state, use a decision box to express alternative event outcomes.

   You may get a long chain of decision boxes.

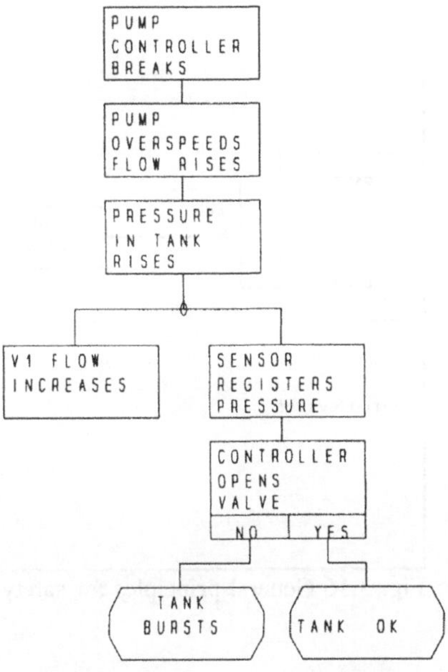

7. Put final states of the plant as consequences.

8. Describe the conditions causing "no" exits to be taken in a decision box by means of a small fault tree.

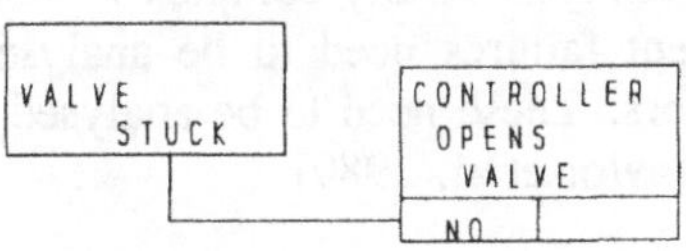

9. Use fault tree methods to find the causes of the critical event.

10. Repeat the procedure for a full range of critical events.

Figure 3.15 Cause consequence analysis

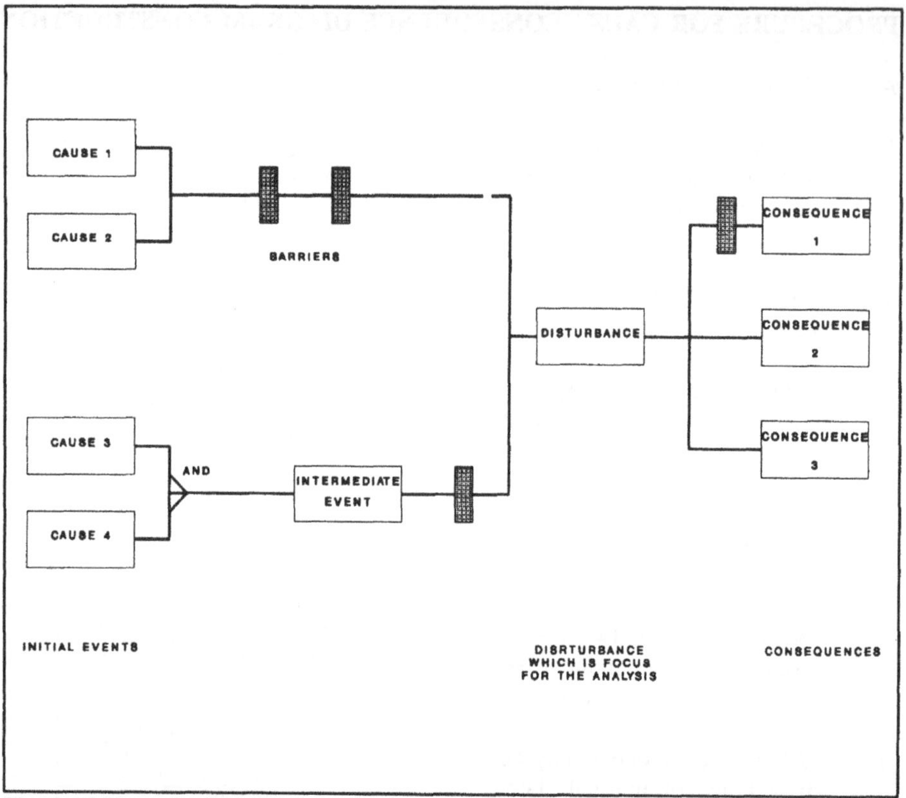

Fig. 3.16 General principles for safety barrier diagram

Safety barrier diagrams

A full safety analysis of any complex system requires a large amount of effort. Not only equipment failures need to be analysed, but also operator, maintenance, and computer errors. These need to be analysed for all the different operational states of the system (Taylor et al, 1989).

It is necessary, in performing this kind of analysis, to limit the amount of work involved. This is necessary not only because of the cost and time scale, but also to keep the results as comprehensible as possible.

The top down form of analysis provides one way of limiting the amount of work needed. The work can be further reduced, and the results made more clear, by focussing attention on the accident, as a starting point for the analysis, and on the safety barriers which surrounding the potentially hazardous object.

A safety barrier is defined as:

- any wall, shield, switch, bolt, interlock, software or operational check which is intended to prevent a signal or activation from reaching a place where it can cause an accident.
- a mechanical barrier which can prevent external influences from causing an accident.
- a mechanical barrier which can prevent a release of energy or poison from having adverse consequences.
- distance from the source of hazard.

A "safety configuration" is defined as a combination of safety barriers.

A rational approach to safety analysis is then to identify concentrations of energy. For example in a chemical reactor the energy sources might be the compressor, the reactor itself, and a fired heater. The various safety barriers surrounding each hazard source are listed. The intended combinations of safety barriers for each operational state are listed.

Once the intended safety configuration for each operational state has been identified, the analysis proceeds, starting at the barriers closest to the hazard, and working backward towards the causes of activation of the hazard. At each barrier, the reliability of the barrier and the possibilities for bypassing the barrier are investigated. This process proceeds from barrier to barrier, until the combined reliability of the barriers analysed is sufficient to meet the acceptability requirements.

The "safety barrier" diagram has been found to give a good visual presentation of this process. Fig 3.16 shows the general principle of the diagrams, and fig. 3.17 gives an example In this kind of diagram the sequences of events from initial trigger to accident consequences is shown on a diagram. The barriers which can prevent the accident are then shown as solid bars, lying across the chain of events.

In analysing systems in terms of safety barriers, it is not sufficient just to ensure that each barrier is present. It is also necessary to check that:

1. The functional capability of the barrier is sufficient to deal with the full range of accidents which it is intended to protect against.

2. Any signal or operator action which is required to activate a barrier can actually be detected, and is detectable at all times when the barrier should be in operation.

3. The barrier meets accepted construction and quality standards.

4. The barrier is fail safe, self testing, or is tested periodically.

5. The barrier, if it is activated, will not cause other accidents.

6. The barrier will not be destroyed or put out of action by the accident which it is intended to prevent, or by a trigger event which can cause the accident.

7. The barrier will not be prevented form working by an external catastrophic event such as flooding or fire.

8. The barrier will not fail as the result of activation of another barrier designed for the same accident type.

9. Two barriers cannot be put out of action by the same event.

10. The demands on the barrier are not changed beyond the capacity of the barrier as a result of changes or improvements in the primary system.

If any of these conditions are not met, it will be necessary to improve the barriers or add further barriers.

Safety barrier diagrams can be made using the backward searching approach described above. The process then is like a simplified form of fault tree analysis. The approach is useful because it allows cut off techniques to be used to limit the amount of effort.

It is also possible to use the diagrams to illustrate the results from hazop analysis. This kind of use makes it possible to relate hazop studies to fault tree and cause consequence analyses, and to combine them onto one diagram. The diagrams can be used as a basis for quantification, making it possible to introduce a quantitative element into hazop analysis. This is difficult when hazop remains in a tabular form, even when special notation such as numbering is introduced to express parallel event sequences in the tables, since the tables do not represent precisely the way in which parallel and alternative branches occur.

Various techniques have been developed for producing safety barrier diagrams automatically, on the basis of computer aided hazard and operability analyses. One such system is described in chapter 16. The hazop tables are filled out normally, with the exception that the consequences and causes of disturbances are structured explicitly as tree structured lists (i.e. with text in heading, subheading, sub sub heading form etc.). Programs then convert the tables to diagrams, and vice versa.

Safety barrier diagrams have begun to be used rather than fault trees or cause consequence diagrams because they have been found to be easier to understand. This is especially important when the diagrams are to be presented to non specialists. Fault trees, in particular, have been found to be confusing to persons who have not trained in their use. The limitation of safety barrier diagrams is that failure causes for safety measures and barriers cannot be expressed easily, if they involve complex patterns of redundancy. In such cases it may be necessary to combine fault tree and safety barrier diagramming techniques.

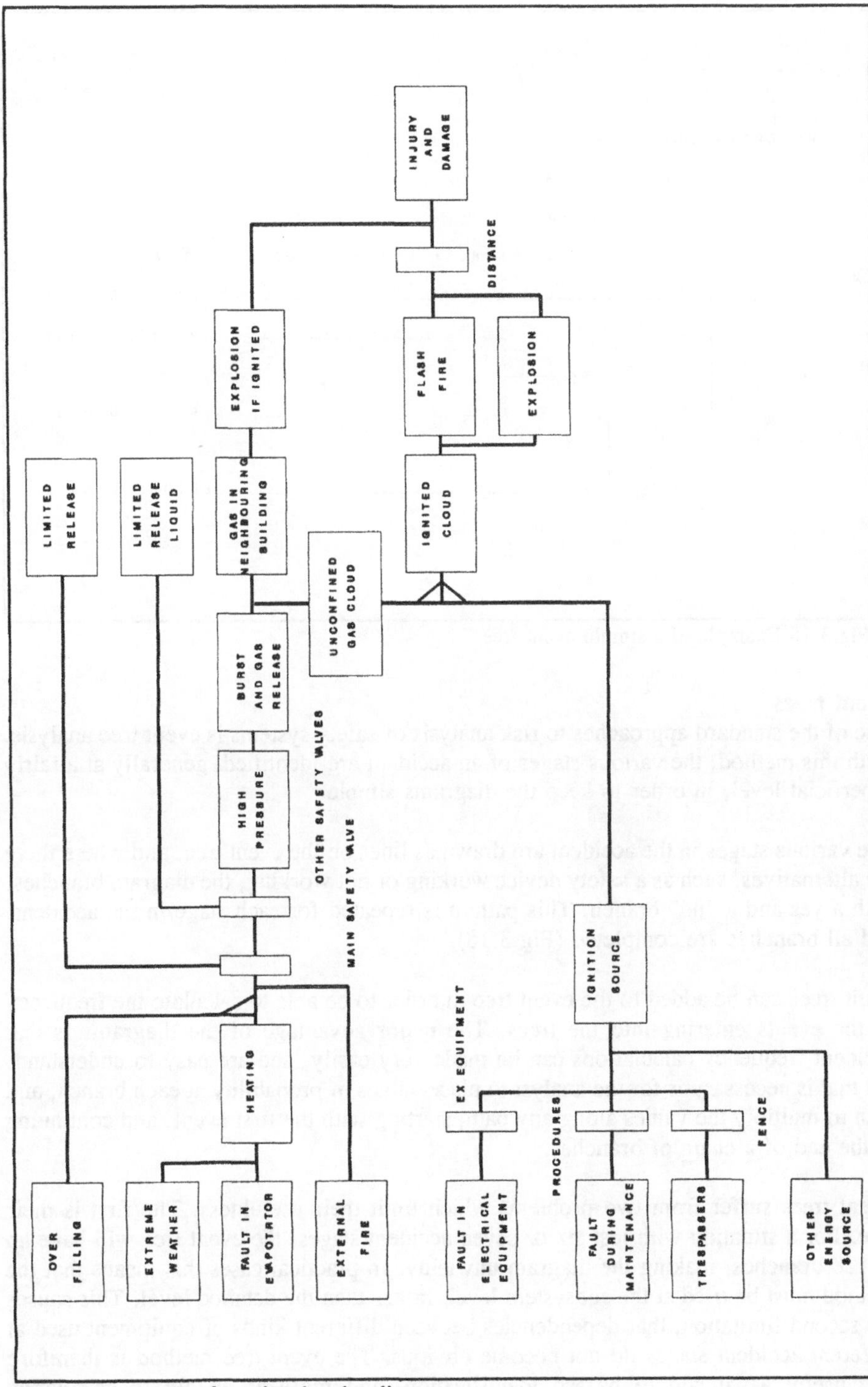

Fig 3.17 Example of a safety barrier diagram

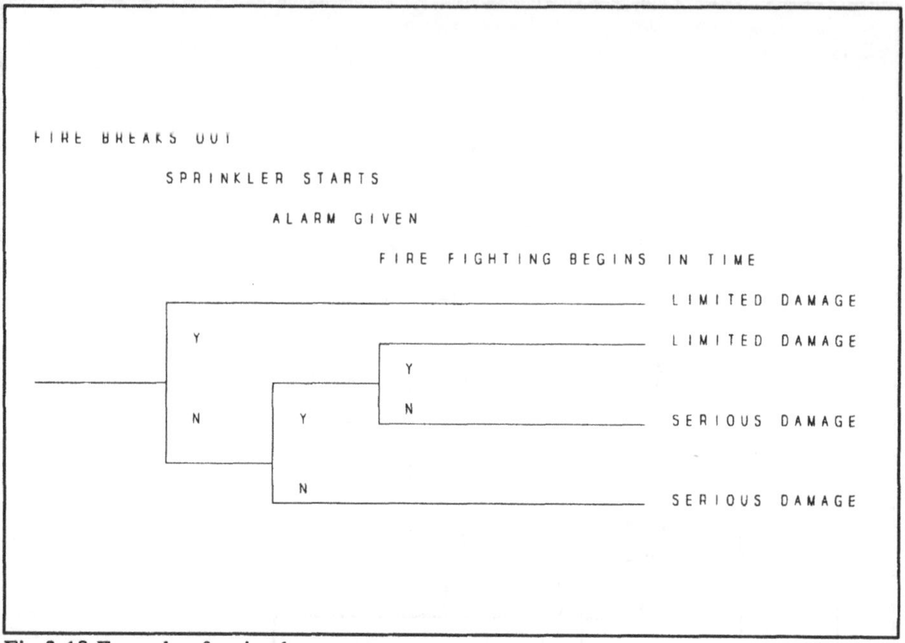

Fig 3.18 Example of a simple event tree

Event trees

One of the standard approaches to risk analysis of safety systems is event tree analysis. With this method, the various stages of an accident are identified, generally at a fairly superficial level, in order to keep the diagrams simple.

The various stages in the accident are drawn as lines on the event tree, and where there are alternatives, such as a safety device working or not working, the diagram branches, with a yes and a "no" branch. This pattern is repeated for each stage in the accident, and all branches are completed (Fig 3.18).

Fault trees can be added to the event tree in order to be able to calculate the frequency of the events entering into the trees. The major advantage of the diagrams is that accident frequency calculations can be made very easily, and are easy to understand. All that is necessary is for the analyst to place values of probability at each branch, and then to multiply the values along any path, starting with the first event, and continuing to the end of a chain of branches.

Event trees suffer from two problems which limit their usefulness. The first is their size. For a situation with just six or seven accident stages, the event tree will have up to 128 branches, making the diagram unwieldy. In practical cases this means that the method must be used at the subsystem level, rather than the detailed level. This causes the second limitation, that dependencies between different kinds of equipment used at different accident stages do not become obvious. The event tree method is therefore less useful as an aid to hazard identification, and more as an aid to calculating probabilities and recording results.

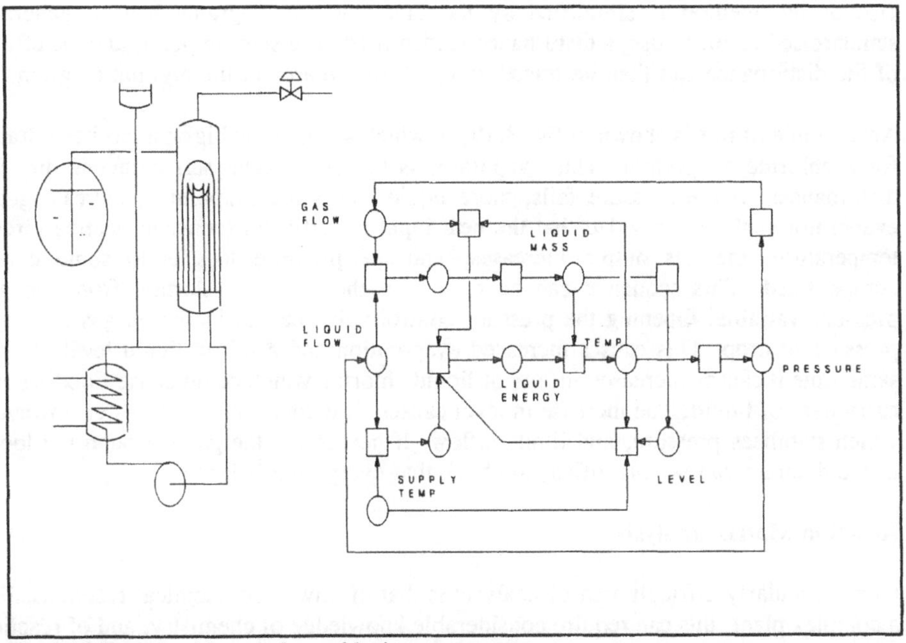

Fig 3.19 Equation bigraph for a chlorine evaporator.

Equation bigraphs

One of the hardest problems in analysis of process plant is to treat systems in which there is a lot of inherent feedback, recycle loops, or complex control actions. It can be a problem to know whether a disturbance will increase or decrease pressure or temperature in evaporators, distillation columns etc. when subject to control loops.

A method to reduce the problems is to use the equation bigraph technique. A diagram is drawn for the plant with a node (square) for each equation, and a circle for each variable describing the plant. The equations will generally be mass and energy conservation equations and constitutional equations for flow rate, heat flow rate, etc. A line is drawn between each equation node and each variable node if the variable enters into the equation.

Arrows expressing causality can then be added to the diagram according to the following rules:

1. When there is a time delay in the equations, cause comes before effect.
2. Each equation determines just one variable.
3. Each variable is determined by just one equation.
4. For integration equations, a change in flow precedes a change in accumulation.

Use of the method is simplified by the fact that the diagrams have a generally standardised form. In use, a disturbance is identified, as a starting point, and the effects of the disturbance can then be traced along all of the arcs in the bigraph diagram.

An example of this is shown in fig. 3.19, in which an equation bigraph has been drawn for a chlorine evaporator. This evaporator is to some extent self compensating for disturbances. If the pressure falls, more liquid enters the evaporator, increasing the evaporator surface. Provided that the heat input is sufficient (constant heating circuit temperature, the gas output increases, and the pressure loss is to some extent compensated. This sequence can be traced on the diagram, starting from the gas pressure variable. Opening the pressure control valve causes increased gas flow and pressure to drop. This causes increased evaporation and a fall in liquid level. At the same time it causes increased inflow of liquid chlorine which could cause flooding and carry over of liquid. The increase in level causes an increase in heat transfer however, which stabilises pressure, and limits inflow. If necessary, the gains around the loops in the diagram can be quantified, to check that everything balances.

Reaction Matrix analysis

One particularly difficult area of analysis is that of unwanted chemical reactions. For a complex plant, this can require considerable knowledge of chemistry, and of reaction kinetics. A technique which can simplify the work involved is reaction matrix analysis. The basic idea here is to form a matrix with rows and columns labelled with the substances present in the plant. The elements of the matrix are then filled out with the reaction potential for each substance pair (a weakness is that three substance reactions are not treated).

	acid	alkali	aldehyde	phenol	water
acid	-				
alkali	Possible violent reaction	-			
aldehyde	Possible explosive condensa-tion	Polymerisa tion	Polymerisa tion		
phenol	-	phenolate formed	Polymerisa tion	-	
water	Possible violent reaction	-	-	-	-

Table 3.9 Example of a reaction matrix.

In order to make a good analysis it is necessary to remember substances which may be added in error (stored in the same area, for example), and substances which may be mixed if failures and leakages occur. It is also necessary to remember that water and air are almost always present, or can become present due to failures. The diagonal elements of the matrix represent condensation or decomposition reactions.

Choosing hazard identification methods

To some extent, the freedom of choice for hazard identification methods is limited by the fact that there are appropriate methods to be used at each stage in project development. As a minimum, any project will require a PHA and hazop analysis. Fault tree and cause consequence analyses, and sneak path analyses (Ch 6), may be added as needed, depending on the level of hazard. A much more significant question is the depth to which the analysis should be taken. The deciding factor is the degree of completeness required. In depth hazop analyses, supplemented by fault tree analyses for control and safety shutdown, and power, air and cooling supply systems will be made when the hazard level justifies it.

The degree of completeness has been investigated by Soukas(1985) and by the author (Taylor 1983, 1987) for a number of analysis sequences, and variants of the methods. The investigations were carried out by comparing different analyses of the same systems, and following up the analyses over a number of years of plant operation. The results show that a typical degree of completeness for industry risk analyses is about 85%, increasing to 97% for the very best analyses. The values here are taken as a percentage of the total number of hazards found by predictive analysis of the plant hardware when results are pooled from the different analyses and operational experience. The last 3% are typically found during operations.

These results concern analysis based on hardware drawings and descriptions. Taking the hardware and process hazards found as a base of 100%, analysis of operator errors has typically added an extra 60%, and plant inspection an extra 12%, of the total number of hazards, when batch plant were concerned. The corresponding figure for continuously operating plants was an extra 30%.

These results appear disappointing but not in themselves critically so, since the omissions may lie within the uncertainties inherent in any risk assessment. The oversights, however, have a tendency to be more probable causes of accidents, and in several cases have been found to dominate the overall risk, when follow up studies have been made. For this reason, it seems appropriate that as much effort as possible is concentrated on the hazard identification aspects of risk analysis.

The best overall procedure which has so far been found involves:

- PHA
- Hazop analyses for all vessels and lines, using check lists
- Fault tree analyses for control and services failures
- Sneak analysis, for selected hazards
- Design review for pressures, temperatures, pump flows, overflow and blowdown capacity, and reaction rates.

- Safety barrier diagramming for accident scenarios
- Action error analyses for operator procedures (Ch 6)
- Inspection and safety review of the working plant (Ch 17)

References

CIA	A Guide to Hazard and Operability Studies. Chemical Industries Association, Alembic House, 93 Albert Embankment London SE1 7TU
Lawley, H.G.	Operability Studies and Hazard Analysis, Chem. Eng. Progress, April 1974
Rogers	Safety Engineering Handbook
Nielsen, D.S	Use of Cause Consequence Diagrams as a Basis for Quantitative Accident Analysis, Risø-M-1374, 1970, Risø National Laboratory, Denmark
Barlow, R.E. Lambert, H.E.	Introduction to Fault Tree Analysis, in *Reliability and Fault tree Analysis,* ed Barlow, Fussel, Singapurwalla, SIAM 1975
Haasl, D.F.	Advanced Concepts in Fault Tree Analysis, *in Proc. System Safety Symp.*, Seattle, 1965
Taylor et al.	Qualitative and Quantitative Criteria for Risk Acceptance, Danish Environmental Agency 1989
Taylor, J.R.	Quality and completeness of risk analyses Vol 1 manual methods, ITSA 1987 Vol 2 automated analysis, ITSA 1991
Taylor J.R.	Burteilung von Kosten, Vollstændigkeit und Nutzen von Risikoanalysenverfahre, in Ermittlung und Bewertung Industrieller Risiken, ed. S Lange Springer Verlag 1984
Suokas, J	On the reliability and validity of safety analysis, Technical Research Centre of Finland, 1985
Taylor, J.R. and Olsen J.V.	A comparison of Automatic Fault Tree Analysis wit Manual Methods of Hazards Analysis. Int Symp. Loss Prevention and Safety Promotion in the Process Industries, Harrogate 1983

4 Failure and Accident Probabilities

The probability of failure of a component such as a pump can be described by its "failure rate" or more correctly, hazard rate. This is the probability per unit of time that it will fail within the next short time interval t - t+dt, given that the component is running at time t.

The "failure probability density function" for the component is the probability that the pump will fail within a short time interval, divided by the time interval. The difference between the failure rate and the failure probability density function is that for the density function we do not assume that the pump is running at the start of the short time interval.

The "cumulative failure distribution function" for the component is the probability that the pump has failed before time t, given that it was working at time t = 0.The "reliability" of the component is defined as the probability that the component will operate according to specification for a defined period.

These quantities are related. The cumulative failure distribution function is found by integrating the failure probability distribution function.

At the beginning of its working life, a component generally has a high failure rate because of the possibility of installation errors, design errors, or poor fabrication. There follows a period of normal operation, in which the failure rate is more or less constant. Finally, there is a period in which the failure rate rises due to wear out of the pump. These are expressed in the form of a "bath tub curve" (fig 4.1). (For some component types, particularly electronic systems, the initial "infant mortality" period overlaps the "wear out" period).

It is unusual to take into account the possibility of wear out or "infant mortality" in everyday risk analysis work. The reasons for this are that:

- components fail and are repaired, so that at any one time, there may be components of many ages in use

- it is generally assumed that components will be overhauled before wear out becomes significant,

- it is very difficult to obtain data for infant mortality and wear out.

For these reasons, a constant failure rate is generally assumed, which is an average value covering all phases of component life. This leads to the so-called exponential distribution for failure rates (Fig 4.2).

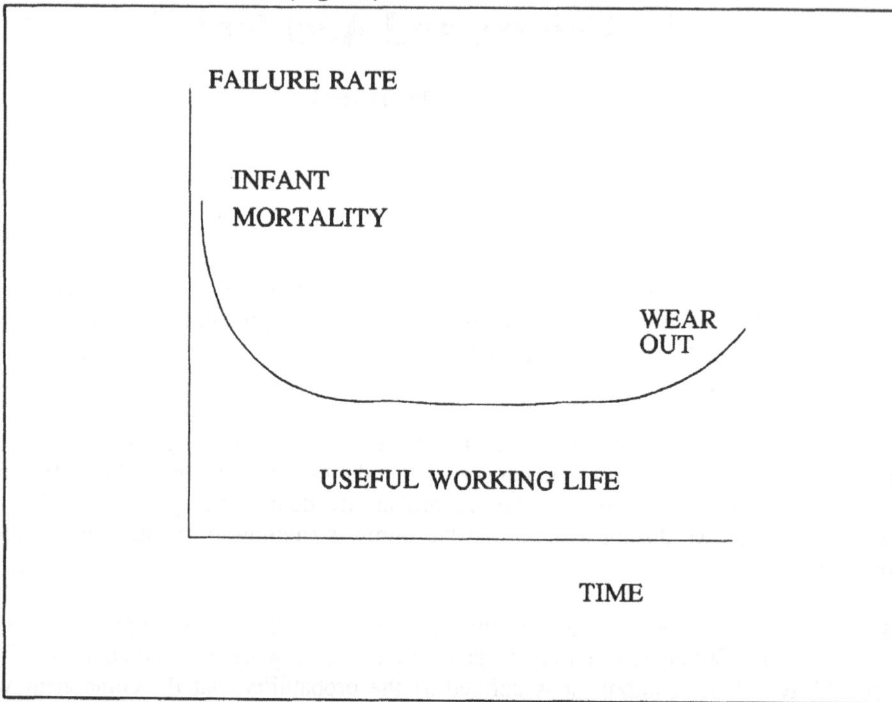

Fig. 4.1 Typical pattern of development for a component failure rate

---

Reliability (R)

Probability that a system performs correctly under specified conditions for a specified period of time (no repair)

Point availability (A(t))

Probability that a system functions correctly at a specified point in time (if no repair A(t) = R(t))

Interval availability ($A(T_1, T_2)$)

The expected fraction of a given time interval that a system is able to perform correctly.

Unavailability = 1 - availability.

Table 4.1 Reliability and Availability Definitions

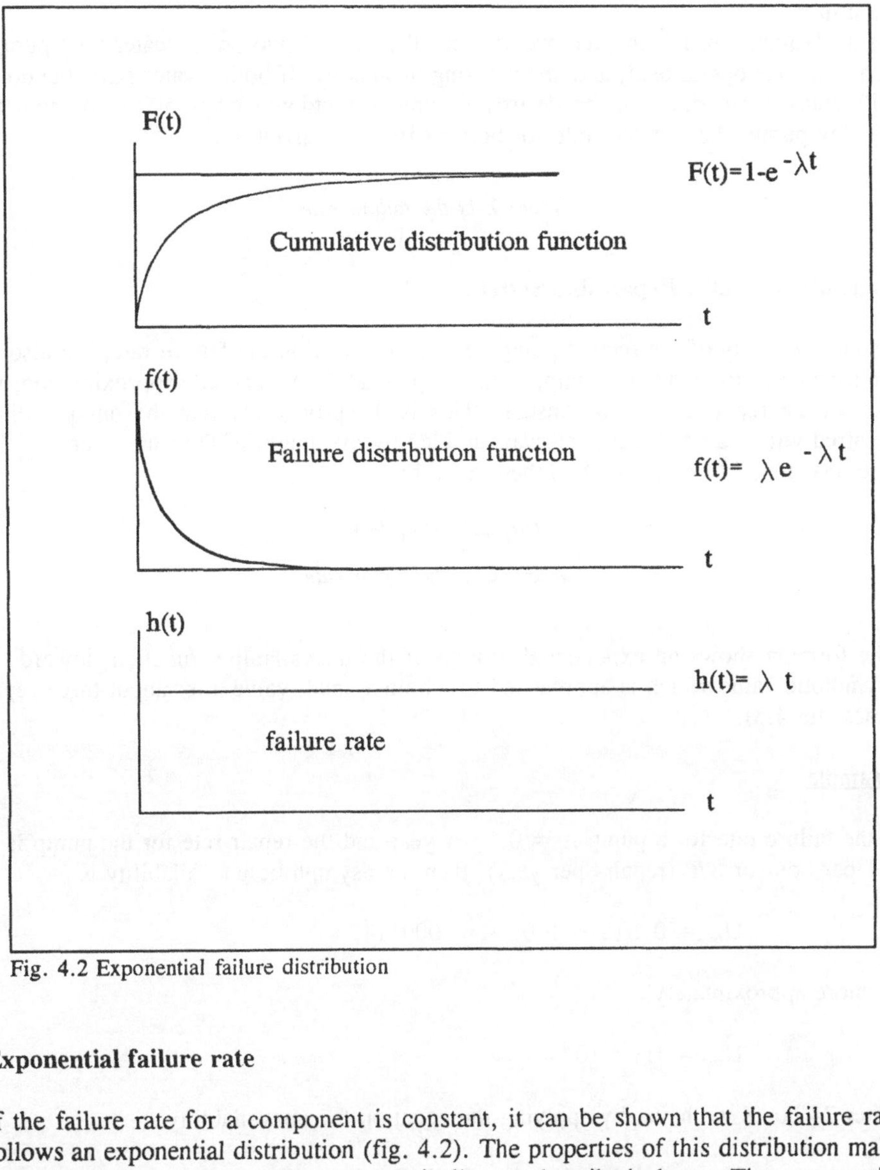

Fig. 4.2 Exponential failure distribution

Exponential failure rate

If the failure rate for a component is constant, it can be shown that the failure rate follows an exponential distribution (fig. 4.2). The properties of this distribution make it particularly easy to handle mathematically, as described above. The exponential failure distribution is often used also because it is simple, requiring only one parameter. If very little data is available, it may be impossible to derive better or more appropriate distributions. In a well run plant, many components will be maintained and repaired, more or less continually. Most components then are in the "useful working life " phase, so that the exponential distribution is also reasonable from an engineering point of view, when no attempt is being made to operate components "to the limit of their useful lives".

Example

As an example in this chapter, we will use the case of two boiler water feed pumps, A and B, one operational, and one working in standby. If boiler water fails, the boiler will quickly run dry, and be destroyed, unless shutdown takes place. Without the standby pump, the accident rate for boiler damage is given by:

$$F = \lambda$$

Where λ is the failure rate

Unavailability of a Repairable System

The unreliability of the second pump depends not just on the failure rate, but also on the time taken to repair the pump, or the "repair rate". As a simple approximation, we can set the repair rate as a constant. This is the probability that the pump will be repaired within a small time interval, divided by the length of the time interval.
The unavailability of pump A is then given by

$$U(t) = \frac{\lambda}{\lambda + \mu}(1 - e^{-(\lambda+\mu)t})$$

Where μ is the repair rate

The formula shows an exponential growth of the unavailability function, towards an asymptotic value which is approached to within a small value after about three repair times. fig 4.3).

Example

If the failure rate for a pump is = 0.1 per year and the repair rate for the pump is = 0.1 per hour or 876 (repairs per year), then the asymptotic unavailability is

$$U_\infty = 0.1/(876 + 0.1) = .0001142$$

or more approximately

$$U_\infty = 1.1 * 10^{-4}$$

After 24 hours we have, by applying the equation for unavailability

$$U_{24} = 1.1 * 10^{-4} * (1 - .090718)$$
$$= 1.1 * 10^{-4} * 0.909$$

This means that the limiting (asymptotic) unavailability is reached within a few days.

The unavailability function can be expressed in different terms since, in the case of exponential distributions.

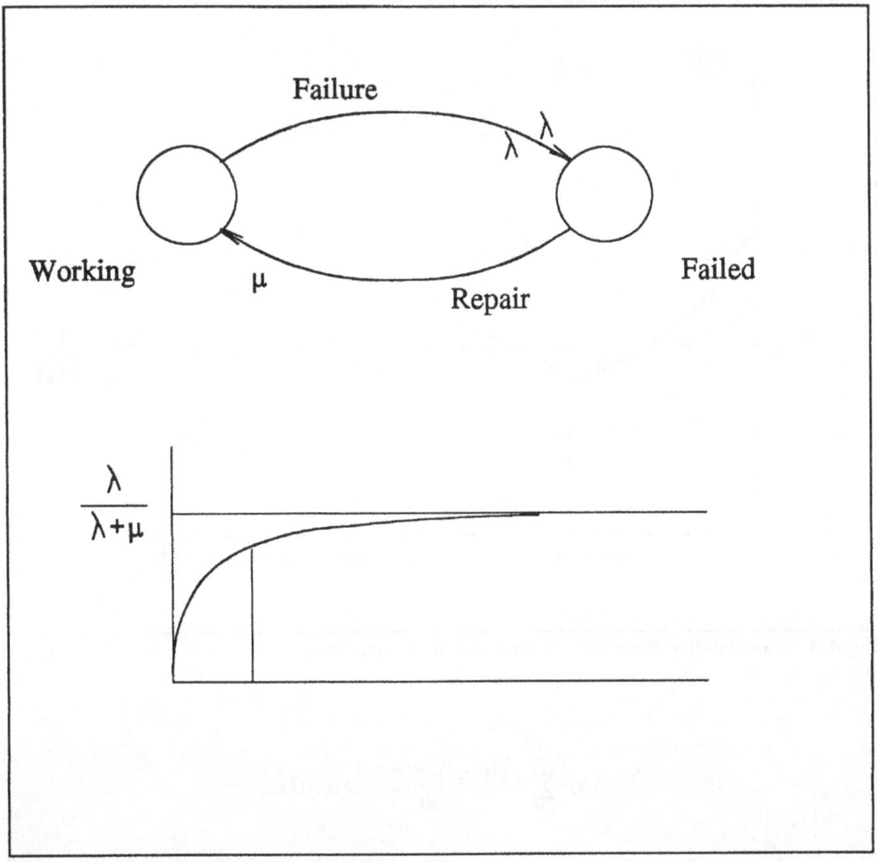

Figure 4.3 Unavailability of a repairable system

$$\textit{mean time to repair} = 1/\mu$$

$$\textit{Mean time between failures} = 1/\lambda$$

$$U_{\infty} = \frac{\lambda}{\lambda + \mu}$$

= MTTR/MTBF + MTTR

A more typical situation is that repair is characterized by a constant failure time rather than constant repair rate. In this case, the derivation of the formulae is much more difficult, but the result is given as follows:

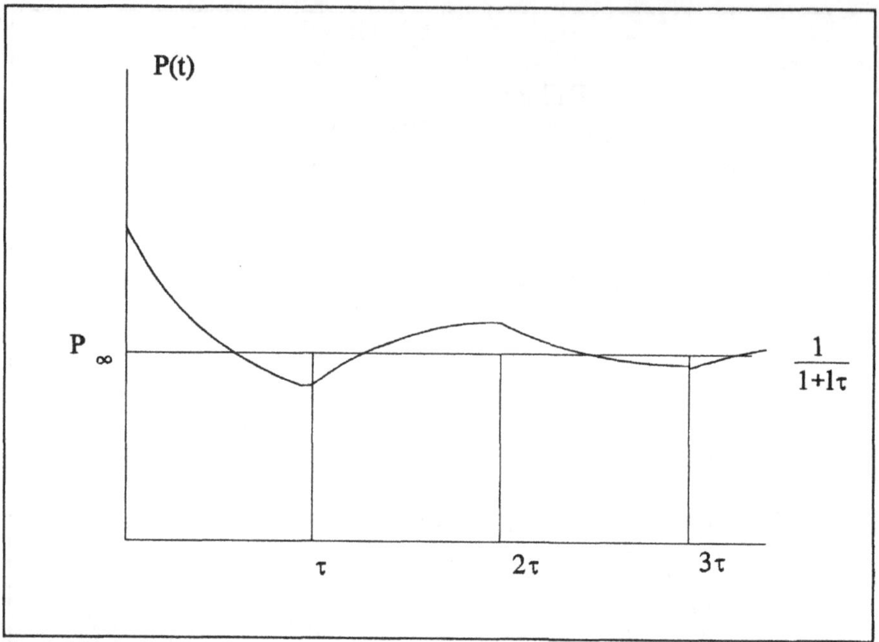

Fig 4.4 Unavailability for a system with constant repair time

$$A(t)=\sum_{n=0}^{\infty}\lambda^n\frac{(t-nT)e^{(1-(t-nT))}}{n!}X(t-nT)$$

Where

X (t-nT) = 1 if t > nT

0 if t < nT

T is here the repair time.

The formula for asymptotic unavailability is then

$$U=\frac{1}{1+\lambda\tau}$$

<u>Example</u>

If as before failure rate = 0.1 per year, and repair time T = 10 hours

then T = 0.1 * 10/8760 = .0001142

$U_\infty = 1.1 * 10^{-4}$

The difference between this and the earlier formula is very small, even at early times, and disappears in the limit.

Repair Times

In the simple example, only a few factors affecting repair times were noted. A more complete check list is

- time to notice the failure
- time to alert repair team (night time/weekend)
- time for repair team to reach equipment (remember bad weather, nighttime, transport difficulties)
- time to obtain tools and equipment
- delays to obtain special tools, equipment (cranes, etc.)
- plant shutdown time
- time to make ready or make safe
- failure diagnosis time
- delay in obtaining spares
- stripping plant or erecting scaffolding to gain access/digging out of piping
- dismantling/machining/replacement/reassembly
- testing
- start up time

All of these factors can be added together to give a fixed repair time estimate. Most of the factors are subject to variation, and it is more accurate to give a repair time distribution. A log normal distribution is generally found to fit the observed data best.

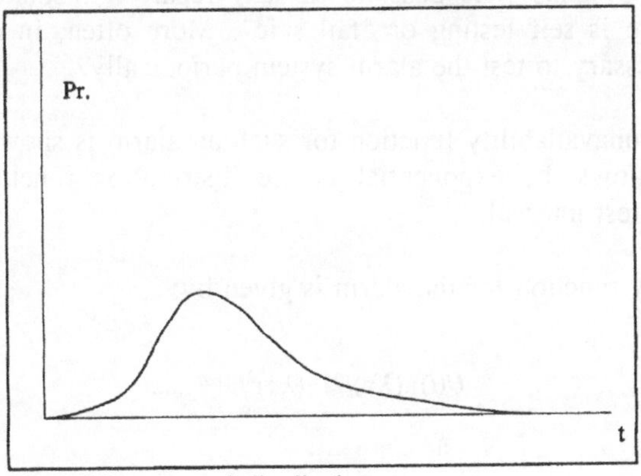

Fig 4.5 Log normal distribution

Example

If the pump has a failure rate of 0.1 per year as before, but the "repair" consists of fitting a new pump, then we may have to wait for delivery of the new pump, with a delivery period of 100 days.

In this case, the unavailability of the standby pump is given by:

$$U \approx \lambda / \mu$$

$$T = 0.1 * 100 * 24/8760 = .0274$$

$$U_{\infty} = .0274/1.0274 = .026$$

The overall frequency of loss of boiler feed water is in this case

$$F_e = 0.1 * .026$$
$$= 2.6 * 10^{-3}$$, or about once every 400 years.

This is too high a rate, under normal circumstances, for serious boiler damage or even explosions. For this reason a level alarm and burner shut down system is fitted.

Alarm Availability

If a safety alarm fails, the system may be severely damaged. The availability for the alarm system is therefore critical. If the alarm fails, repair or replace is necessary.

For some alarm systems it is possible to start repair immediately after a failure, because the alarm is self testing or "fail safe". More often, in conventional alarm system, it is necessary to test the alarm system periodically.

The form of the unavailability function for such an alarm is shown in Fig. 4.5. The unavailability follows the exponential failure distribution function until testing is performed, after test interval.

The unavailability function for the alarm is given by:

$$U(t) = (\lambda\tau)/2! - (\lambda\tau)^2/3! + ...$$

Usually, the test interval is much shorter than the failure rate, because otherwise the unavailability of the alarm becomes rather high. In this case, we have

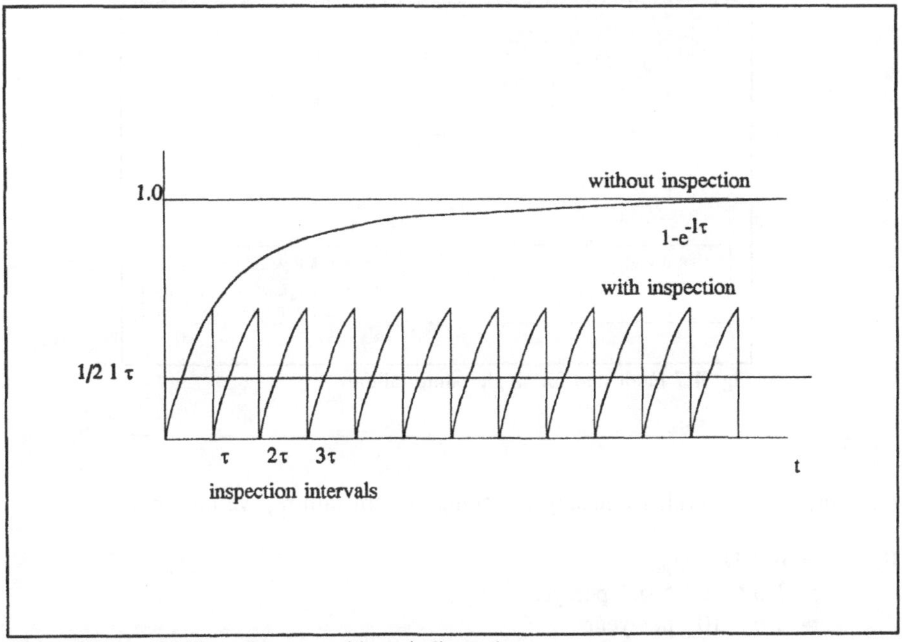

Fig. 4.6 Unavailability of system with periodic testing

$$U_a = \frac{1}{2}(\lambda\tau)$$

Where U_a is the average unavailability.

<u>Example</u>

A typical failure rate for an alarm system, including sensors, register, reset, alarm lamps. and audible alarms, is 0.2 per year, or one failure every five years.
If the alarm is tested every year, the unavailability is given by

$U_a = 1/2 * 0.2 * 1 = 0.1$

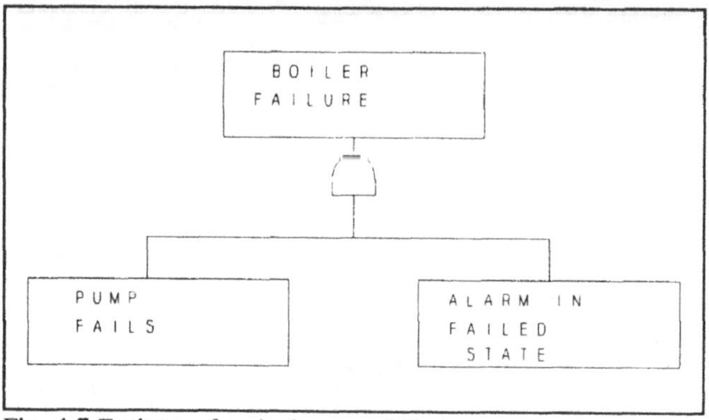

Fig. 4.7 Fault tree for single pump failure.

Looking at the overall frequency for pump system failure, we now get

$$F_e = F * U * U_{alarm}$$
$$= 2.6 * 10^{-3} * 0.1 \text{ per year}$$
$$= 2.6 * 10^{-4} \text{ per year}$$

Comparing with "acceptable risk" levels, this frequency would usually be too high if it could lead to fatalities. Regarding the economic aspects of the risk, the main source of major damage for a plant of this type in case of an explosion in the boiler would be loss of production time.

The net rate of return of the (plant yield - cost of production) might be for example $10,000 per day. A typical "down time" for an accident of this kind would be about 25 days, provided that the damage was not too serious. This gives a "loss rate" of

$$2.6 * 10^{-4} * 25 * 10{,}000 \text{ dollars per year}$$
$$= \$\ 65 \text{ per year.}$$

Taking the "present worth" of this rate of loss as 3 times the yearly loss rate, this gives a value of $195.

This is hardly sufficient to pay the cost of additional safety equipment, unless human life is also at risk.

The failure costs may be sufficient to pay for additional safety equipment if other failures, different from the pump failures, are also taken into account. The $65 per year may also be sufficiently high to make additional testing worthwhile. For example, testing twice per year reduces the yearly cost of pump failure to $32, and may cost in practice an additional $20 per year.

More Realistic Calculations of Alarm Availability

The calculation above assumed that testing removed all failures. In practice, this is not so. Some failures are too difficult to detect, and some will be overseen by the maintenance team, and maintenance may introduce new failures.

The unavailability of the plant is then given by

$$U=\frac{1}{2}\lambda\tau+\alpha(1-e^{-\lambda t})+P$$

where α is the fraction of failures which cannot be detected by testing, and P is the probability of introducing a failure during testing.

A further factor to be taken into account is that testing may result in tripping the plant, due to errors during testing. Also, if testing is performed while the plant is on stream, then accidents can occur at this time.

The "cost" of a trip system does not just include the problem of failing to trip. Failures may cause the trip to shut down the plant at the wrong time.The complete calculation of the net value of a trip is given by:

V =	$F * 1/2(1-\alpha)\ D * K$	cost of failure on demand
	$+ F * (1 - e^{-\alpha Dt}) * K$	cost of untestable failure on demand
	+ F.Pc	cost of failures introduced by testing
	- C	rent on capital cost
	- A	cost of testing
	- BS	cost of fail safe trips
	- FT	cost of failure to trip due to testing
	$- B(1-e^{-It})$	cost of shutting down if a failure is found

Where

V	is the net value per year
F	is the frequency of demand for the trip
t	is time
D	is the failure rate for dangerous failures
S	is the failure rate for safe failures
C	is the rent on the capital cost of the trip
A	is the cost of labour for a test, and the cost of any necessary shutdown
B	is the cost of a shutdown
I	is the test interval
T	is the time taken for testing.

The alarm/shutdown should preferably have a positive net value. This will not always be the case. In some cases safety devices which have negative value from the point of view of production economics are required by law or by standards.

The above calculation is given as a cost per year. The capital cost needs in this case to be expressed as a "rate of return on capital". The average rate of interest on investment capital in the company is the usual basis for the calculation. With present day rates of interest, it is usually sufficiently accurate simply to divide the capital cost by three, representing the net value over three years.

Common Cause Failure

One of the problems of redundant systems such as the two pumps is that the components may fail at the same time, due to the same cause. The pumps can fail for example due to a blockage in feed water supply, due to a fire, etc. Such failures are called "common cause failures". A common cause failure is defined as a failure which cause redundant components to be in the failed state at the same time, due to the same event or effect.

The best way to treat these problems is to separate out "common causes" from the description of failure, and to calculate unavailabilities and failure frequencies from these causes separately, for example:

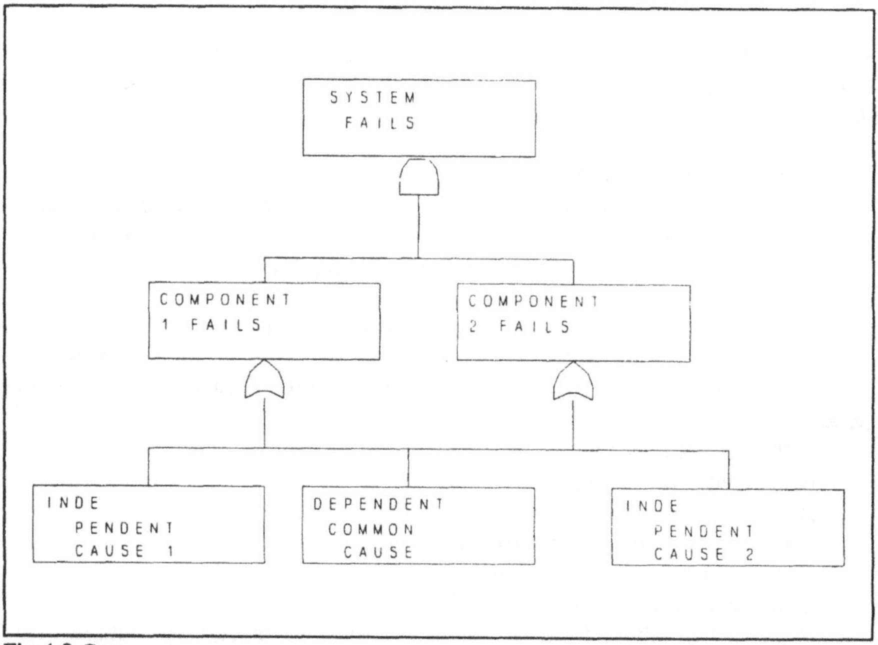

Fig 4.8 Common cause

Unfortunately, it is not always so easy to identify all possible common causes, so treatment by separating them out in this way may be impossible. A more general approach is the so-called "beta factor" method. The beta factor (ß) is defined as the fraction of failures (in a doubly redundant system) which are common cause failures. For unavailability calculations for two identical components, this gives:

$$P(A\&B) = (1-\beta)^2\, P(A) \,.\, P(B) + \beta . P(A)$$

For the frequency of the double failure, we have without common cause failure

$$Fr(A\&B) = 1 * Fr(A) * U(B)$$

With common cause failure we have

$$Fr(A\&B) = 2 * (1 - \beta)^2 Fr(A) . U(B) + \beta Fr(A)$$

Common cause failure may affect redundant equipment of any kind, but is typically a problem for identical equipment. Typical values of ß lie between 0.3 to 0.01 for similar equipment types.

Common cause failure is a major problem for high reliability safety systems which depend on redundancy. A typical unavailability for a pressure operated blow down valve is, for example 0.01. A doubly redundant system should in principle have an unavailability value of 10^{-4} but common cause failure may well increase this to 10^{-3}. For this reason, safety systems are often made with diverse (i.e. different) redundant systems.

For redundant safety systems, it is also usual to perform a "common cause analysis". This analysis looks for specific root causes of component failure. A typical analysis looks at the following problems:

- external effects (fire, flooding, dropped objects, (zonal analysis inspects the areas which these causes may affect).
- design
- installation
- testing and calibration
- damage, corrosion, vibration, foreign objects, blockage due to substances arising in the process.

Identifying Common Cause Failure

Several "methods" have been developed for common cause failure identification. Most of these however correspond to a straightforward checklist applied to pairs of redundant components. A suitable check list is given in table 4.1.

The most probable areas for common causes are those with similar redundant items. For this reason redundant safety systems are often deliberately made with different component types, so called diverse redundancy.

To apply the check list, list those components which are redundant, and which areas or "zones" they lie in. Then consider which of the common failure causes apply to each redundant group.

Table 4.2 Common cause failure cause types

Environment	Function
overheating	Operation outside design envelope
freezing	Inadequate instrumentation
corrosive atmosphere	Inadequate capacity
dust, dirt	
vibration	**External events**
contamination	
Humidity	Fire
	Flood
Design	Missiles
	Earthquake
Underdimensioning	Freezing
Unsuitable material	Explosion
Overstressing	Missiles
Undesired interaction	Sabotage
Overlooked side effect	Physical damage, crash
Overlooked precondition for operation	Falling objects
	Installation and Maintenance
Operation	
	Omission of component
Common error in start up	Assembly error wrong way
Calibration	Assembly error, jamming
Switch off A instead of B	Foreign object
	Over or under lubrication, painting, sticking
	Forget to switch on, or align valves

Many Different Causes of Failure

If there are two or more possible failure causes, these can be represented on a fault tree by an OR gate (see e.g. Fig 4.9).

If an OR gate is used to express failure frequency, the formula for the output frequency of the gate is

$$Fr(C) = U(A) + U(B) - U(A) \,.\, U(B)$$

The product term in this formula is included to avoid counting twice the case in which both components are failed at the same time.

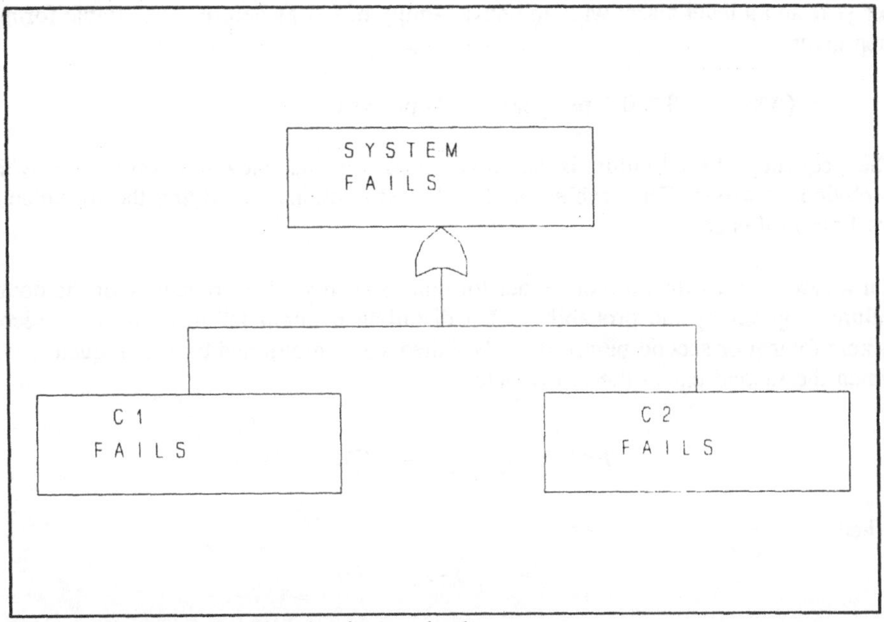

Fig.4.9 OR gate relationship, showing overlap between cases.

Example

The unavailability of a pump is calculated as .026. The valve following the pump, may fail closed as a result of breakage of the piston or seat, or as a result of blockage. The repair time is two hours, and the level alarm gives a good indication so repair starts quickly. The asymptotic availability is then given by

$$U_x = T = .02 * 2/8760 = 4.56 * 10^{-6}$$

The total unavailability for the pump and valve together is then

$$U_x = .026 + 4.56 * 10^{-6} - .026 * 4.56 * 10^{-6}$$
$$= .02600462$$

The valve is seen to have very little effect on overall availability.

A Pitfall

The formula for frequency of failures in one pump while another is failed, or while the alarm system is failed, was given as

$$F(A\&B) = F(A) * U(B)$$

These formulae apply so long as $t >> T$ for an immediately repaired system and

$$\lambda\tau < 1, \ t > 1/\lambda$$

for a periodically tested system.

Consider the problem of a single pump with a very high failure rate, say three failures per year and a level alarm with an unavailability of 0.1 as before. The simple formula then gives

$$Fr(A\&C) = 3 * 0.1 \text{ per year} = 0.3 \text{ per year}$$

The frequency of total failure is then higher than the frequency of alarm failure, which is obviously wrong. The problem is that the assumptions underlying the formulae do not hold in this case.

For a case such as this, a more exact formula is needed. The frequency of the double failure is given by the probability of a disturbance pump failure) while the second system (alarm or second pump) is in the failed state, multiplied by the frequency with which the second enters the failed state.

$$Pr(C \wedge D)_{at\ time\ t} = (1 - e^{-\lambda\tau/2})(1 - e^{-Dt})$$

When

$$\lambda\tau \approx 1$$

the frequency of the double failure is given by

$$1/\lambda$$

Calculations for gates in a fault tree

Fault trees provide an ideal way of describing very many faults and failure combinations.

For a simple fault tree, the calculations are very simple.

AND GATE

If a component has two supplies, or is provided with support, from two subsystems, only one of which is necessary, the system is redundant. It will fail if supply 1 and supply 2 fail. This is described by an AND gate. We can say that the two supplies are in parallel.

The probability of total system failure is then given by the product law for probability, provided the failures of supply 1 and supply 2 are statistically independent.

Pr (System fails) = Pr (Supply 1 fails) . Pr (Supply 2 fails)
or more generally

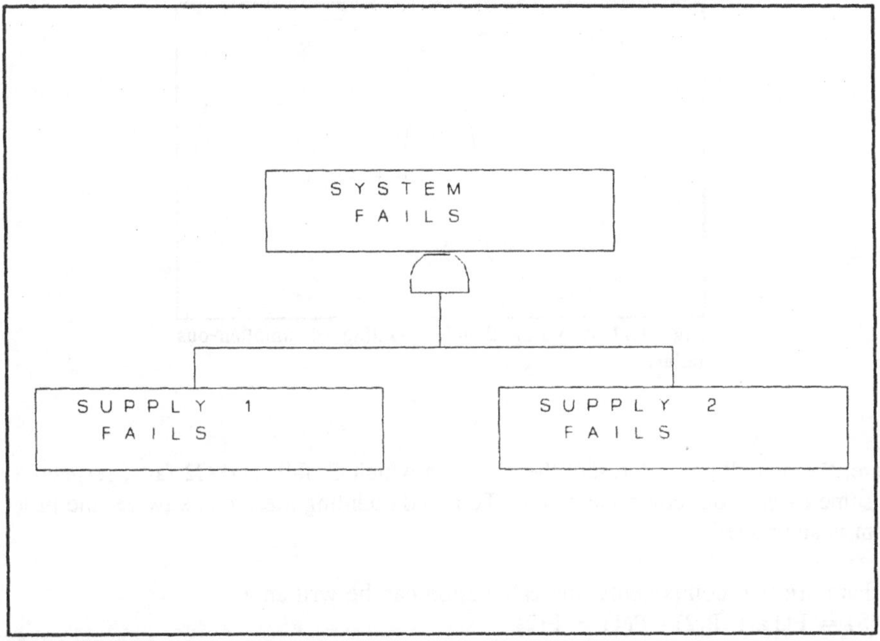

Fig. 4.10 AND gate

Pr (Event) = Pr (E_1) . Pr (E_2).
This can be generalized further for three or more input events e.g.
Pr (Event) = Pr (E_1) . Pr (E_2) . Pr(E_3).

OR GATE

If a system has two components which are connected in series, failure of either the first OR second component will cause the system to fail. This can be described by an OR gate in a fault tree.

Provided that failure of C_1 and C_2 are statistically independent, the probability of system failure is given by

Pr (System = Pr (C_1) + (C_2) - Pr (C_1) . Pr (C_2).

The importance of the last term in the calculation here can be seen from the following diagram

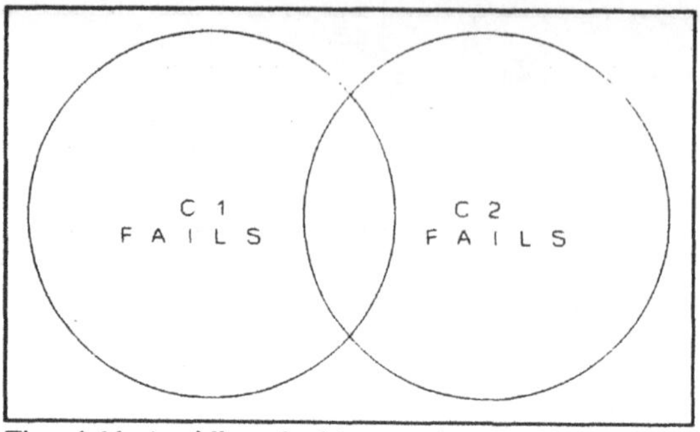

Fig. 4.11 Avoiding double counting of simultaneous failure

Here, the two ellipses represent the events in which C_1 fails and C2 fails, respectiv
In some cases, both components fail. To avoid counting these cases twice, the proc
term is subtracted.

if there are two components, the calculation can be written as
P.(S) = P(1) + P(2) - P(1) = P(2).

With three components, the calculation can be written as

$$
\begin{aligned}
P(S) &= P(1\ \text{OR}\ 2) + P(3) - P(1\ \text{OR}\ 2)\ .\ P(3) \\
&= P(1) + P(2) - P(1)\ .\ P(2) + P(3) \\
&\qquad - P(1)\ .\ P(3) - P(2)\ .\ P(3) + P(1)\ .\ P(2)\ .\ P(3) \\
&= P(1) + P(2) + P(3) \\
&\qquad - P(1)\ .\ P(2) - P(2)\ .\ P(3) - P(3)\ .\ P(1) \\
&\qquad + P(1)\ .\ P(2)\ .\ P(3).
\end{aligned}
$$

<u>Exercise</u>

Evaluate the probability formula for a four input OR gate.

In general for a N input Or gate, the probability is given by

$$
P(system) = \sum_{components} P(i) - \sum_{pairs} P(i).P(j) + \sum_{triples} P(i).P(j).P(k) + ... + \prod_{alli} P(i)
$$

Events which are not Independent

The calculations given above apply if the events, such as failure of component 1 and 2, are statistically independent by definition if:

$$P(1 \text{ and } 2) = P(1) . P(2).$$

What does it mean to say that two events are not statistically independent?

One way in which two events can be dependent is that one causes the other, or both have the same cause. The events then occur together. This is total dependence

$$P(1 \text{ and } 2) = P(1)$$
$$P(1 \text{ or } 2) = P(1) \qquad \text{total dependence}$$

Another way in which dependence can arise is that the events are mutually exclusive. This is not often a logical condition, for example a valve cannot both fail open and fail closed. Another possibility is that one event prevents another, for example fire and flood are in some cases mutually exclusive.

In this case, the probabilities are given by

$$P(1 \text{ and } 2) = 0 \qquad \text{1 and 2 mutually exclusive}$$

$$P(1 \text{ or } 2) = P(1) + P(2) \qquad \text{1 and 2 mutually exclusive}$$

In more general cases, there is some statistical dependence between events, but the dependence is not total. There may be some causes which affect both components at the same time. This can be drawn on a fault tree, and the techniques described later can be used for evaluation.

In other cases it may be impossible to identify a cause, only a statistical relationship being found . In this case, a "correlation coefficient" may be used to represent the dependency.

Assigning Probabilities to Events and Conditions in a Fault tree

One of the most difficult aspects of fault tree calculation is finding the proper probability or frequency model for events in the fault tree. Most textbooks are of very little help in this respect, because they do not distinguish the different cases very clearly.

The first and most important distinction is between events and conditions.

An event is a change of state in the plant, or an action by a person.

A condition is a description of the state of a component or of part of the plant.

Example

A valve opens (event).
If there is water pressure at the valve inlet (condition) and the outlet pipe is open (condition).
Then water begins to flow (event).

Events are characterized by a frequency (usually given as so many occurrences per year, per million hours, or per ten thousand years in risk analysis work).

If an event leads directly to an accident, or to the top event in the fault tree, then it is the frequency of this event which is required. For risk analysis work, it is most often the frequency of the top event of the fault tree which is needed. The calculation of fault tree frequency then for the most part consists of small calculations of the following kind.

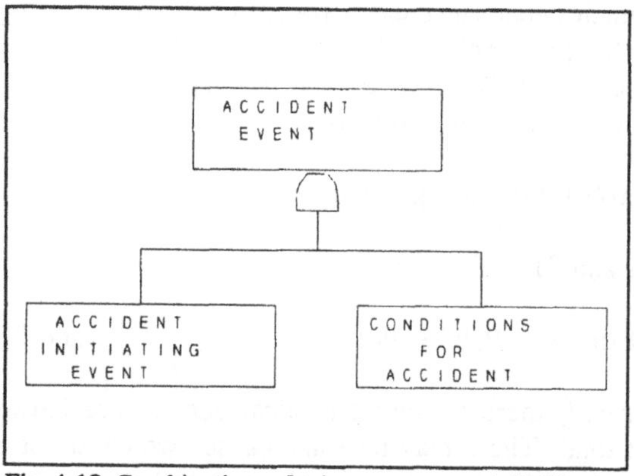

Fig 4.12 Combination of trigger event and facilitating condition

The frequency of the accident event is then given by

$$Fr(accident) = Fr(initiating\ event)\ .\ Pr(condition).$$

The condition probability is in practice normally a product of a series of subconditions. These are typically:

- the probability that the necessary substances or energy for the accident are present

- the probabilities that the various pemanent safety measures are in a failed state or are being tested or repaired.

- the probability that ther are persons or property exposed to the accident

Procedure for assigning Event Probabilities and Frequencies in a Fault Tree

1. Does the failure lead immediately to an accident (or to the top event)?
   If yes, class 1
   Event frequency = F (F is failure rate)

2. Does the failure lead to a dangerous or less reliable system state? If yes, go to 3. If no - check that the event should be in the fault tree - if yes go to 6.

3. Does the failure show itself immediately?
   If so does repair process start immediately?
   If yes U = FT (T is repair time)
   (class 2)

4. If the failure does not show itself immediately, can it be detected by inspection? If so, how often does the inspection (or monitoring or supervision) occur?
   U = 1/2FI (I is inspection interval)
   (class 3)

5. If the failure cannot be inspected, or there is no inspector, what is the resulting time interval in which the failure can occur?
   $U \approx F\,t$
   (t is the time interval in which failure can occur)
   or $U = 1 - e^{-Ft}$ if t is >> 1/F
   (class 4)

6. Is the event an operator error or an environment condition, or an error rate per operation? If so, use the pure probability P (class 5).

<u>Example</u>

A major fire may break out if an ignition occurs (event), and the fire fighting system is in a failed state (condition).

Fr. (fire) = Fr (ignition) . Pr (firefighting system failed).

If there are several events which can lead to an accident, then an OR gate will be used. If one is calculating frequencies, the frequency of the top event is the simple sum of the frequencies of the contributing events.

Fr. (accident) = Fr.(1) + Fr.(2) + Fr(3).

<u>Example</u>

A fire may be caused by ignition by lightning, by a runaway reaction, or by ignition by static electricity. The frequency of ignition overall is the sum of the frequencies for each ignition type.

Once the frequency of initiating events has been decided, the probabilities for conditions needs to be determined. Probabilities will generally be concerned with the unavailability of safety equipment, and similar problems. The calculations from the section on component reliability come into use.

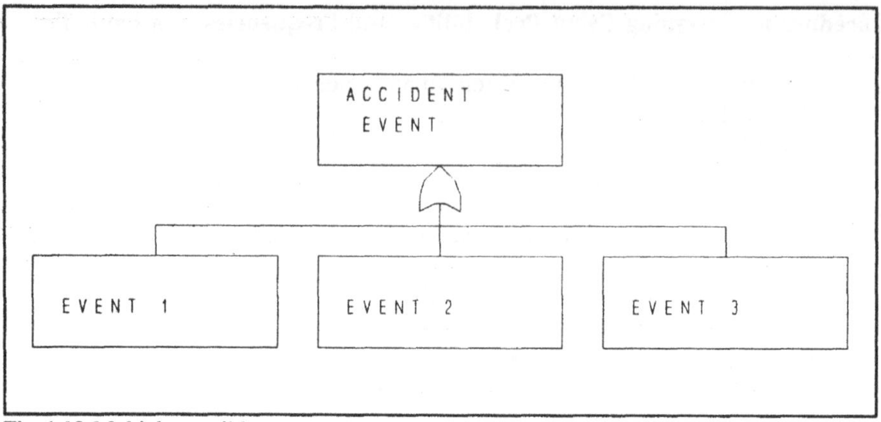

Fig 4.13 Multiple possible cause events.

For some components, the failure will show itself immediately, and the unavailability is then dominated by the repair time. The average unavailability is then given by the equation:

$$U=\lambda T$$

Example
A cooling pump is necessary to prevent overheating in a reaction. The correct functioning is monitored and there is an alarm. In many such cases, if there is only one pump, the operator will shut down the plant. In this case, the time T in the above equation will be the time taken for shutdown rather than the time needed for repair. If there is redundancy in the system, then the repair unavailability will be the relevant factor for the fault tree calculation, with T being the repair time. There may, though, still be a rule requiring shutdown of the plant if two cooling pumps fail. In this case, using the repair time in the calculation will produce unduly pessimistic results for the accident rate calculation.

Some types of failure do not show themselves immediately. For example, a failure of a safety valve in the form of a blockage cannot be seen immediately. In this case, the appropriate calculation formula is:

$$U=\frac{1}{2}\lambda\tau$$

Example
The failure rate for a safety valve is 10^{-4} per year. The inspection interval is once per year. The unavailability is $U = 1/2 * 10^{-4} * 1 = 0.5 * 10^{-4}$.

If the failure is of a kind which cannot be inspected, for example corrosion in an inaccessible pipe, the appropriate equation is:

$$U = 1 - e^{-\lambda t}$$

When

$$t < 1/\lambda$$

As is generally the case, the formula

$$U = \lambda t$$

gives a very good approximation.

In some cases, accident initiating event can lead to an accident if a safety action fails. The success of the safety action may depend, not on the state of the plant, but may simply be a constant probability. The appropriate equation is then

$$U = P$$

Example

A relay may fail in the stuck closed mode once for every ten thousand operation.

Example

An operator should respond to a cooling water failure alarm by starting the standby pump. He fails to start the pump with a probability of 0.001.

Example

Figure 4.14 shows a diagram of a pressure vessel, together with a safety valve. The diagram also shows a fault tree for the safety valve. Many causes of safety valve failure have been included in the tree).

The frequency of overpressuring of the vessel is given as 50 times per year. Table 4.3 shows the way in which the failure rates are allocated to the tree. Note that some of the failure modes can readily be identified and failure rate data can be supplied, using the failure rate data collection given in these notes. Others require a considerable degree of judgement and the use of other analysis techniques, as is indicated in table 4.3.

A combination of fault tree analysis and failure mode and effects analysis can be useful in cases such as this in order to ensure that a full range of causes, detection measures, and test and repair activities are taken into account.

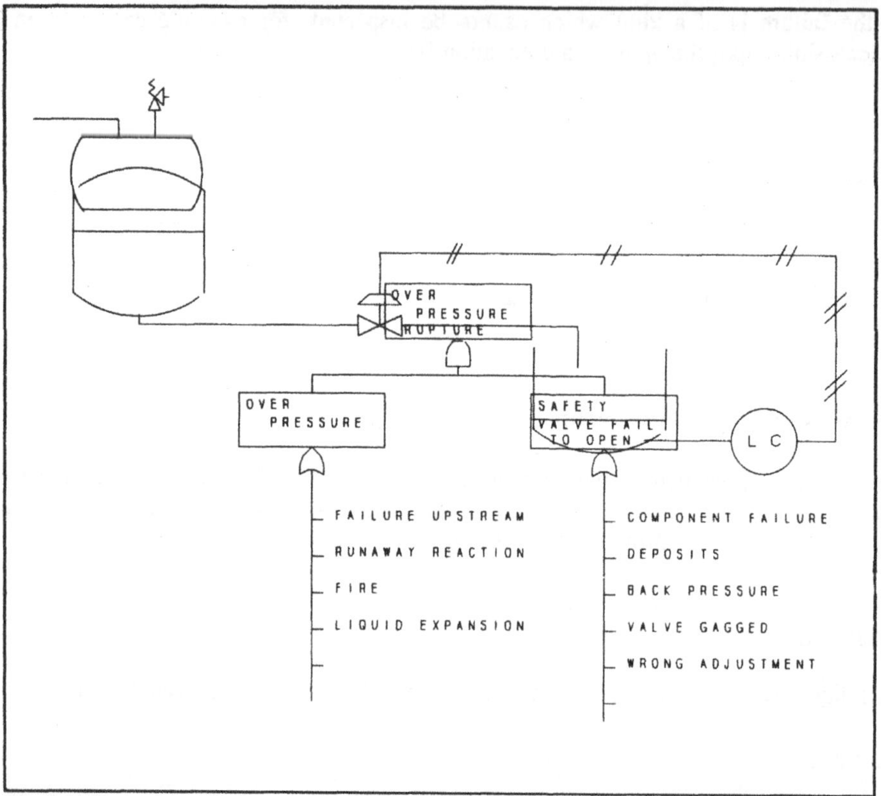

Fig 4.14 Example of fault tree with failure mode details

Table 4.3 Analysis methods for failure and disturbance causes

Failure cause	**Assessment**
Failure upstream	Fault tree analysis upstream
Runaway reaction	Fault tree analysis for runaway reaction
Fire	Fire analysis
Liquid expansion	Requires overfilling
Component failure	Data base value, see ch 5
Deposits	Depends on operating conditions
Back pressure in blowdown line	FTA of downstream pipelines
Valve gagged	Maintenance error, about once per thousand to once per ten thousand. See Ch. 6

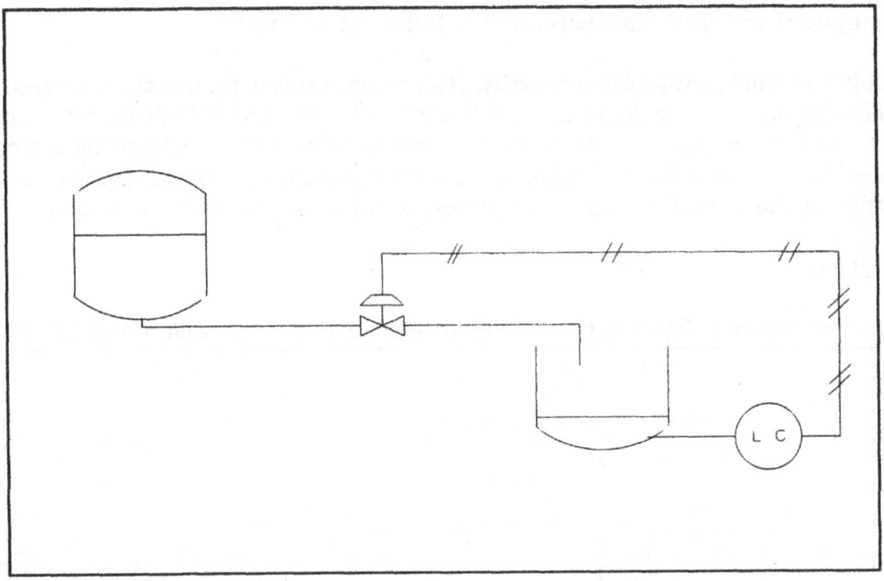

Fig. 4.15 Buffer tank example

Exercise

Fig. 4.15 shows a level controller for a buffer tank between two production units. The objective of the level controller is to ensure that there is sufficient feed to the downstream processing, and to avoid the need to shut down the downstream unit. The downstream unit requires a continuous flow, but the flow amount may be varied from 20% up to 100% of nominal flow.

1. Try to draw the fault tree with failure modes.

2. Assign probabilities to the various events.

3. Calculate the probability of level-controller unavailability, and the probability of tank overflow.

4. How would the situation change if there were a high level and a low level alarm?

Complete Fault Tree Calculations with Independent Events

Fault trees can be evaluated very easily, if all of the events in the tree are independent. The principle is to find the gates at the lowest level of the tree and calculate the output probabilities for these events. Then each lowest level gate is replaced by a single "super event". with the corresponding calculated probability. The process is repeated until only the top event is left. This process is called "series/parallel reduction".

Example

The tree shown in figure 4.16 has 7 basic events A to G, each with probability 0.1.

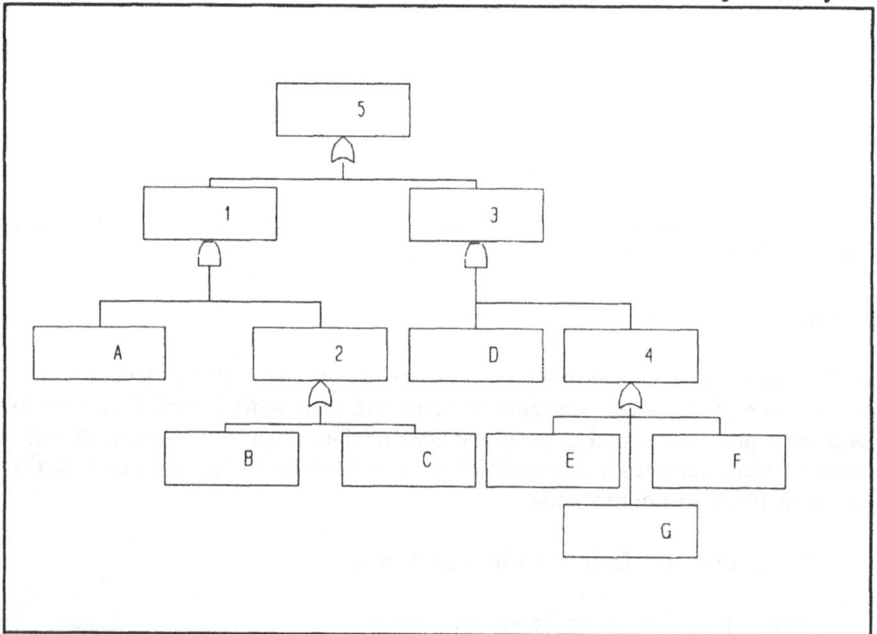

Fig 4.16 Example tree

Gate 2 is reduced first. It has events B and C as inputs, so
Pr(2) = 0.1 + 0.1 - 0.01 = 0.19.
Gate 4, when reduced, gives
Pr(4) = 0.1 + 0.1 + 0.1 - 0.01 - 0.01 - 0.01 + 0.001 = 0.271
The tree now appears as in Fig. 4.17

Gates 1 and 3 can now be reduced as follows:

Pr(1) = Pr(A) . Pr(2) = 0.019
Pr(3) = Pr(D) . Pr(4) = 0.0271.
Gates 1 and 3 are replaced by events 1 and 3.

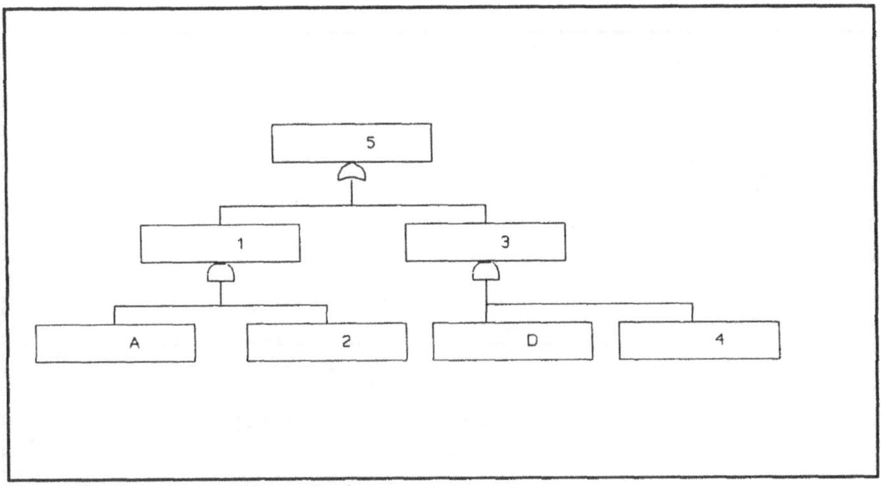

Fig. 4.17 Example tree after reduction

The next step is to reduce gate 5, which gives

$$Pr(5) = Pr(1) + Pr(3) - Pr(1) . Pr(3)$$
$$= 0.019 + 0.0271 - 0.019 * 0.0271$$

Series parallel reduction may be applied throughout a fault tree if the events in the tree are not repeated i.e. do not appear more than once. Series parallel reduction may also be applied to more complex trees, with repeated events, provided that it is only applied in those parts of the tree (subtrees) in which there are no repeated events. This can be very useful in reducing the complexity of the trees.

Note that common cause failure almost always introduces repeated events into fault trees at the lowest level, and this often prevents the series/parallel reduction technique from being applied. A way around this, which is equivalent to the more formal techniques described later in this chapter, is to analyse the effects of common cause failure separately. The system is analysed as though common cause failure did not exist. Then branches are added to the fault tree specifically for the common cause failure effects, at the point in the tree where an AND gate appears to indicate the presence of redundancy. It can be helpful if you include, alongside any AND gate, an alternative failure "mode" describes as "Failure of redundancy".

An alternative procedur to this is to analyse common case failure separately, using FMEA, event tree, or cause consequence analysis techniques. For very large analyses, with many redundant components, this may be the only practical approach, because of the very large number of repeated events. This kind of thing occurs for nuclear power plant safety systems, for example when flooding or earthquake are taken into account. For chemical plant, where redundancy of safety equipment is less common, fire and explosion represent a common causal event which can damage many items at once. The explosion can however be analysed as an event in itself, using cause consequence analysis.

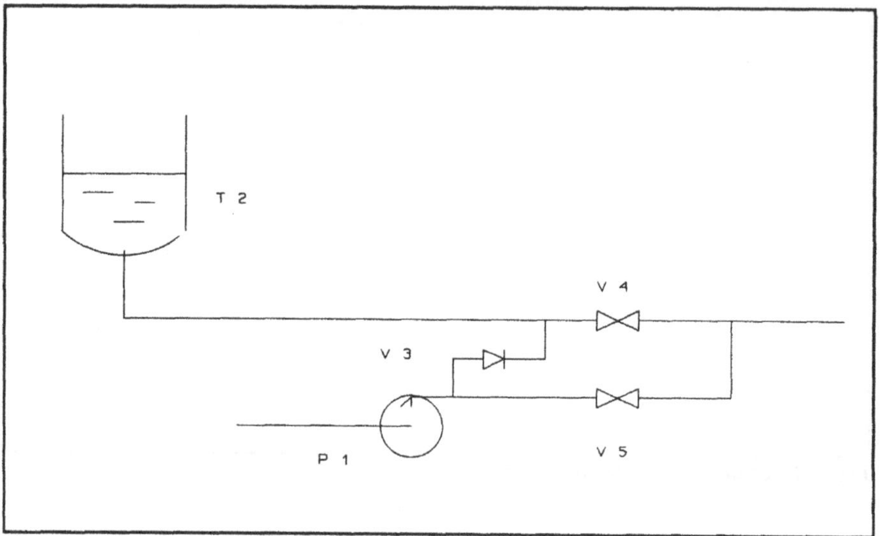

Fig. 4.18 Dual water supply.

Calculations for Fault Trees with Repeated Events

The simple procedure given above cannot be applied if there are repeated events in a fault tree. The reason is that the events in the tree are not independent, so that the equations for gate probabilities do not apply.

As an example, consider the water supply plant shown in fig. 4.18. A fault tree for this system is shown in Fig. 4.19 for the top event "No flow from system".

The probability for valve blockage is .001, for pump stoppage .1, for an empty tank is .1.

By simple fault tree reduction, this gives

P(2) = .1009
P(5) = .1009
P(4) = .01009
P(3) = .01098
P(1) = .001098 (which is wrong).

To see that this result is wrong, note that failure of the pump and tank together is sufficient to give loss of flow. The probability of this occurring is P(Pump and Tank) = 0.1 * 0.1 = 0.01.
The reduction technique has given a result which is almost ten times too small. The reason is that we have multiplied by the failure probability for pump failure twice, in a hidden way.

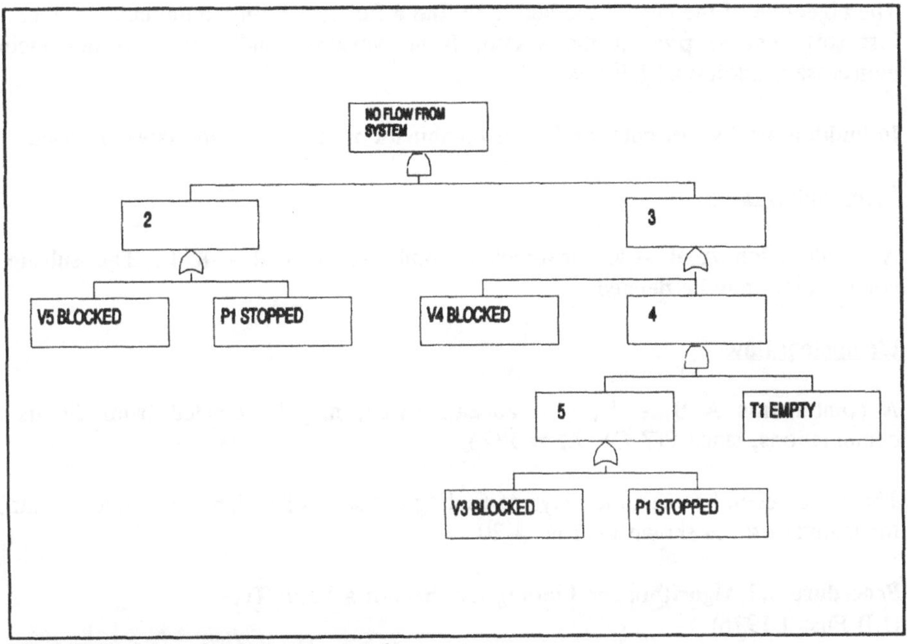

Fig 4.19 Fault tree for dual water supply

A way of avoiding this problem is to take each combination of failures separately.

Then it can be seen that the following failure combinations give loss of flow

1. T_2 and P_1
2. P_1 and V_4 and V_5
3. T_2 and V_3 and V_5
4. T_2 and V_4 and V_5
5. V_4 and V_5
6. P_1 and V_4.

Of these combinations note that 2, 4 and 5 all include failure of valves V_4 and V_5. Failure of these two valves alone will cause loss of supply. The failure of a further component such as P_1 in combination 2 is therefore irrelevant. Combinations 2 and 4 can be deleted.

The probability for each combination can now be obtained as an OR function of each of the combinations.

This approach to fault tree analysis is called the "cut set" method, because each of the fault combinations represents a cut across a network representing the system.

The objective of the cut set method is to find a full set of "minimum cuts", which are just sufficient to prevent the system from working, and which do not include unnecessary additional failures.

In building up lists of cuts or failure combinations, the following laws are used

Subsumption Law

A combination A B ---K subsumes a combination A B ---K L. The subsumed combination may be deleted.

Idempotent Law

A combination A B --- K, if it appears twice, may be deleted from the list of combinations, since,P(Z OR Z) = P(Z).

There are several systematic ways of finding cut sets, of which the simplest, suitable for manual use, is shown in Fig. 4.20

Procedure 4.2 Algorithm for Finding Cut Sets of a Fault Tree
(J.B.Fussel 1976)

1. Place the highest level gate on the cut set list

2. Take a cut set which contains a gate from the cut set list, and "expand" it as follows

   a. If the cut set contains an OR gate, replace it by new entries on the list, one for each input. The entries consist of the original cut set, but with the selected gates replaced by one of its inputs.

   b. If the cut set contains an AND gate, replace it on the list with a new cut set, consisting of the original cut set with the selected gate replaced by all of its inputs.

3. If a cut set contains a repeated event, delete all but one occurrence of the event.

4. If a cut set is a superset of another, delete the cut set with the largest set of events.

5. Repeat steps 2 and 4 until there are no cut sets sets left which contain gates.

<u>Example</u> Using the fault tree from Fig.4.19 as starting point.

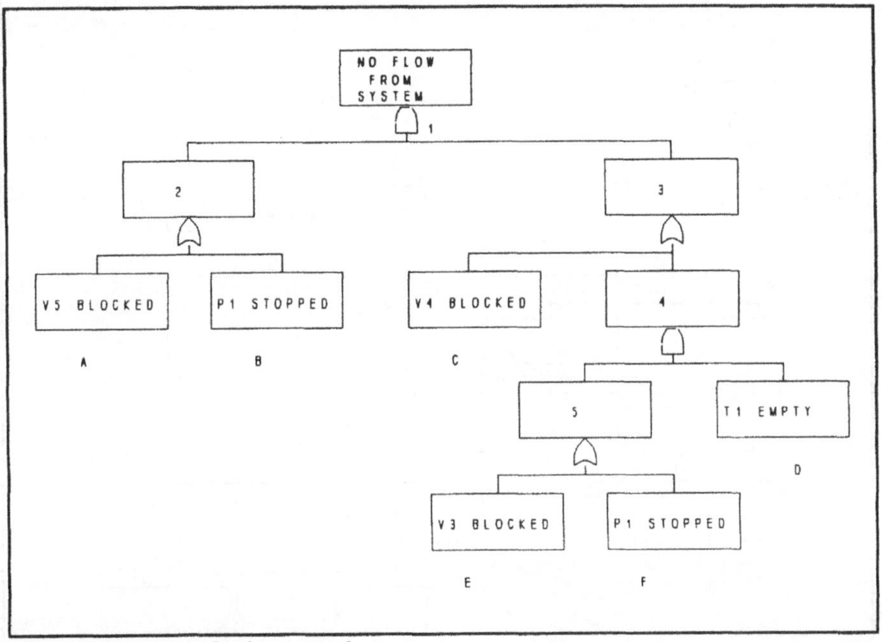

Fig. 4.20 Cut set evaluation example.

Cut set List		Expanded Cut Set List
1		AC
23		AED
A3		ABD
B3		BC
AC	BC	BED
A4	B4	BBD = BD
A5D	B5D	BD subsumes ABD, BED
AED	BED	
ABD	BBD	

<u>Reduced cut sets</u>

AC	V4, V5
BC	P1, V4
BD	P1, T1
AED	V4, V3, T1

Example

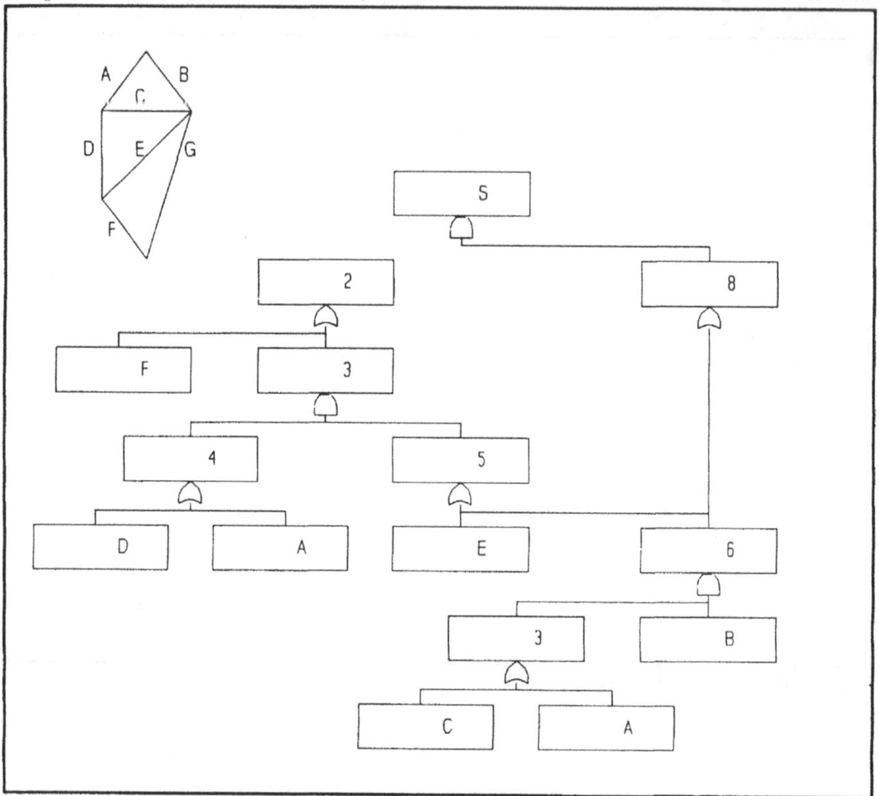

Fig. 4.21 Fault tree for cut set reduction.

Cut set list

S	4,5,8	A,E,8	A,E,G
1	D,5,8	A,6,8	A,E,6
2,8	A,5,8		
3,8	D,E,8	D,E,G	A,G,6
F,8	D,6,8	D,E,6	A,6,6
F,G			
F,6	D,6,9	A,6	
F,7,B	D,6,6		
F,C,B		A,7,B	
F,A,B	D,6		
	D,7,B	A,C,B	
	D,C,B		A,A,B
	D,A,B		A,B

Probability Calculations for Cut Sets - The Inclusion - Exclusion Algorithm

Once a fault tree has been reduced to cut set form, calculation is fairly straight forward.

First, the probability for each cut set is calculated, as the product of the probabilities for each of the cut sets.

The next step involves regarding the fault tree as one large OR gate, with the top event as output, and each of the cut sets as an input. The equation for OR gate probabilities applies.

The first step is then to take the sum of all of the cut set probabilities, then to subtract the sum of the products, taking two at a time, then to add the sum of the products taken three at a time, and so on. In symbols this is

$$P_{TOPEVENT} = \sum_{all\,cut\,sets\,i} P_i - \sum_{all\,pairs\,i,j} P_i . P_j + \sum_{all\,triples\,i,j,k} P_i . P_j . P_k + \ldots\ldots\ldots\ldots$$

In practice, an algorithm to do this can be programmed very simply. It is almost never necessary to go beyond fourfold combinations of cut sets, since the product terms become very small after this.

<u>Example</u>

Evaluate the unavailability for Fig. 25 (27) with
P(A) = P(C) = P(C) = 0.001 and P(B) = P(D) = 0.1.
The cut sets are AC, BC, and BD.
The cut set probabilities are
P(AC) = .000001
P(BC) = .0001
P(BD) = .01
The top event probability is then
P_{TOP} = .010101
- .000001 - some very small terms
+ some very very small terms
= .01

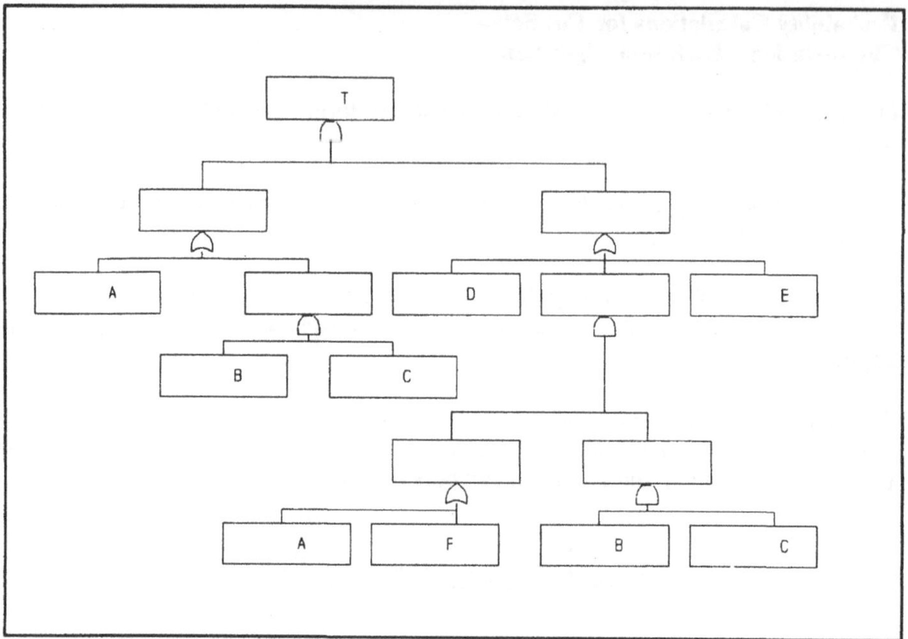

Fig.4.22 Modularisation example

Modularisation

Evaluation cut sets for large fault trees is time-consuming. For very large fault trees, computer methods are required. Some fault trees for military and nuclear power plant systems are too large for any but the largest computers. Some chemical plant fault trees are too large for any computer. The solution to this problem is to build smaller fault trees, and then to combine them. This is only possible, though, if the separate small fault trees do not contain the same events, so-called repeated events.

Another, similar, approach is to make a large fault tree, and then to reduce it by combining parts of the fault tree and replacing them by "super-events". A part of a fault tree may be replaced by a "super-event" if

- all of the events in the selected part of the fault tree are unique and do not appear elsewhere.

OR

- the events which appear in the fault tree appear in other places, but always in the same pattern.

Example

The parts of the tree with events B and C can be replaced by super-events, because the events appear in two places, but in the same pattern. Events D and E can be replaced by a super-event, because they only appear once in the tree. Events A and F cannot be combined as a super-event, because event A appears twice in the tree, and in different patterns of events. The reduced tree then appears.

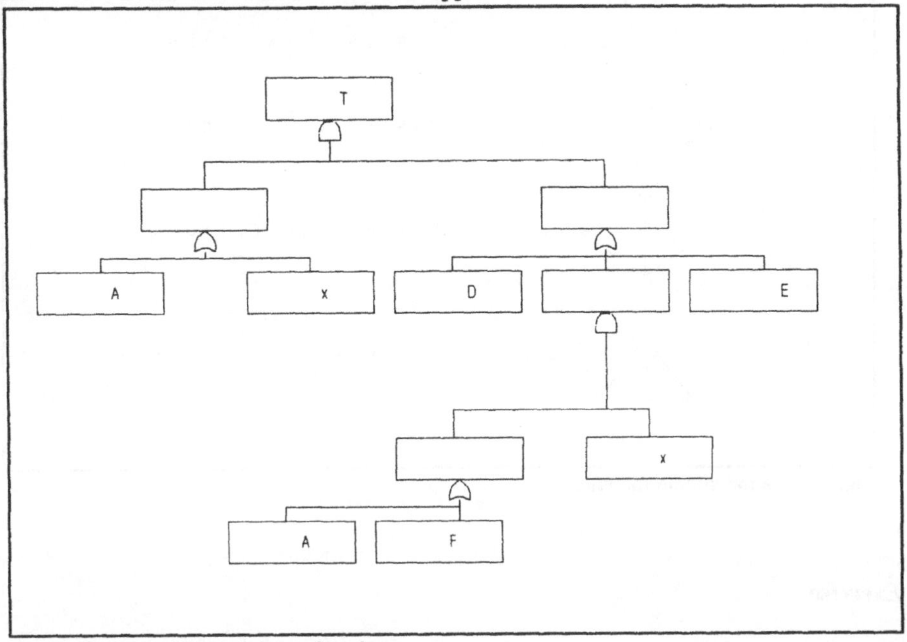

Fig. 4.23 Tree with modules reduced.

Parts of the tree which can be reduced in this way are called "modules". In a network representation of a plant, a module corresponds to a part of the network with a unique label for each arc in the network, and a single input and output. Fig. 4.24 shows some typical modules.

Modules correspond to parts of a system with a single "input" and a single "output". This means that the module may enter into the system in several different ways, but that in each case it enter into the system in the same way. As a result, the module can be regarded as a single component. Not that the "inputs" and "outputs" referred to here concern all possible causal effects. If a part of a system causes side effects in several different ways, it is unlikely that it will be a candidate to be a module, and so be reducible in the tree.

As an example, consider a DC power supply. If this has just two voltage terminals, and if its switching is not made to depend on changes in the circuit supplied, it will be a candidate for consideration as a module. If the power supply can cause fire however, and if this fire affects other systems in a variety of ways, and if these fire effects enter into the fault tree, then it will not be possible to regard the power supply as a single component.

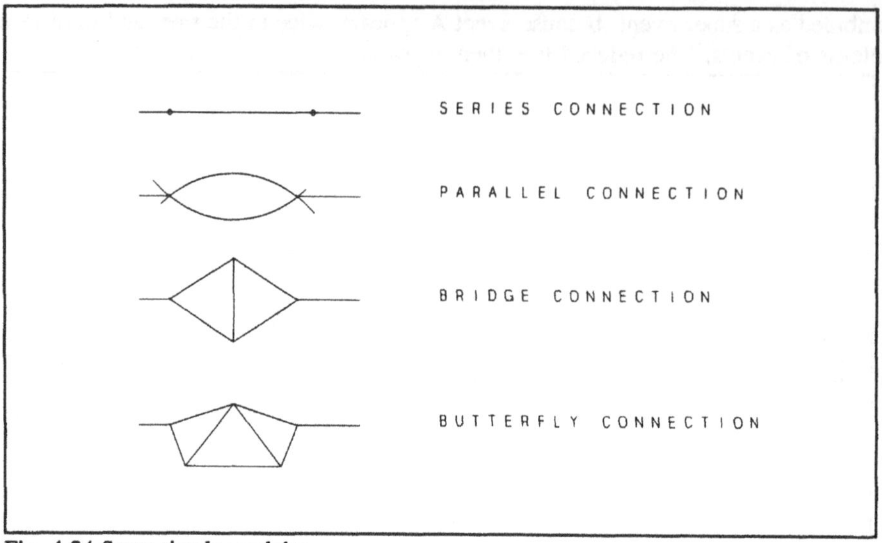

Fig. 4.24 Some simple modules

Exercise

Make fault trees for each of the modules in Fig. 4.24

Simplification of Fault Trees -Pivoting

If there are only a few repeated events in a fault tree, it is possible to manipulate the fault tree so that the event only appears once.

The pivoting method is based on the following theory

$$P(A) = P(A/B)\,P(B) + P(A/B)\,P(B).$$

In fault tree terms, this can be written

$$\begin{aligned} &P(\text{top event}) \\ &= P(\text{top event given B}) * P(B) + P(\text{top event given not B}) * P(\text{not B}). \end{aligned}$$

The procedure for simplifying the tree is then that given.

1. Select the repeated event which is to be simplified (the "pivot")

2. Evaluate the top event probability with the pivot event set to true, P(pivot) = 1, with result P_1.

3. Evaluate the top event probability with the pivot event set to false, P(pivot) = 0, giving result P_2.

4. Get the final result by calculating

$$P_{TOP} = P_1 * P(pivot) + P_2 * (1.0 - P(pivot)).$$

Procedure 4.3 Algorithm for Pivoting

The method can be used symbolically, too, to change the structure of the fault tree. It is especially useful in separating out common power supply, cooling and water supply, and control failure events.

Analysing common cause failures is a particularly difficult aspect of fault tree analysis. The common causes introduce repeated events into the fault tree which are at a very low level, and there are usually very many of them. The result is generally an increase in the number of minimal cut sets, and often a very large increase in the number of cut sets before minimisation. By performing pivotting, this problem can be reduced, and at the same time, the tree is restructured so that common causes can be seen more clearly.

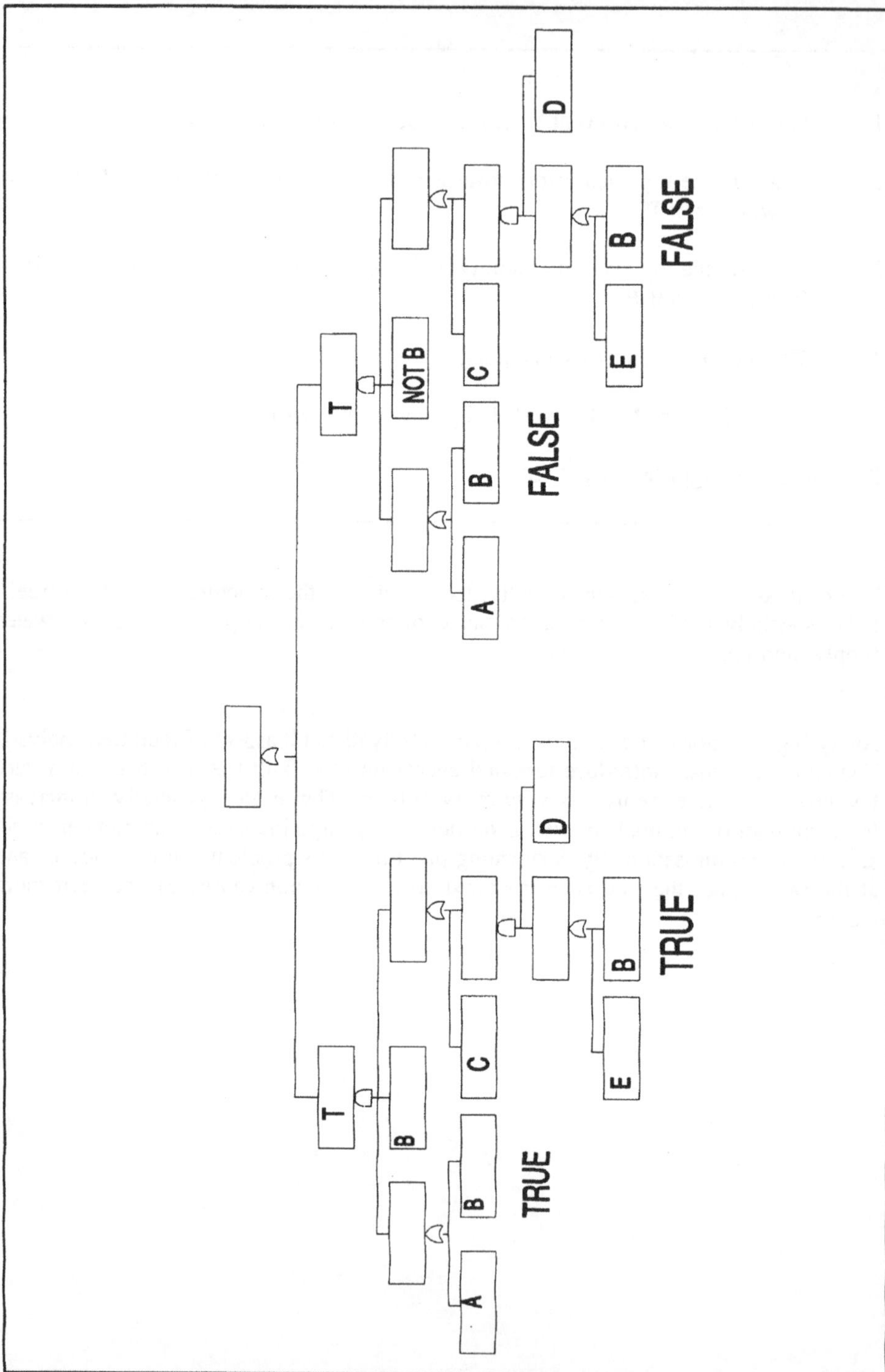

Fig. 4.26 Fault tree after pivoting

Example

The fault tree of Fig. 4.25 has the structure

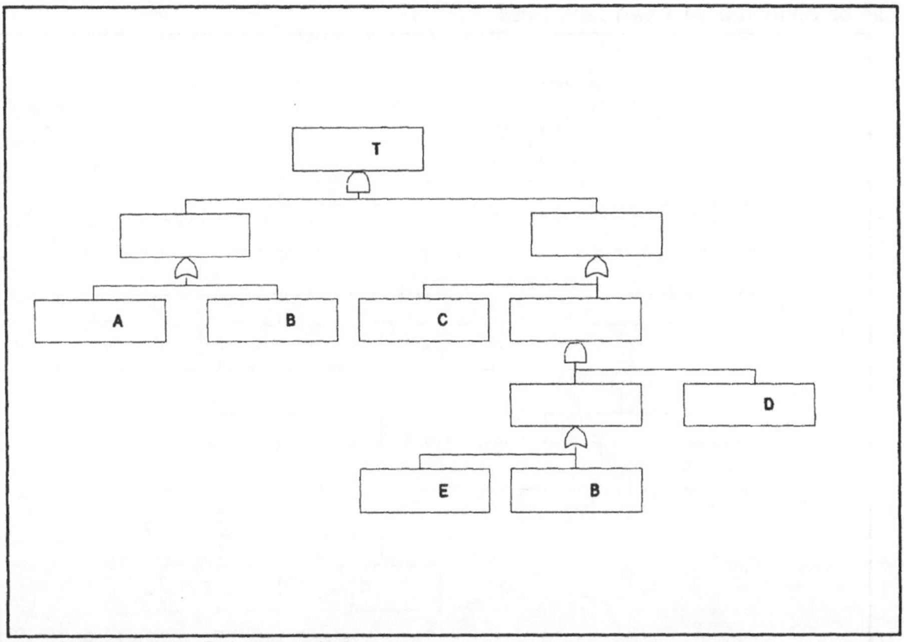

Fig. 4.25 Fault tree before pivoting

The event B is an obvious candidate pivot event. The fault tree is drawn twice with B and not B, as follows

The simplifying rules can now be used, as follows

X AND TRUE = X

X OR TRUE = TRUE

X AND FALSE = FALSE

X OR FALSE = X

The result is given in fig. 4.26 and 4.27.

The advantage here is that the two inputs to the mutually exclusive OR gate can be evaluated independently since there are not repeated events.

Pivoting can be looked at from a systems point of view. If there are two causal paths leading from a component to a top event, the event will appear two times in the fault tree. Dividing the system up into two parts, with the component on two causal paths as one part, the fault tree construction can be divided across two fault tree construction can be divided across two fault trees.

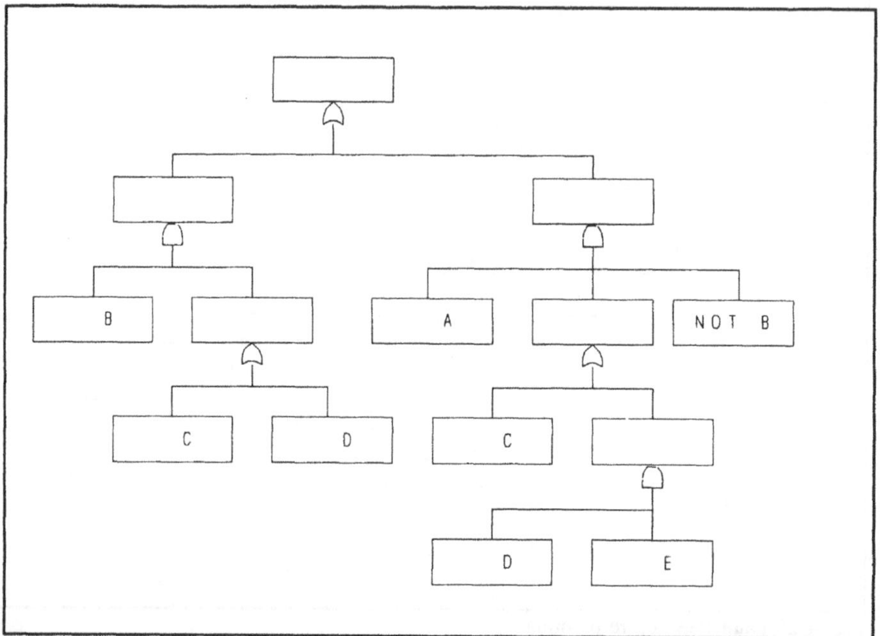

Fig. 4.27 Simplified reduced tree

References

Fussel, J.B. A collection of Methods for Reliability and Safety Engineering, ANCR-1273, 1976

Further Reading

E.J.Henley, Kumamoto Reliability Engineering and Risk Assessment, Prentice Hall 1981

5 Failure Data for Specific Components

Use of generic data in risk analysis is always dangerous. There is the ever present problem that in the actual plant, components may perform more poorly than a general industrial average. The result may be an overly optimistic calculation. On the other hand, it is in practice rare to obtain a significantly pessimistic calculation. For combinations of failures, for component such as pump and valves if average figures are used, since it is difficult to operate components so that their failure rates differ significantly from average. Specially susceptible components are found quickly, and replaced.

During plant design, however, generic failure rate data is all that is available. Some calculations will need to be made even before equipment suppliers are chosen. For this reason, a set of typical data is given here. The data have been obtained by taking fairly representative components which have been studied by the author or his colleagues, and recording their failure modes and failure rates. The results have been cross checked with available data base figures. Note that the reader should contact the author for the most recent data, since estimates are under constant revision.

Only single failure rate figures have been given, since it is generally impossible, with the amount of data available on a single component type, to give statistically significant values for spread. Normally, failure rates given will be accurate to within a factor of three for mechanical components. There is a certain guarantee for this in the behaviour of maintenance teams. With failure rates from one failure per year to one failure per ten years any deviation larger than a factor of three will be fairly quickly noticed, and have a reasonably good chance of correction.

There are special problems for components for which failure rates depend on frequency of use or activation. Examples are infrequently opened valves, or standby pumps. An attempt has been made to relate failure rates to degree of use. But uncertainties here must be at least a factor of 10 in each direction. This is due to the additional uncertainty in determining frequency of use, and uncertainty as to whether some failure mechanisms are operating continuously.

5.1 Valves

There are many different kinds of regulating and shut off valves, used for a very wide range of purposes. The most usual regulating valves though, are globe valves, or needle valves when regulating very low flows, and butterfly valves for regulating flows in very large pipelines. Usual shut off valves are ball valves or plug valves used for small pipes and gate valves or ball valves used for larger pipes (4" +), with a good deal of overlap in the ranges.

It is important when studying reference data to consider whether failure rates include valve actuators or motors. The primary failure modes for valves are:

- The valve leaks to the environment through flanges, through seals on the valve body (split body valves) or especially, through valve stem packing as a result of drying out, corrosion, wear, or inadequately tightened packing nuts.

- The valve cracks and leaks to the environment

- The valve leaks over its seat. Typical causes are corrosion, erosion, trapping of foreign objects on closure such as wild spatter, crud, crystals or metal turnings, or a broken valve trim or seat.

- The valve can jam open because of a foreign body across the seat (anything from large crystals in process liquors to beer bottles and welding rods).

- The valve cannot open or close (sticking). Typical causes are drying out or overtightening of packing, breaks or distortion (temperature expansion or sideload, people standing on valves, hitting them with ladders or cranehooks) on valve stems, installation error, corrosion products, or crud, around the stem, crud or frozen liquor around the valve seat. For valves regulating gas or flashing liquids, the problems of freezing are especially significant.

- The valve opens or closes too slowly, as a result of overtight packing, corrosion or drying out of lubrication.

- The valve opens or closes suddenly usually as a result of a broken valve stem.

- The valve becomes blocked as a result of foreign bodies, crud or freezing of product, sand etc.

Some of these failure modes relate to the fluid passing through a valve rather than the valve itself. Failure rates due to these causes cannot be determined for the valve alone. They must be derived by studying the plant as a whole. For such failure modes, a good deal of engineering judgement is needed.

Failures that prevent opening and shutting of valves are dependent on the frequency with which a valve moves. Infrequent movement gives greater possibility for corrosion or crud to build up.

Tables of failure mode probabilities for hand operated gate valves are given as table 5.1 to 5.3. An assumption is that the flowing material is not especially corrosive, nor at a temperature over 200° C.

Large manually operated valves require gear or worm drivers, and these introduce additional possibilities for failure. Particular problems are objects which lodge in the gearing preventing movement, and pins or keys which work loose or break, resulting in no transfer of movement to the valve.

Motor controls on valve introduce new failure possibilities:

- The motor can burn out, especially if it is overloaded because the valve is too stiff, or jammed.

- Ordinary motor problems can arise, such as brittle insulation, vibration damage, worn out brushes, worn out bearings, loss of cooling due to fan breakage or thick layers of dust, foreign bodies or water in the motor, all of which can lead to short circuits or open circuits.

- Faults in adjustment of end stops or position switches so that valves do not open or close completely, or are overdriven by the motor.

- Excessively tight closure can prevent later opening. Adjustment errors are dependent on conditions especially on whether the actuator is factory or site mounted.

- Faults due to actuator levers fouling other equipment, being bent, or coming loose (especially on butterfly valves).

Failure probabilities for a motor operated gate valve are given in table 5.4.

By far the majority of regulating valves in sizes up to 8" are actuated by compressed air. The actuator introduces a number of new failure modes.

- The airways in the actuator can be blocked, resulting in a problem that the valve can either open spontaneously, or close spontaneously or may refuse to open or refuse to close. Water in the air lines can freeze.

- The framework or harness supporting the actuator can be distorted (physical damage) so that the valve stem jams and cannot move.

- The valve spring can break, or bind, or be badly adjusted. Spring breakage normally leads to total opening or total closure depending on whether the valve is "fail open" or "fail close". Springs are normally installed in such a way as to cause the valve to fail open or fail closed, when air pressure is lost (spring breakage leads to non fail safe action). Spring binding leads to inability to move valve.

- When a valve is used for regulation, it is important that the valve can respond to small signal changes. Overtight, dried out or leaking packings or bearings can mean that the valve sticks slightly, and does not respond to small signal changes, later giving a violent change when signals become large enough (Stick slip failure).

- The membrane in the actuator can leak, pneumatic pipes can break, fire can cause loss of air. This results in opening or closing, depending on whether the valve is designed to be fail open or fail closed.

Table 5.5 shows failure rates which should apply for globe valves in the 3" to 8" range with pneumatic actuators.

Beyond the failure modes so far described there is a series of rare failure modes, which do not play a statistically important role, but which should be checked during inspection, failure diagnosis, or commissioning.

- The valve can be installed the wrong way round. This is not uncommon and should be checked prior to commissioning. The result can be control difficulties, that liquid pressure can affect valve opening and that excessive corrosion or vibration can take place. For three-way valves the result can be misdirection of flow.

- The wrong valve with respect to size valve trim, etc. can be installed which changes the control characteristic.

- The wrong valve with respect to material of construction can be installed. This is especially serious if the material used corrodes quickly e.g. brass fittings on ammonia service.

- Valves should be, but are not always, installed so that a broken valve stem does not give a dangerous opening or closure.

- Failure rates for flange leaks and valve body cracking can be much higher than given here, if the standard of piping designs is low, with excessive stresses on the valve.

	Failure mode	Failure rate
+)	Leak via packing	1 per 10^6 hr.
+) x)	Crack in valve	0.1 per 10^6 hr.
+) x)	Break, burst	0.001 per 10^6 hr.
Normally open valve		
	Will not close	
	often used	0.1 per 10^6 hr.
x)	seldom used	0.3 per 10^6 hr.
		$3*10^{-3}$ per activation
	Leak over valve seat	0.1 per 10^6 hr.
	Unwanted opening	0.01 per 10^6 hr.
	Foreign body in seat	depends on system
	blockage	depends on system
Normally closed valve		
	Will not open	
	often used	0.1 per 10^6 hr.
x)	seldom used	0.3 per 10^6 hr.
		$3*10^{-3}$ per operation
	Leak over valve seat	0.03 per 10^6 hr.
	blockage	depends on system
		higher than for open valve
x)		at most once per month

x) pooled figures for all valves due to scarcity of data

+) For high pressure equipment designed e.g. to API standards and operating at ambient temperature with no significant temperature cycling, these values may by reduced by a factor of 10, or, if indicated by relevant experience. For equipment which undergoes temperature cycling, such as low pressure steam service the values may be much higher.

Table 5.1 Failure rates for hand operated gate valve (8")

	Failure Mode	Failure rate
+)	Leak through stem packing	2 per 10^6 hr.
+)x)	Crack in valve	0.01 per 10^6 hr.
+)x)	Total breakage of valve	0.001 per 10^6 hr.
	Normally open valve	
	will not open	
	often used	.1 per 10^6 hr.
	seldom used	.3 per 10^6 hr.
	Leak across valve seat	.3 per 10^6 hr.
	Foreign body in valve seat	
	Blockage of flow	depends on system
	Normally closed valve	
	Will not open	.3 per 10^6 hr.
	leak over valve seat	.1 per 10^6 hr.

x) figures derived from pooled data for all valves due to scarcity of data.

+) For high pressure equipment designed e.g. to API standards and operation at ambient temperature with no significant temperature cycling these values may be reduced by a factor of 10.

Table 5.2 Failure rates for manually operated globe valve (2")

Failure Mode	Failure rate
Leak across seat, or will not close	6 per 10^6 hr.
Leak through stem packing	1 per 10^6 hr.
Other data unavailable assumed similar to that for 2" valve.	

Table 5.3 Failure rate data for globe valve manually operated (4")

Failure Mode	Failure rate
Will not open	8 per 10^6 hr.
Will not close	
Crack	0.1 per 10^6 hr.
Burst, broken	0.005 per 10^6 hr.
Will not close tightly	2 per 10^6 hr.
For other data see earlier	

Table 5.4 Failure rate data for 8" motor operated gate valve

Failure Mode	Failure rate
Leak across stem packing	30 per 10^6 hr.
Leak across seat when closed	6 per 10^6 hr. (depending on use)
Slow response	6 per 10^6 hr.
Will not open) Will not close)	4 per 10^6 hr.
Crack	0.1 per 10^6 hr.
Break, burst	0.005 per 10^6 hr.
Unwanted opening or closure in fail safe direction	1 per 10^6 hr.
Unwanted opening or closure in non fail safe direction	0.03 per 10^6 hr.

High pressure drop, corrosive or erosive flows increase failure rates by a factor 2-3. Control valves are especially sensitive to crud, foreign bodies, crystals etc, which can raise failure rates by a large factor, even to several failure per year.

Table 5.5 Failure rates for a globe control valve, 4" with pneumatic actuator.

5.2 Check Valves

The purpose of check valves is to prevent reversal of fluid flow. They are often used at the output of centrifugal pumps, so that when the pump stops back flows from the output through to the input of the pump is prevented. A fairly common accident mechanism arises when a pump is stopped, and the following check valve leaks. Reverse flow through the pump can then cause overpressuring or overflow in feed tanks from which the pump draws it feed.

There are many different kinds of check valves, from those with a simple caged ball, or a simple flapper, which closes over a seat, through to quite complex valves which not only prevent reverse flow, but also allow automatic closure.

The failure modes for the simpler valves are

- The valve fails to close when pressure gradients reverse. This happens especially if the valve trim is stuck, through corrosion or through collection of deposits, or if the trim is very worn. Pulsing flows can in some cases cause excessive wear on flapper type check valve hinges, so that the flap fails to seat properly. Foreign bodies in the valve mechanism can also cause failure to close.

  The result of this failure mode is that either a small leakage or even a large flow, may be transferred backwards through the valve.

- The valve may fail to open after it has been standing for a period with no flow. This is caused typically as a result of corrosion, deposits, crud, or simple blockage with foreign bodies or crystalline or pitchy masses.

- Corrosion may cause the flapper or ball to leak

Table 5.6 gives failure rates for check valves.

Failure Mode	**Failure rate**
Failure to close, small leak, over seat	
- depends on nature of flow and on amount of solid material in flow, typically	2 per 10^6 hr.
with vibrating flows, up to	100 per 10^6 hr.
with very clean, non corroding environment	0.2 per 10^6 hr.
Failure to close, catastrophic leak over seat	one third of the above
Blockage - depends entirely on the flow type and possibility for freezing or crystallisation	
Failure to open	
- typically	0.2 per 10^6 hr.
- in corrosive environment	10 per 10^6 hr.
Burst	0.005 per 10^6 hr.
Leak to environment	0.1 per 10^6 hr.

Table 5.6 Failure rates for 4" check valve, flap type

5.3 Piping

The failure modes for pipe sections can be described as either blockage, leak, or break. The frequency of blockage will normally be very dependent on the flow type, particularly freezing or solidification of substances or to crystallisation from solution, due to heat tracing failure or especially cold weather conditions are a cause. Only in extreme cases will corrosion be a cause of blockage. Mechanical damage, as a result of crashes, or explosions, can pinch or bend pipes sufficiently to prevent flow. Otherwise, the main causes are deposits, crud, or foreign bodies, especially at low points in a pipe section, and with still standing flows and freezing of liquids.

Leak or break of a pipe can be caused by many things

- Corrosion or erosion
- Crashing vehicles, cranes
- Overstressing, when pipework is also used as structural support (especially during construction or rigging)
- Damage occurring during installation (this can remain hidden for a long time)
- Expansion stresses, especially if flexible joints are jammed
- Movement in structures or foundations
- Underground pipes may be inadvertently excavated
- Explosion or overpressure, both internally and externally
- Overtemperature due to process runaway or to fire, causing weakening of pipe walls
- Vibration causing fatigue
- For very high temperature pipe work, creep.
- Fire

Failure rates for all these causes are impossible to give - they depend far too much on circumstances. The more generally applicable causes can be estimated as a general average. API failure data have been used as a basis, supplemented by data from direct experience, and from other literature sources. Causes such as crashes, explosions and overpressure bursts, which lead to secondary failures in pipes, have been excluded from the data. The interpretation of data is further complicated by the fact that the term "leak" is not generally well defined.

A leak here is defined as a small crack or pinhole in the range 0.01 mm^2 to 10 mm^2. A large leak is in the range 10 mm^2 to 1000 mm^2. Anything above this is regarded as a break.

Data for small pipes, from 1/2" to 2", and for medium size pipes for general service, in ranges 2" go 6" and pressures up to 30 atm. are described. for larger and higher pressure piping, the fracture mechanics methods described in the section or pressure vessels can perhaps be used.

Failure Mode	Failure Rate
Blockage	depends on system
Small leak	0.001 per 10^6 hr. m = 10^{-9} per hr. m.
Large leak	0.0001 per 106 hr. m. = $1*10^{-10}$ per hr. m.
Break	0.00003 per 10^6 hr. m. = $3*10^{-11}$ per hr. m.

Plus rates for secondary failures e.g. catastrophic corrosion, overpressuring.

Table 5.7 Failure rates for pipes in 2"-6" class up to 30 atm.

Failure Mode	Failure Rate
Blockage	depends on system
Small leak	0.001 per 10^6 hr. m.
Break	0.00003 per 10^6 hr. m.
Plus failure rates due to secondary failures	

Table 5.8 Failure rates for pipes in 1/2" to 2" class.

5.4 Flanges

The most important failure modes for flanges are leak or breakdown.
The causes may be many

- The flange can be too loose due to the fact that screws or bolts are not tightened, or because heating, overpressuring, vibration, mechanical loads have lengthened bolts or opened the flange .

- Overstressing of bolts can cause breakage. Pipe expansion which has not been allowed for, or jammed hangers, can bow pipes and cause opening of flanges

- Sometimes too few bolts will be fitted or some will be left untightened, during installation or maintenance, as a result of error. The remainder may be overstressed.

- Corrosion products in the flange opening itself can force flanges apart.

- Corrosion or aging can weaken packings.

- Corrosion under a packing, particularly if the packing is scored, can lead to a leakage route around the packing.

- Foreign bodies, mishandling, dirt, or burrs on the flanges can cause damage to seals under installation.

- Bolts can break through overstressing, fatigue or corrosion.

- Sometimes flanges are simply unbolted in error by maintenance staff while the system is under pressure, or remains undrained.

- Flanges can fail as the result of other failures. For example packings can be blown out by overpressure (unless recessed) or be destroyed by fire.

Table 5.9 gives failure rates for failure modes which do not involve human errors or are not the result of secondary failures.

Failure Mode	Failure Rate
Small leak +) (0.1-10 mm^2 hole)	0.4 per 10^6 hrs.
Packing blown out, or partially destroyed, or flange opens several mm (not recessed flanges)	0.03 per 10^6 hrs.
Flange breaks open +)	0.01 per 10^6 hrs.

+) for high pressure piping designed to high standards, and with no temperature cycling, the valves may be divided by 10.

Table 5.9 Failure rates for flanges in 2-6" piping.

Small leaks of corrosive liquors such as hot, salt laden water under pressure, can corrode or erode paths over flange faces, so that a large leak is formed in the course of a few hours.

5.5 Pipe Welds

Welds represent one of the weakest points in piping, due to the problems of weld inclusions, material differences, pits and holes, and of unavoidable structural changes to material under the heat of welding. It is probable that many of the failures in piping described earlier in fact occur at welds. Nevertheless, more accurate results are obtained if failure rates specifically for welds are added to the overall failure rates for pipes, at least in the cases where there are many welds. Failure modes concern leakage and breakage. WASH 1400 data, which concerns high pressure, moderately high temperature steam and water service, is reflected in table 5.10.

Failure Mode	**Failure Rate**
Leakage (less than 10% of weld area)	0.03 per 10^6 hr.
Breakage	0.003 per 10^6 hr.

Table 5.10 Weld failure rates for larger pipe sizes (1 1/2" to 24")

5.6 Pressure Vessels

There is an enormous number of potential causes of pressure vessel failure, and of these, only a few can be evaluated in terms of a generic failure rate. For example, airplane crash is a potential cause of tank failure, but it depends more on tank size, distance to airports, and air traffic rate, than on the tank itself.

A list of more common causes of pressure vessel failure is:

- Design errors, including underdimensioning, specification of inadequate materials, specification of wrong welding procedures, overloading of supports and designs in which there are weak points or stress raisers.

- Some pressure vessels require cooling, and multiwall or wound vessel require drains in the walls to protect against leakage - These present special problems.

- Overload, due to too high pressure or temperature, as a result of equipment failures, operation or administrative errors or fire.

- Material faults, welding faults, and faults due to errors in heat treatment.

- Corrosion (especially stress corrosion)
  Corrosion can attack both internally and externally. A frequent cause is improper water treatment, or contamination of water supplies. Stainless steel is especially vulnerable to chloride contamination. For vessels holding liquids other than water high water concentrations are often a problem (e. g. in liquid ammonia).

- Ageing (creep or fatigue)

- Excessive vibration (can cause fatigue or direct overload, usually at flange or weld attachment to vessels.

- Foundation collapse

- Frost heave

- Earthquakes

- Crashes (aircraft, ground vehicles, cranes, missiles from explosions.

- Internal explosions and runaway reactions.

- Structural overload of vessels due to external stresses, especially pipe expansion or contraction.

- Liquid expansion when a vessel is completely full of liquid, and is also shut off.

Some failures are the result of several causes combined.

Of these causes, overload, crash, explosion and vibration are properties of the application and not of the pressure vessel. They must be separated out in any treatment of statistical data. In risk analysis, risk from these causes should be assessed separately.

Similar arguments might also be applied to corrosion, since this depends on the environment in which tanks operate. So far, however, this has not been the normal practice in risk analysis, because of the difficulty of obtaining data.

Failure probabilities will be very dependent on the frequency with which pressure tests and inspections are carried out on tanks. Many of the direct causes of failure are small cracks or pits, which develop over a longer period until a critical size is reached. Non destructive testing, using ultrasonic or X-ray or gamma-ray photography can reveal many such flaws, as can surface inspection in some cases, For some applications pressure vessels are inspected regularly, every two or four years (especially transport vessels).

There have been several thorough studies of pressure vessel failure rate data. Phillips and Warwick (PHI 68) studied pressure vessels built to very high standards, and in a following study (SMI 74) the number of "tank years" was brought up to 105402. These studies showed probability for catastrophic failure of pressure vessels of $4.4 * 10^{-5}$ per year. Of these, catastrophic failures were such a small part that a failure rate for catastrophic damage of $3 * 10^{-6}$ could be given. In all, the study covered 1,700,000 tank years. Table 5.10 shows the distribution of errors.

Boesbeck (1975) undertook an evaluation of these and other data, to find a probability of catastrophic failure which is somewhere between 10^{-5} and 10^{-6}. These studies lead to the results in table 5.11.

A. M. Thomas of Rolls Royce Ltd. has, in a series of articles, used the available data for tanks together with data from pressure vessel testing, to build up a model for pressure vessel reliability. This uses a theory of fracture mechanics which is not universally accepted, but which seems to give a reasonably good correlation with the available data.

Failure Mode	Failure Rate
Catastrophic failure	$3 * 10^{-6}$ per annum $2.5 * 10^{-4}$ per 10^6 hr.
Small leak or small break	$3 * 10^{-5}$ per annum $2.5 * 10^{-3}$ per 10^6 hr.

Table 5.11 Failure rates for pressure vessels.

The theory is based on the conditions which are necessary for a built in flow to reach a critical size. The constants in the theory are adjusted to fit data from 700 well studied tank failures.

5.7 Storage Tanks

For small storage tanks, almost the only failure modes are leaks, and blockage of input or, more especially, output pipes. Bursting, as a failure mode, is generally caused by outside events such as overpressuring, crashes, or fires. Cover filling or filling with wrong substance and foundation or support collapse are also possibilities.

The probability of leaks is a function of tank age, substances stored, how big a corrosion margin was applied etc. A minimum value of 1 per 10^6 hours is reasonable, it being hard to come under this with ordinary mild steel tanks, unless there is frequent inspection (every 2 years or so).

The leak size will normally be at most a few square millimeters initially.

There are several things which can occur to tanks as a result of failures and errors in pumping procedure. Tanks can be overpressured because they are fully filled with liquid, with no overflow or expansion possibility. Similarly, they can be sucked in, if that are drained while vent lines are closed or blocked (depends on wall thickness, and tank size and form). Vent lines are sometimes closed to keep out rain, or to prevent ingress of dirt, during repairs or installation, and this has led to the destruction of many tanks by over or underpressuring. Blockage with ice or birds nests has also been a problem. Note that even the pressure drop due to flow down through a drain pipe can cause a thin walled tank to be "sucked in". Vent valves are often fitted, or vacuum safety valves. These can jam, with a reliability of the same order as low pressure safety valves.

Large storage tanks have many and varied designs of which the principal ones are cone roofed tanks and floating roofed tanks. Corrosion is a possibility here, and leads to a need for periodic inspection. Failure rates depend very much on the material stored.

Apart from various valves, the most likely causes of leaks are either overfilling (this can cause tank roofs to raise) and corrosion attacks around welds. A leak probability of 1 per 10^6 hours for a few square millimeter leak, and 0.1 per 10^6 hours for a large leak, has been given by Davidson.

Catastrophic failures generally arise from overfilling, as a result of operator error, administrative error, or instrument failure, and internal fires or explosions in oil tanks. The reliability of inerting systems used in hydrocarbon storage tanks to prevent explosive atmospheres arising requires a special study in itself. A typical pattern involves overfilling or overpressuring, which then results in a release from a weak, improperly welded, or corroded tank seam. Japanese data gives values of 10^{-5} per year for such failures.

Floating roof tanks have some special failure modes, In some (generally earlier) designs it was possible for the floating roof to jam and then to be damaged as its full weight was unevenly supported. Seals against air ingress can also fail. A special failure mode is the possibility of the roof sinking, if sufficient blockage of drain pipe).

An unusual but relevant failure mode for large tanks in that they can float, and break free from pipe work, in floods, or if the area within bund walls or berms becomes flooded.

Fire in large oil storage tanks has a historical probability of about 0.05 per 10^6 hr, (API).

5.8 Centrifugal Pumps

Single stage centrifugal pumps are used almost everywhere in the process industry to transport water, oil or reagent from tank to tank, or between process units, or to circulate heating and cooling fluids or reagents.

By far the most frequent failure with such pumps is leak through the seals. Such a failure is typically caused by :

- Wear and correspondingly inadequate maintenance.
- Drying out of packings, and repeated starts with empty or unlubricated pump seals. This is especially a problem if pumps standing still can allow crystallization of process liquors in and around seals.
- Hard substances, grit, or crystals forced onto seals due to overpressure, or due to failure of seal lubrication flows on larger pumps.
- Wrong seal types fitted, or seals fitted wrongly.

As indicated above, some pumps are fitted with special cooling, seal washing, and bearing lubrication systems, though this is more common on multistage pumps. A failure in this system can quickly lead to seal failure. Analysis of such systems needs to be carried out as a separate exercise. If seal and lubrication systems are common to several pumps, the possibility of simultaneous failure, or of one pump contaminating seal liquids for all, needs to be taken into account.

In cases where particularly flammable or toxic materials are pumped, double seals are often fitted. In this case it is possible to fit drains between the seals which lead to safe disposal, and to fit pressure monitoring between the seals.

Other failure modes for simple centrifugal pumps are:

- Blockage of pumps, especially as a result of dried crud, crystallisation, or solidification of product or freezing, so that the pump will not start.

- Blockage of the pump during operation, as a result of foreign bodies, or slugs of solid material.

- Blockage of the pump during operation, as a result of seizure of bearings (overheating, lack of lubrication).

- Breakage of pump axles, or decoupling of flexible joints between motor and pump. This can arise because of fatigue, corrosion, misaligned bearings, or, in the case of flexible couplings, loose joints.

- Impellers can break off, as a result of corrosion, erosion, fatigue. Erosion occurs especially if the net positive suction head (NPSH), that is, the pressure of the input to the pump is too low,so that cavitation (small vapour bubbles forming and collapsing) occurs. The cause of such low NPSH is typically blockage of feed lines to a pump, particularly at filters, or fitting of too many pumps to the same feed line. High temperature in feed flows can be a problem too, because cavitation occurs when pressure in certain parts of the pump suction falls to the vapour pressure of the liquid.

- If there is blockage at the input to the pump, it is possible for air to be sucked into the flow.

- Some pumps are self priming, that is, they do not need to be filled with liquid before starting. But most pumps need to be primed, and loss of priming due to other events means that no liquid will be pumped. Operation without priming can cause bearings and seals to fail and either leak, or seize.

  A sudden pressure fall at pump input, such as when another pump starts up on the same feed line, can cause loss of prime.Depending on motor type, loss of prime can prevent pump start or cause overspeeding.

- Operation of pumps against too high a head, for example as a result of closure of valves in output lines, can cause damage to seals, by forcing liquid into them. Lack of flow through a pump can lead to overheating, and resulting seal damage or even seal leakage and fire in the case of flammable liquids.

For these reasons, many pumps are fitted with temperature or flow trips, and they may in extreme cases be fitted with safety valves or spill lines which ensure a minimum flow through the valve.

- When a centrifugal pump stops, it is possible for liquid to run backwards through the pump. For this reason check valves are generally fitted in output lines. Back flow can lead to overpressuring of supply lines and tanks. Back flows can also, in some cases, cause the pump to work like a turbine. Overspeeding can as a result cause pump damage. With some pump motor types, the reverse rotation can cause generation of electricity. There have been cases of electrocution of maintenance workers as a result of such generation.

- Pumps can leak or burst, from overpressure, fatigue, corrosion, fire, etc.

Beyond the pump itself the drive motors are a significant cause of failure. The most common failures here are short circuits due to overheating or mechanical damage. These can in turn be caused by

- Blockage in the pump, or operation against too high a back pressure (closed valves) can cause motor power to rise leading to overheating, and possible short circuit or fire.

- Bearing lubrication can be inadequate leading to overheating or seizure.

- Dust or dirt covering motors can prevent cooling, and lead to short circuits or fire.

- Ageing or vibration can cause motor parts to loosen and break free. These can then cause physical damage, stopping the motor or short circuit.

- Bearing problems, and mechanical damage, can prevent a pump starting.

- An especial problem with some pump motors is that they can be misconnected during installation or maintenance, so that they run the wrong way round.

- Breakage of leads, or winding open circuits, can cause motors to stop, or prevent starting.

- Most motors are fitted with temperature trips, and many with current trips. Failures in these can cause motors to stop, or can prevent starting.

- Some motors have gears or drive belts. Broken drive belts, especially as a result of wear, foreign bodies in the drivers, or simple breakage can stop pump or prevent starting.

- For motors with brushes (direct current, on three phase induction motors) brush wear can cause overheating, short circuits or open circuits.

It is virtually impossible to break down pump or motor failure data for all these failure causes. In table 5.12, data are given for pump failure modes, which includes both single stage centrifugal pump, and drive motor failures.

Failure Mode	Failure Rate
Leak through seals	30 per 10^6 hr.
Pump stops, or fails to pumps, during operation (internal failure rate, excluding suction head or electrical supply problems)	20 per 10^6 hr. 300 per 10^6 hr. in poor (dusty, corrosive) environments
Pump will not start	0.001 per activation (frequent activation once per day) 0.01 per activation (infrequent activation) 30 per 10^6 hr.
Destruction of seals	Depends on system and maintenance
Leak or burst of pump housing	0.03 per 10^6 hr.

Table 5.12 Failure rates for single stage centrifugal pumps

An important question is the size of leaks from pumps. The size of possible seal leaks can often be read from the pump drawings. A reasonable minimum figure and a 1 mm leak over the whole seal circumference a maximum. Maximum leakage occurs in 1/5 to 1/10 of leakages and involves breakage of a section of the seal. The wear modes of leakage often occur slowly at first, but accelerate, so that frequent inspections may reveal leaks before they become serious.

5.9 Safety Valves and Burst Discs

The most usual form for failure in safety valves is that they open, and then do not close completely. Unwanted opening as a result of vibration or waterhammer is another reasonably frequent failure mode. Failure to open on demand is on the other hand a fairly rare failure mode.

Safety valve systems can suffer from many problems.

- If the safety valve lies too far from the vessel it should protect, there is a possibility for vibration to arise, or oscillation. The vibration can shake flange bolts loose, can cause pipework to break, can cause valve internals to jam open, and reduce the capacity of the safety valve to relieve the overpressure.

- If several safety valves are installed on the same vessel, and with similar opening pressures, there is again a good chance of vibration due to flow interactions.

- If the valve has a setpoint close to the operating pressure of a plant unit, it is possible for the valve to open and close frequently, or to chatter. This can cause wear on seats. If the vessel contains solutions, or liquids which can freeze or polymerize, then deposits can build up in the valve, or in its outlet pipe, which can block the valve. For this reason, metal burst discs are sometimes fitted in front of safety valves, with precautions being taken to prevent overpressuring between the disc and valve. Graphite burst discs are not fitted here because of the danger of valve blockage with pieces of graphite.

- Liquids and deposits can gather in the blowdown pipe following the safety valve. These can cause blockage, can change the pressure at which the valve opens, and can reduce the flow rate through the valve. They can also lead to water hammer effects when the valve opens. For these reasons, blowdown pipes leading to a blowdown tank should generally be fitted with a drain, and the drain needs to be inspected regularly.

- Valves may be "gagged", that is locked closed, in order to carry out pressure tests. If the gag is left installed by error, the valve will not open.

- For a host of different reasons, safety valves may be adjusted to the wrong pressure.

- When a safety valve opens, there is a strong jet of vapour expelled. The reaction from this jet creates significant force. A surprisingly frequent failure mode arise if the valve supports are not designed to take this force. The result is that the valve pipe work bends. It can then pinch off the flow through the valve, or can break.

Failure rates for safety valves themselves, assuming proper installation and calibration given in table 5.13. The data are reasonably typical for 1 to 4 inch valves. Large valves, and those that work at very high pressures (> 100 atm) have an apparently higher failure rate.

Failure Mode	Failure Rate
Unwanted opening	0.1 to 1000 per 10^6hr. depending on setting and operating conditions
Leak over seat	1.0 per 10^6 hr. 10.0 per 10^6 hr. if the valve opens frequently
Valve does not open on high pressure	10^{-5} per activation
Valve does not close after opening	3 * 10^{-3} per activation

Table 5.13 Failure rates for safety valves, 2" to 6" in gas and vapour service

Burst discs have a very simple construction, so that failure to open in a properly constructed and installed burst disc is almost unheard of. Reverse buckling discs can fail to open sufficiently quickly to ensure full cutting, if they have been damaged during installation. Beyond this, almost the only cause of failure to open is fitting of the wrong discs type or wrong pressure level discs, and blockage due to deposits. In some cases dished discs can be fitted the wrong way round.

There are two other significant failure modes for burst discs - leakage due to corrosion, and unwanted bursting. Leakage occurs in corrosive environments and can lead to blockage or blowdown lines if the leaking substance can give rise to deposits. When burst discs are used in combination with a safety valve, pinhole leaks become a danger too because pressure can build up behind the disc, insufficient to open the safety valve, but sufficient to prevent the burst disc from bursting. Breakage and unwanted opening generally occurs as a result of problems in installation, uneven seatings etc, or due to pulsating pressure causing fatigue. Table 5.14 gives some reasonable reliability estimates, based on accident case stories.

5.10 Positive Displacement Pumps

Positive displacement pumps for liquids are used to achieve high pressures with relatively low flows, and in some cases, when the amount of liquid needs to be closely metered. There is an enormous number of types, but the type considered here is that where there is one or more reciprocating piston.

The failure modes are then

- Leaks, especially around pistons, often due to wear or scoring. In many pump designs, there are additional seals around piston rods or pumps, and leakage may require failure of both piston rings and seals.

Failure Mode	**Failure Rate**
Unwanted opening	Depends on installation, and on degree of pressure pulsing, but generally >0.1 per 10^6 hr. up to 100 per 10^6 hr. in extreme circumstances
Leak	Depends on substances and materials used V. Low for graphite discs, usually due to corrosion or pressure pulsing fatigue.
Fails to open as a result of high pressure (depends on installation error)	10^{-4} per demand 10^{-3} per demand unless specific rules and procedures are given for installation.

Table 5.14 Failure rates for burst discs.

- Overheating, especially due to blockage in cooling systems. Overheating can cause bearing and piston damage, and can, in high power pumps, lead to boiling, or overheating of process fluid.

- Mechanical damage, such as broken pistons, broken piston rods, cam shafts or crank axle, and pump blockage. Some of these failures can cause gross leakage or just failure to the pump.

- Valve failures, leading to leakage back across the valves, and loss of pumping efficiency. Double failures can lead high pressure liquids back to feed lines.

- Various motor failures, as described for centrifugal pumps.

- If the output of a positive displacement pump is blocked, for example by erroneous closure of a valve, output pressures can rise dramatically and in some cases can cause piping to burst (positive displacement pumps with powerful motors approximate closely a constant flow device). Many pumps have built in safety valves which can spill liquid from output back to input, in case of overpressure. In such cases, the analyst must take account of three other failure modes - failure of the safety valve to open, failure to reclose, and overheating in the liquid recirculated through the safety valve.

- Dirt in the liquid being pumped can cause piston ring or seal failure, and can block valves.

Failure rates for these modes are shown in table 5.15

Failure Mode	Failure Rate
Leak (packings, piston)	60 per 10^6 hr.
Will not pump (motor failure, valve failure)	40 per 10^6 hr.
Refuses to start	30 per 10^6 hr. for pumps in standby
Cracking, or burst	5 per 10^6 hr.

Table 5.15 Failure rates for piston pumps 0.3 up to 2 HP, pumping liquids

5.11 Filters

Filters for straining solids, grit, and foreign bodies from liquid flows can take many forms. The actual straining unit consists usually of a layer of metal mesh, plastic polymer mesh or felt, or cellulose felt or paper. Many filters take the form of a pleated "cartridge" which can be removed from the filter holder, and replaced.

The three main failure modes for filters are as follows.

- The filter can become blocked by solids, sludge, crystals, etc., preventing flow or causing excessive pressure drop.
- The filter can be holed so that solid material can pass through it.
- The filter can break up (usually due to corrosion) and release "foreign bodies" in the form of filter material, into the flow.

The frequency of blockage is a function of the flowing fluid, rather than the filter, although the filter mesh size is relevant. The probability of blockage also depends on how often the filter is changed or inspected. Hole formation depends very much on the compatibility between the flowing liquid and the filter material. A frequency of hole formation of 3 per 10^6 hr. has been observed for steel mesh filters in pump cooling water flows.

5.12 Compressors

Reciprocating compressors, for air or other gases give failure possibilities very much like those for reciprocating liquid pumps. The most common failures are those which cause compressors to stop, to pump ineffectively, or to fail to start.

Faults in cooling systems, and leaks, are important reasons for requiring the compressor to be stopped, even though these failures do not necessarily in themselves stop the compressor.

For oil lubricated compressors, and even for some oil free compressors, there is an unusual and specific failure mode. That is, that oil film deposits build up on the outlet piping. Subsequent overheating can thereafter cause explosion in the piping, an effect which has been seen in several accidents.

Failure rates for the simpler failure modes are given in table 5.16.

Failure Mode	Failure Rate
Compressor stops, or fails to deliver air	200 pr 10^6 hr.
Compressor will not start	10^{-3} per activation (activated more frequently than once per week) 100 per 10^6 hr. (infrequently started)
Leak (cooling water system, or air/gas systems)	20 per 10^6 hr.
Burst cylinder block or pistons	0.3 per 10^6 hr.

Table 5.16 Failure rate for air compressor - plant air supply (4 bar)

5.13 Sensor Systems

All sensor systems have installation, and calibration problems, which can lead to failure. During installation it is necessary to ensure

- That instruments are placed correctly, so that they measure what they are supposed to measure at all times.
- That they are not affected by other equipment, vibrations, or temperature so as to give false readings.
- That hold down clips or screws have been removed.
- That instruments are correctly wired up, and especially, are wired to the right place and in the right sense.
- That correct calibration valves, correct calibration instruments and procedures have been used.
- That instruments are reconnected properly to the plant after calibration.

The Action error method, described in Ch. 6, can be used for analysis of the simpler kinds of errors in building up an instrumentation system. Beyond this, each type of instrument shows its own failure modes and rates.

5.14 Level Sensors

A simple type of level sensor is the float gauge, with a float, arm, and potentiometer read out, or a switch to indicate low or high level. These devices have many failure modes.

- The float can stick due to corrosion or deposits resulting in too high or too low a reading.
- The float can leak, so that it sinks, giving a false low reading.
- The float arm can break off, giving an unchanging or constant reading.
- The float or arm can come in contact with tank walls, so that it sticks.
- The arm can break off, giving an unchanging or constant reading.
- The potentiometer (on a measuring instrument) can break, open circuit, drift, or short circuit.
- Contacts on a switch instrument can weld together, corrode, become covered in dirt. Results may be permanent open circuit or short circuit. Typical failure rates are given in table 5.17.
- For differential transformer sensors, jamming of the transformer is sometimes a problem.

Failure Mode	Failure Rate
Sticking	50 per 10^6 hr.
Fails with low indication	5 per 10^6 hr.
Failure with high indication	0.1 per 10^6 hr.
Drift more than 5%	2 per 10^6 h.

Table 5.17 Failure rates for level sensors.

More reliable sensors are based on floats on a central induction sensing rod, or on resistance or capacitance measurements in an insulated tube. Such sensors are nevertheless subject to failures due to deposits. Sticking of rod and float type sensors depends very much on fluid and possibility of deposits, but is generally less than for arm type sensors. High and low signals in capacitance type sensors depend on both deposit formation and on possibilities of open and short circuits. Failure rates up to 50 per 10^6 hrs. have been observed (chlorine service) but 20 per 10^6 hrs. seems to be more normal.

Apart from these types of instruments pressure sensors such as diaphragm types are often used to measure level, even though the measurement is sensitive to changes in density. Such sensors may be used directly, or may be used with bubble types in which a slow stream of air is bubbled through the liquid, and the air pressure in the supply measured. Failure rates for the pressure sensors are described in the next section.

Bourdon tube and bellows sensors also have higher failure rates, according to up to 150 per 10^6 hours.

For sensors which must measure small pressure changes at high pressures (over 100 atm.), drift failures have been observed at higher rates, nearer 20 per 10^6 hr.

Failure Mode	Failure Rate
Drift greater than 5% from set point	5 per 10^6 hr.
Will not make contact (normally open contacts)	3 per 10^6 hr.
	10^{-4} per activation (Wash 1400)
Will not break contact or short circuit (normally closed contacts)	2 per 10^6 hr.

Table 5.18 Pressure switch failure rates when used for level measurement

5.15 Flow Sensors

The most usual form of flow sensor in the process industries is that which uses a precisely dimensioned orifice plate, and a device to measure pressure drop across the plate. Diaphragm or bellows type instruments are often used. Failure possibilities are primarily in the pressure measurement, or in the piping to the pressure instrument. For the orifice plate itself, the failure possibilities are.

- The wrong orifice plate can erode, again affecting calibration
- The orifice place can become blocked affecting both calibration and flow.
- The flanges at the orifice plate can leak

Orifice plate failures are fairly rare and completely dominated by pressure sensor problems. Flange leak probabilities are generally similar to flange failure given earlier.

Another form of flow sensor is the paddle types, especially used for flow switches. A paddle projects into the flow, counted eccentrically on a shaft. The flow turns the paddle against the force of a spring, and the turning causes contacts to open or close. In addition to the failures common to most mechanical instruments (loose parts, wear on bearings, breakage of levers, open circuits, short circuit welded contacts) paddle switches suffer from breakage of the paddle due to flow forces, and blockage.

Orifice plate failures are fairly rare and completely dominated by pressure sensor problems. Drift may occur due to erosion of the orifice plate, if the flowing liquid is of a kind likely to produce erosion, or if there is any flashing or serious pressure reduction across the orifice plate. Flange leak probabilities are generally similar to flange failure given earlier.

Other types of flow sensor are small turbines, in the flow itself, sensed mechanically or magnetically, and vortex meters which measure the rate or vortex formation.

Lees gives rates of failure for such devices up to 250 per 10^6 hr. or less for some types. Modern instruments have failure rates closer to 40 pr. 10^6 hr.

Failure Mode	Failure Rate
Pneumatic, Gas flow	
Drift greater than 5% from calibration	12 per 10^6 hr.
No output signal	4 per 10^6 hr.
No signal gain	4 per 10^6 hr.
Output signal high	1 per 10^6 hr.
Pneumatic, liquid flow	
Drift greater than 5% from calibration	5 per 10^6 hr.
No output signal	4 per 10^6 hr.
No signal gain	3 per 10^6 hr.
Output signal high	1 per 10^6 hr.

Table 5.19 Flow sensor failure rates

5.16 Temperature Sensors

The most common temperature sensor in the process industry is the thermocouple. To evaluate failure probabilities, one must look at a couple temperature sensing system, including the thermocouple pocket, the thermocouple itself, cold junction, and amplifier.

Thermocouple pockets are vulnerable to failure through corrosion or direct physical damage, and can leak. Corrosive products can affect the thermocouple itself, and give erroneous readings. Deposits on the thermocouple pocket can cause false readings.

The thermocouple can become open circuit, or can short circuit to ground or at some point along the circuit. The result is either no temperature signal, or a misleading signal.

The cold junction can fail also open circuit or short circuit, and additionally can have the wrong temperature, leading to erroneous readings.

The Thermocouple amplifier can in general change calibration, can fail giving no signal, or can give maximum signal. Failure rates are shown in table 5.20.

Failure Mode	Failure Rate
Short circuit to earth	3 per 10^6 hr.
Amplifier no signal	2 per 10^6 hr.
Open circuit	3 per 10^6 hr.
Drift > 5%	10 per 10^6 hr.
Amplifier maximum signal	1 per 10^6 hr.
Leak of thermocouple pocket	0.3 per 10^6 hr.
Cold junction failure	2 per 10^6 hr.
Wear from flow (where relevant)	100 pr 10^6 hr.

Table 5.20 Failure rates for thermocouple systems (apart from high temperature applications)

The analyst should note some system design problems which can lead to errors and failures. First, it is difficult to place the measurement in precisely the right place to observe the temperature required. Temperatures near pipe or tank walls are measured. If the flow is not homogeneous then temperature readings can be misleading. Temperature in vessels cannot be inferred by reading temperatures in attached pipes.

Secondly, if flow is low, the heat conducted to the thermocouple may be insufficient to maintain its temperature, as heat is conducted away. A similar problem when thermocouples are affected by heat tracing.

The third problem is that a thermocouple pocket has an appreciable heat capacity, which takes time to warm up. This lengthen the response time of thermocouples up to 5-10 secs.

Resistance thermometers, which measure the resistance of alloy wires, have similar failure modes to thermocouples, and similar overall failure rates. Cold junctions are not necessary in resistance thermometer systems.

5.16 Pressure Sensors

There are many types of pressure sensors for process plant use, but the most common ones are bellows and bourdon tube types, in which a bellows or tube changes shape, and diaphragm and strain guage types. For the bellows and tube types, the mechanical distortion must be transformed to a pneumatic or electrical signal, while for the strain guage type, the output signal is already electrical. The conversion of mechanical displacement to an electrical signal is performed for example by microswitch contacts, for an alarm, or by for example a differential transformer.

For pressure sensors, typical failure modes ar:

- leakage of the tube or bellows, so that the device releases a possibly hazardous substance. Leakage of hazardous substances along instrument wiring, back to control room fittings, have occurred
- leakage so that the sensor fails to register, or registers wrongly, due to pressure build up in the sensor housing
- jamming, or physical damage, so that the device fails to register
- open and short circuits, or pneumatic amplifier leaks, so that signals fail to register, or register wrongly.
- blockage of the connections to the process. This is an insidious failure mode, because the pressure guage will often pass calibration tests even when this fault is present.

Table 5.21 and 5.22 give some failure rates for pressure sensors

Failure mode	Failure Rate
Fail to switch on demand	6 per 10^6 hours
Spurious trip	32 per 10^6 hours
Leakage	.1 per 10^6 hours

Table 5.21 Failure rates for pressure switches (bourdon and bellows)

Failure mode	Failure rate
Erroneous output, drift	3 per 10^6 hours
No output	1 per 10^6 hours
High output	.5 per 10^6 hours

Table 5.22 Failure rates for pressure indicators (strain guage)

5.18 Transmitters

The signals from various instruments must generally be converted, from low levels up to some standard level for example 2-20 volt, or 2-20 milliamps, before being transmitted to a recorder or control room. In some cases, the transmitter is an integral part of the instrument. In all cases, a simple amplification, and a power supply line, are required. Power supplies are often taken from the controller or indicator which receives the signal.

Failure Mode	Failure Rate
Short circuit/no signal	.5 per 10^6 hr.
Fail high signal/short circuit	.2 per 10^6 hr.
Drift greater than 5%	1 per 10^6 hr.

Table 5.23 Failure rates for simple electronic transmitters.

In the other end of a control loop, an electrical to pneumatic transmitter is often necessary. Failure modes similar to those for the electronic transmitter are generally possible. But in addition, the fine mechanical parts of a pneumatic amplifier can stick, work loose or wear, and pneumatic jets can become blocked (damp, corrosion, dust).

Failure Mode	Failure Rate
Fail high signal	1 per 10^6 hr.
Fail low, no signal	.5 per 10^6 hr.
Drift greater than 5%	2 per 10^6 hr.
Stuck, no movement	.5 per 10^6 hr.
Blocked jet, no output	2 per 10^6 hr.

Table 5.24 Failure rates for electronic to pneumatic converters

5.18 Trip Amplifiers and Alarms

In many alarm systems, an analogue (continuously varying) signal must be compared with a constant reference value and a signal generated if the limit is exceeded. Trip units usually have a state register to record when a limit has been exceeded, and reset buttons to reinitialise the amplifier. Depending on the system design, the amplifier may have built in alarm lights and audible alarms.

Such amplifiers have problems of calibration, open and short circuit, component failure which may lead to no signal or spurious signal.

Modern circuit techniques can ensure a fail safe design, in that any single failure will be immediately registered as an alarm. In the failure rates in table 5.23 though, a simple, non fail safe design is assumed.

Failure Mode	Failure Rate
Input short circuit	.2 per 10^6 hr.
Input open circuit	.2 per 10^6 hr.
Drift > 5%	5 per 10^6 hr.
Output signal high/spurious trip	1 per 10^6 hr.
Output no signal	2 per 10^6 hr.
Output cannot shift	.3 per 10^6 hr.

(Data inferred from observed failures in fail safe amplifiers using integrated circuit techniques)

Table 5.25 Failure rates for trip amplifiers and electronic alarm circuits.

Additionally, calibration errors and errors of forgetting to switch back form calibration to on line mode must be remembered.

For switch instruments such as pressure switches, similar circuits are necessary to "hold" start signal pulses, and to drive audible alarms and lamps.

Audible alarm generators consisting of peizoelectric crystals have been observed to have a failure rate (no signal) of less than 2 per 10^6 hr.

For overall circuits, failure rates for one of signal registers are somewhat less than for trip amplifiers, the "no-signal" and "signal high" failure rates being about halved.

Alarm lamps are built into systems normally as double lamps, or with a test button which can light a complete set of lamps for test purposes and alarm circuits have often in addition a test circuit and an alarm acknowledgement button which allows a shift from blink to steady light. The variations in design do not permit a single figure to be given for an alarm system.

Single filament lamps for alarm circuits have failure rates in the region of 2 per 10^6 hr. but this depends on age, vibration and the current level. Lamps are often derated to give longer life. Small LED devices (light emitting diodes) have failure rates of about 0.5 per 10^6 hr.

5.19 Controllers, Indicators, and Recorders

Regulator instruments usually have much more equipment in them than is necessary for regulation alone. Modern instruments have, for example many regulating function options set point circuits, interfaces for remote set point control, manual/automatic switches, manual inputs, and set point and measured signal indicators. The overall failure rate is in the region of 4 per 10^6 hr. (IEEE 500) with about one tenth being indicator failures, and one fourth as intermittent failures. Drift is given in the IEEE 500 tabulations as one twelfth of the overall failures, but the authors observations give a significantly higher proportion for this failure mode.

Lees, Cox and Skala give higher values, with a high proportion being "no signal" or "high signal".

Some indicator instruments have contacts installed on the indicator, so that the indicator can produce alarm signals. These are often troublesome in that mechanical indicators can stick. Interpretation of the importance of such failure modes is difficult. If an indicator sticks on a contract, an alarm is generated anyway. Moving coil instruments can stick due to wear or vibration and refuse to move, but such "still standing" instruments will often be noticed by operators. Alarms from such instruments will generally have a reliability of about 10^{-3} pr activation.

Chart recorders are notorious source of reliability problems, especially because of movements sticking, and running out of paper or ink. If a recorder has many channels, it is not always easy to see that one has failed. Many such "failures" are not recorded, because they can readily be cured by the operator e.g. by tapping the instrument gently, some sources give failure rates for these instruments as high as 60 per 10^6 hr., but in practice intermittent failures are primarily the cause.

5.20 Control Computers

Computers and programmed logic controllers have failure problems which are unique when compared with other components. In particular, they suffer relatively severely from programming errors (over 50% of functional failures in practice). They can also be self monitoring, so that up to about 95% of failures can be detected and reported by the computer itself, and instrument failures can be monitored.

The technology of control computers is changing rapidly, and most smaller installations have switched to micro-computer control. Failure rates (ceasing to operate) once every two to three years (2 - 300 per 10^6 hr. are currently typical. During the first year of introduction of a new device type, failure rates are often much higher. The data in table 5.24 are therefore for well tried components. As stated above, the ratio of software to hardware errors is high, often with a ration 2:1 and for software systems over 50 kilobytes in size, and for new special purpose software, the ratio is more typically 1 to 10 (hardware to software).

Causes of failures of hardware are often in contacts (edge connectors) in backplane wiring, and in power supplies, as well as in individual circuits.

Failure Mode	Failure Rate
Micro computer with 32 kilobyte of store (stops)	40 per 10^6 hr.
Binary interface card (no signal)	3 per 10^6 hr.
Analogue to digital input	6 per 10^6 hr.
Digital to analogue output	6 per 10^6 hr.
Indicator lamp + interface	5 per 10^6 hr.
Auto/manual interface	
- stuck	2 per 10^6 hr.
- no signal	10 per 10^6 hr.
Power supply failure (no power/low voltage) (remember fuse blowing)	20 per 10^6 hr.
Display screen (no display, meaningless display)	15 per 10^6 hr.
PC, 640 K byte (malfunction)	30 per 10^6 hour
Hard disc 100 Mbyte	20 per 10^6 hour

Table 5.26 Failure rates for control computers

References

Anyakora, S.N. and Lees, F. — Some data on the Reliability of Instruments in the Chemical Plant Environment, Chemical Engineer No 225, Nov 1971

Lees, F.P. — Some Data on the Failure Modes of Instruments in the chemical Plant Environment, Chemical Engineer No 227, 1973

Phillips, C.A.G. and Warwick, R.G. — A survey of Defects in Pressure Vessels, AHSB (s) R 162 1968 UKAEA

Thomas, H.M. — A probability Model for the Failure of Pressure Containing parts, Third National conf reliability, UK, 1981

Pipe and Pressure Vessel Failure Probability, Reliability engineering 2 (1981) 83-124

Kellerman, O. and Siepel, H.G. — Analysis of the Improvement in Safety Obtained by a Containment, in *Containment and Siting of Nuclear Power Plants*, IAEA 1967

Further Reading

There are a number of standard data bases for many types of equipment, the most useful of which are recorded here:

IEEE	Reliability of Electrical Equipment, IEEE Std 500
OREDA	Offshore Reliability Data, Veritec/Pennwell Books, 1984
DOD	Reliability Prediction of Electronic Equipment, MIL-HDBK-217E, 1986 Rome Air Development Centre

For high reliability trip and shutdown systems, the following reference is useful:

RKS/SKI	Reliability Data Book Swedish Nuclear Safety Board

For general reliability data for chemical plant, by far the most directly relevant and thorough source is:

Doberstein, H., Hauptmans,U., Hoemke, P., Verstegen, C., Yllera, J. — Ermittlung von Zuveralssigkeits-kenngrössen Fur Chemianlagen, GRS-A-1500, Gesellschaft für Reaktorsicherheit, 1988

This reference gives data for a wide range of components, with a thorough statistical and failure mode analysis for each type. (for some component types, only a few failures are recorded.

6 Human Error in Plant Operation

Introduction

Human error can occur on virtually every stage in a process plant's life cycle, in

- planning
- design
- construction
- maintenance
- repair
- close down and dismantling.

Errors in design and planning are treated elsewhere (Taylor 1979). The material in this chapter covers errors in operation, maintenance and repair.

Operator error is inevitable in system design. The operator, like any other component in a process plant has a certain capacity.. When demands on the operator exceed his capacity, the result will be failure or error. The operator also inevitably has a stochastic aspect to his behavior. If the system cannot tolerate the occasional randomness in his responses, the result will be an error.

While it may seem somewhat inhumane to regard the operator as "component in the plant", it is even more inhumane to demand that he should perform perfectly all the time. We must accept that the operator has certain environmental requirements, that tasks must be adapted to his capabilities, and that even under the best circumstances, errors will occur.

As will be seen, operator error is a significant contribution to safety-related incidents. It tends to be a prevalent aspect of the more serious incidents.

The various places where operator and maintenance error can enter into plant failure behaviour are shown in Fig. 6.2

During maintenance, testing, or repair, an error can lead to an active disturbance in the plant, or more usually, to a latent failure. The latent failure can be revealed during normal operation and so lead to a disturbance.

An alternative pattern of disturbance starts with an erroneous operation under normal operation. Disturbances can also arise due to technical failures, and the disturbance brings the plant into an abnormal state. When a plant is brought into an abnormal state, there are normally possibilities for either automatic or manual safety actions, there is the possibility of omission or incorrect execution, so that the correct safety result is not achieved, or the disturbance is made worse.

Reactor Plant

Cause	Frequency %
Design error	35
Random component failure	18
Operator error	12
Error in procedure	10
Maintenance or installation error	12
Fabrication error	1
Unknown	12

Fig 6.1 Causes of incidents in nuclear plant (Taylor 1975)

Human error plays a role in performance of automatic safety systems. Errors in maintenance or repair can prevent safety action. Human error in testing can also mean that failures are not revealed. A possibility, which has been recognized as significant after the Three-Mile-Island incident, is that the operator can interfere with safety system operation, if he thinks the safety action is dangerous.

Finally, there is the possibility of extraneous activities, unconnected with the plant, or unconnected with operation, to interfere with performance. The fitter who drills a hole in a wall, and then through instrument cables. The welder who starts a fire with sparks. The electrician who drops a spanner into a relay cabinet. All these can give rise to plant disturbances.

From this description, it can be seen that analysis of operation and maintenance error must cover a wide range of circumstances and types of error.

Analysis of operator error has developed gradually over some twentyfive years, starting with basic ergonomic studies during the 1950's, aimed at determining just what an operator could do. A tradition grew up around the study of aircraft control, to determine the limitations of persons in directly controlling plant. This work tends to be mathematical and experimental. Study of general industrial accident mechanisms for example by Surry and Leplat gives useful background. The main radiation of process plant operator error analysis is based on the work of Meister and of Swain and his associates in the 1960's and early 1970's. This tradition has reached its current summit in the handbook of operator error analysis by A. Swain and Guttman. /NUREG CR 78/

The work presented here is based on the line of development initiated by J. Rasmussen in the early 1960's and continuing until the present day in studying the full range or operator control room behaviour. In particular, it attempts to include the aspects of operator decision making, to the extent that this can be analysed at all.

- Active error in plant maintenance
- Error in plant maintenance leading to latent failure
- Oversight in plant testing
- Active error in plant operation
- Active error in safety system maintenance
- Error giving latent failure in safety systems
- Oversight in safety system testing
- Omission or inadequacy in manual safety action leading to larger or new disturbances
- Error in disturbance response, leading to interference with safety system
- Extraneous error or act, which interferes with part of plant not connected with current activity.

The most recent work on this area deals with the problems the operator has in understanding plant performance an in diagnosing disturbances. However, this work has hardly gone far enough yet to be directly useful in risk analysis.

The plan of this chapter is to give first a definition of operator error as a background. Then to describe a model of error, and a classification of error types. This is followed by a number of practical methods for identifying error possibilities. After this, some of the more complex error mechanisms are described and examples given.

Human reliability analysis has been sadly neglected in chemical plant safety, in spite of the fact that human error is an important cause of accidents. The reader will find that the action error mode analysis described here is a very effective way of analysing error possibilities, and improving plant safety. No attempt has been made here to give methods for quantifying human error possibility. The simple procedures currently available give results which are too uncertain and variable. More precise methods are outside the scope of this book. The error cause analysis methods described are, however a necessary precursor to quantification.

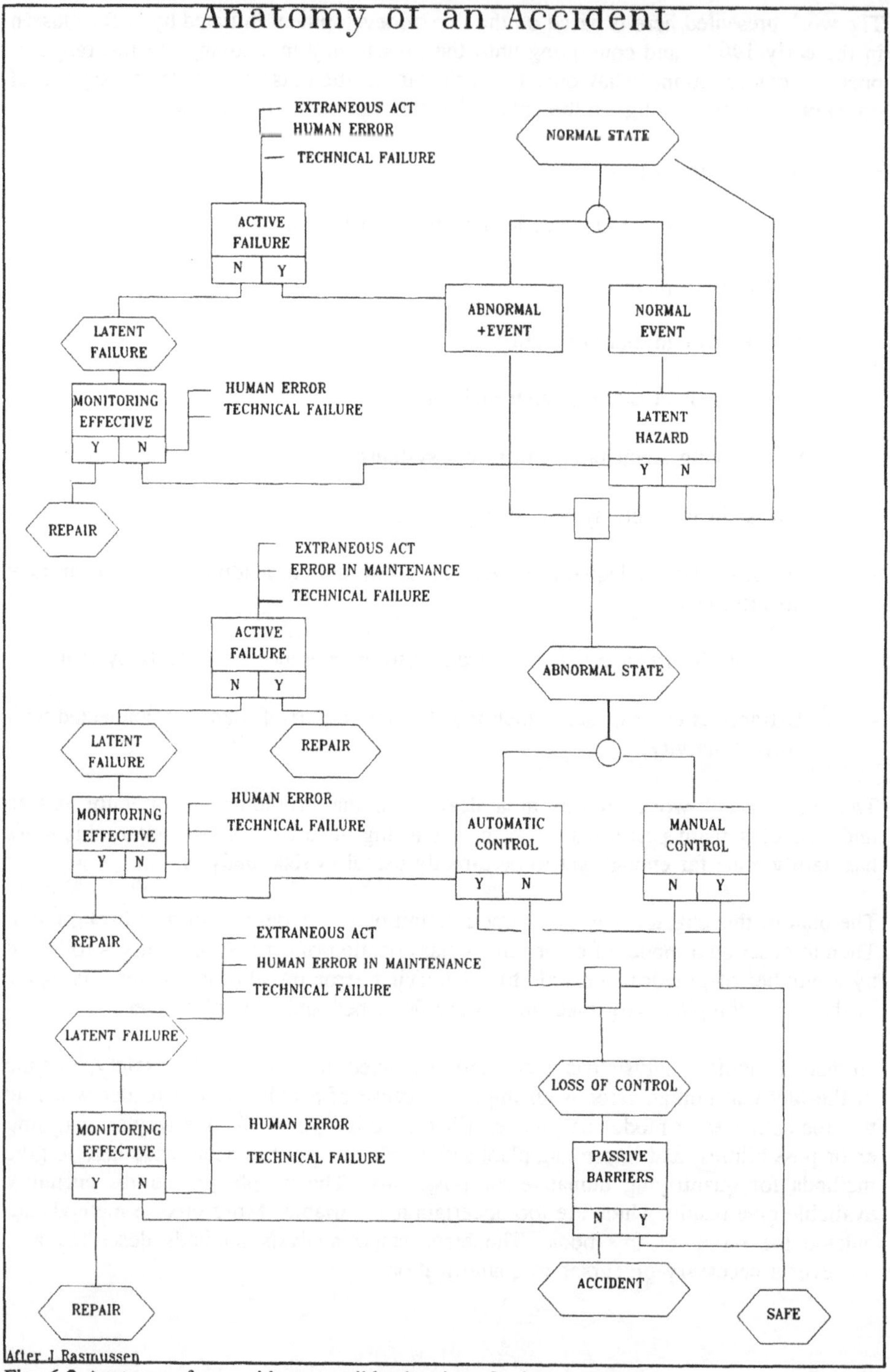

Fig. 6.2 Anatomy of an accident, possible situations for human error

Definition of Operator Error

A simple definition of operator error is "an action which is not in accordance with planned procedures".

This definition is only useful in dealing with the most banal errors. Fixed procedures are often lacking, or do not deal with the unusual situations the operator meets. This is especially true in the critical cases in which the operator is to take safety action.

Another related problem is that the operator's function is often to deal with situations which the designer cannot foresee. In this case, just which action would be correct can often only be determined by hindsight.

A third problem with this definition is that one of the functions of an operator is (often) to optimize plant operation. It is obvious that he can only do this if he has some freedom of action. Some safety incidents can, therefore, be regarded as unsuccessful experiments in optimization of plant operation.

In some cases so-called "operator errors" are the direct result of overloading or misleading the operator. An example is the situation in the power plant control room with many hundreds of alarms. Under normal operation, the operator can respond reasonably , if there are only a few "red alarms" at one time. In a disturbance in which many hundreds of alarm lamps are lit, the operator cannot be expected to cope.

Here, the "error" can be characterized as a mismatch between demands on the operator, and his capabilities. Such "errors" are more correctly regarded as design errors. But however classified, the engineer needs to take them into account and to take steps to minimize resulting risk.

A still more problematic example is the case where a procedure is erroneous or has omissions. If the operator obeys the procedure instructions, and this results in an accident, is this an error?

One must accept that a single simple definition of operator error does not exist. What we have is a range of error phenomena, each of which must be taken into account in process plant design.

A Model of Operator Activities

Any model of operator (or maintenance) behavior must take account of some important features of this behaviour. Some operator tasks are carried out almost intuitively, with a minimum of conscious thought, as for example steering of a crane to a particular position. Other operator tasks require that the operator "pays attention", for example, steering the same crane in an area where there are people and vehicles moving. (This conscious thought may make use of unconscious routines, as for example, writing down the results of a planning activity). And there may be unconscious association between the conscious activity and potential solutions of problems. Nevertheless, conscious thought is characterized by one particular feature.

Only one line of conscious thought may be followed at once. The model must also account for "associations" between data received and task to be performed.

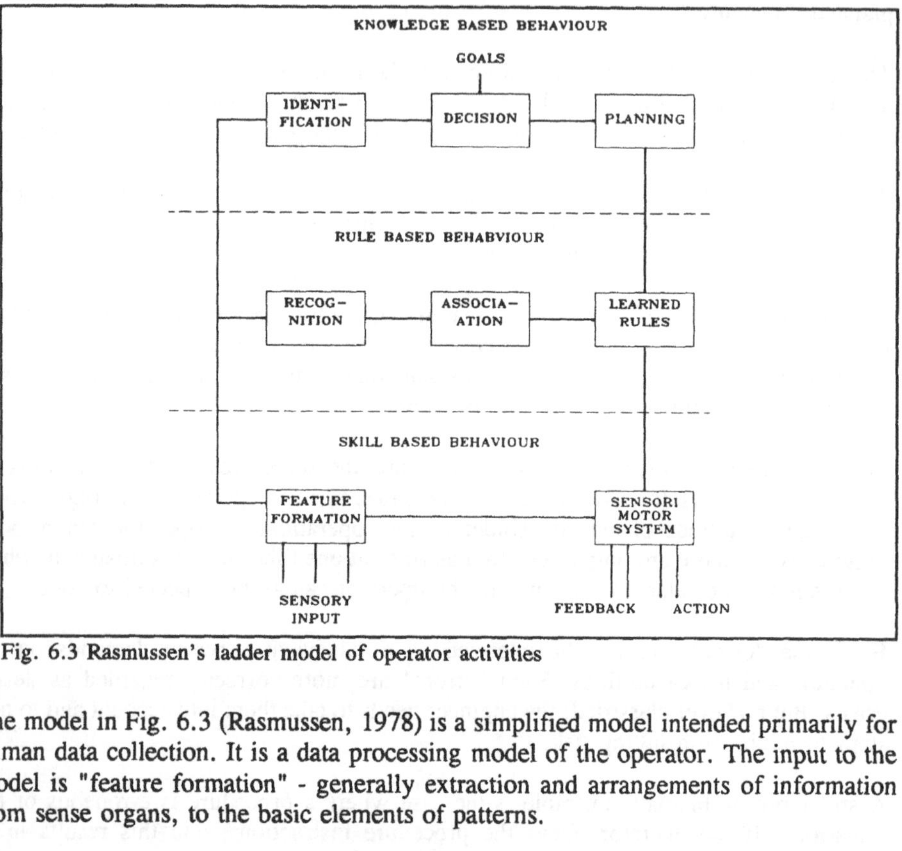

Fig. 6.3 Rasmussen's ladder model of operator activities

The model in Fig. 6.3 (Rasmussen, 1978) is a simplified model intended primarily for human data collection. It is a data processing model of the operator. The input to the model is "feature formation" - generally extraction and arrangements of information from sense organs, to the basic elements of patterns.

The next area of processing is recognition of the patterns formed which can lead directly to an association to a standard task to be carried out.

An alternative to direct recognition of required action is to identify the plant situation leading to a decision concerning some goal to be achieved, and further, to planning a sequence of tasks which can be executed to achieve the goal.

Task execution is considered to be completion of well trained sub-routines (e.g. pressing a button) and is essentially direct activation of sensori-motor control of limbs, vision, voice etc. The subroutines may be activated through direct recognition of the situation, or through conscious planning.

It is possible to take "short cuts" through this sequence. One has been mentioned - direct association between a recognized plant state and corresponding task. Another is a direct connection between feature formation and sensori-motor actions. This is the area of reflex responses and skilled behaviour for example in steering.

The model allows a classification of the way different jobs are carried out - as trained skills, as trained rule based behaviour or as planned tasks.

Much more detailed physiological and psychological models are possible. (More complete models similar in spirit to the one referenced here can be found in J. Rasmussen 1974). The one described here does not explain how attention is switched between tasks, nor how conscious planning makes use of sub-tasks and subconscious associations. It does not explicitly account for learning. It does not take account of communication in co-operative tasks. But the advantage of the model in this simple form is that it allows a classification of operator errors which covers by far the majority of errors occurring in practice, which is fairly simple to apply, and which can be used as a basis for prediction of some error rates with some hope of success. It serves also as a framework into which more detailed models can be placed.

Sensori-motor Errors

Starting with area 1 of the model - the feature extraction/sensori-motor control path, this is the area primarily associated with manual skills and steering.

The error types associated with this area of the model are typically the random errors associated with stochastic variations in the sensori-motor system and errors because the system to be controlled changes character.

Taking stochastic variation errors, examples are placement errors, such as those of threading a needle, adjustment errors for control system set points, errors of threading a nut on a machine screw, steering errors for vehicles, cranes.

Care needs to be taken in this area, in using the model for examples. During initial learning, tasks such as steering may be carried out "consciously" and during the learning process, different steering strategies may be used at different stages, each with its own error pattern.

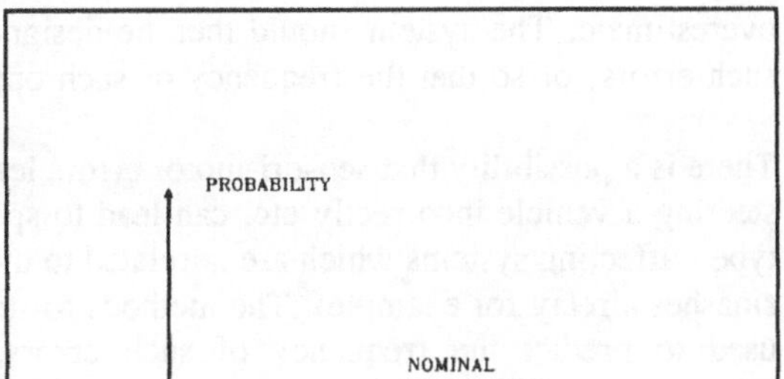

Fig. 6.4 Errors due to variability of performance

Errors of this kind can be predicted, using methods which regard the operator as a control system component.

In order to be able to use such methods, data is needed concerning particular operators in particular control situations - data can be gathered from actual systems or from simulators. Account needs to be taken of different steering strategies, which may be used by different operators. The results of data collection can in simple cases be represented by diagrams of the kind shown in fig.6.4. Errors arise when variable operator performance passed outside limits of acceptability.

The other major type of "operator error" at this level arises when the properties of the system change, without this coming directly to the attention of the operator with sufficient warning for him to respond safely. An archetype for this is oil on a motor clutch, changing the friction properties of the clutch. Given warning, a driver can adapt to a poorly working clutch, but the probability of error in doing so is quite high. Other examples arise when tools are changed, or designs are changed, etc. to change control properties of plants.

Whether problems arising from such changes in control properties can truly be called operator error is a fair question. What is certain though, is that a risk analysis needs to take account of "error" possibilities of this kind. At present, practical quantitative methods for this kind of error are not available. Qualitative prediction of possibilities of this kind of error is at present possible only for restricted classes of problems - a survey of the types of phenomena which can give rise to this type of error is required.

Given that some kinds of "error" of this type can be identified though, the most practical approach to treating them seems to be to assume that if the opportunity for such an error arises then the error will follow. The predicted error rate will be an overestimate. The system should then be designed either to prevent opportunities for such errors, or so that the frequency of such opportunities is acceptably low.

There is a possibility that sensori-motor errors leading to such things as dropping tools, steering a vehicle incorrectly etc, can lead to sporadic errors of the extraneous action type - affecting systems which are unrelated to the task being carried out (a falling tool smashes a relay for example). The methods for treating sensori-motor can probably be used to predict the frequency of such errors, but if these are to be related to consequences of the errors the topographic (layout) information will be required. For practical purposes it will often be easier to protect equipment which is important for safety from such "extraneous act", for example by enclosures, than to try to predict the probability of such acts.

Errors in Well trained Procedures

Errors in the second area of operation activity are confined (by definition) to inaccurate execution of well learned procedures, described as sequences of stereotyped "actions".

Possibilities here include:

- carrying out an action too early
- reading an instrument wrongly
- not carrying out an action
- omitting to check a precondition for an action
- carrying out a correct action on the wrong object
- reading the wrong instrument
- carrying out an action in the wrong sequence
- carrying out an action in the wrong direction (turning left instead of right etc.)
- a combination of the above error types.

What the "actions" are in such well-learned procedures is an open question. An operator may have learned to regard "starting a pump" as a single action. The analyst is free, though, to define his own "actions" in a way which allows straightforward data collection and simple analysis. The results will be valid provided that error rates in performing such actions do not vary too much from task to task, or that the main factors causing variation or error rates can be identified.

The classic method for reliability analysis of well-trained procedures is THERP (Technique for Human Error Rate Prediction) [Swain 1977] (see later section).

For this method the "actions" are such things as "pressing buttons". Whether error rates are reasonably constant can be checked to some extent by comparing the "error rate data bases" established to support the methods. But it would be desirable to check these error rates (gathered from production line work and from electronic equipment operation and maintenance for example) with error rates from industrial process plant.

In order to be able to use such methods certain criteria must be fulfilled in the way that tasks are carried out.

- The task must be well trained so that in practice only small (single action) variations from the defined sequence occur, or the sequence of actions must be cued by the system so that modification of the procedure does not take place. Major ad hoc improvements in procedures must be impossible.

- The task can be broken down into sub-tasks or actions for which failure data can be obtained, and for which failure rates do not vary too widely.

One point which helps <u>reliability</u> analyses of errors of this type is that it is much easier to obtain data concerning the probability that <u>a</u> wrong action will be performed, than it is to predict just <u>which</u> actions may be performed. This same fact makes safety and risk analyses more difficult.

Errors in Planning and Decision Making

In this area of the operator performance model, the opportunity for error is very wide. Decisions can be made on the basis of incomplete diagnosis of plant state, insufficient information on present plant state, and insufficient information on plant performance in general. The prediction of results from carrying out a plan can be in error (wrong result predicted), or the prediction can ignore some important side effects of carrying out a plan. The goal of a plan can be incorrectly set, for example so as to be incompatible with safety requirements.

Examples of such planning/decision making errors are given in Rasmussen & Taylor (1977) and Rasmussen(1979). This class of errors poses some special problems. One is that it is almost impossible to predict the result of an operator's planning - many different plans can lead to the same result and operators will typically find different ways of carrying out the same job. The other problem is that a single decision making error can lead to many separate "action errors". This presents a problem in risk analysis of complex process plant because the consequences of error are evaluated by considering individual actions at the man-machine interface. If there are many possible erroneous actions, then there will be more combinations of such actions.

The facts that planning and decision can result in many different plans, and can result in large numbers of combinations of errors, leads to difficulty. It seems doubtful that a practical consequence analysis of planning and decision errors can be made at present. Rasmussen (1987) has pointed out one way of accounting for the reliability of tasks carried out as a result of planning. This depends on the fact that planning is goal-oriented.

Whatever way a task is planned and carried out, the goal of the task will generally be clear cut (hypothesis). At some stage the task will be completed and the goal achieved.

Checking the achievement of the goal will generally also involve a fairly clear cut method, being directly limited by the instrumentation and direct observation possibilities (hypothesis).

If the goal of the planned task is not achieved, then there is a good probability that the task will be repeated. An analogy might be drawn to a child completing a jigsaw puzzle. The method may be varied, but in the end, something will come out, even if this means changing the method (fetching an adult) after the first few attempts. The criterion for success is clear, and the effort continues until the result is correct.

For this kind of task, reliability will be determined just as much by the reliability of the final checking as by the probability of error in the task. If the checking can be made very reliable, then a good upper bound can be set on the reliability of the overall task.

It is usually much simpler to determine the reliability of a simple checking process than it is to determine the probability that an error will be committed in a complex planning process. The value can be used as an upper limit on the error rate of the overall process. The limit is generally a good one, in that it is not too conservative. The value can generally be determined by using methods such as THERP, or by direct observation.

The objective of such a reliability analysis of planning and decision making tasks will generally not be one of reliability, but one of risk. If an error can lead to a major accident during the execution of the task, then the probability of error in checking results is pretty well irrelevant.

In order for the above method to be applicable, the system must therefore fulfill several criteria:

- The errors which can be committed in the task must lead to reversible effects.
- The probability of detection of error (or its results) must be susceptible to analysis.
- The time available for correction of the error must be sufficient for a high probability of success.

If these criteria are not fulfilled, there seems little hope at present of quantitative prediction of risk from freely planned tasks. Some effort at qualitative prediction, though, may be worthwhile. In order to fulfill the design criteria given above, the effect of individual errors must be predicted (to check that the errors are reversible and observable). This can be done for individual action errors (fairly simply for small plants, with difficulty for large ones). But an analysis of combinations of erroneous actions which could in principle result from an erroneous decision is often impossible. An analysis of typical patterns of errors resulting from particular types of decision error is an approach which may help, and is at least a promising line of research.

Errors in Identification of Plant State

In the upper area 4 on Fig. 6.3, three functions are shown which are concerned with noticing features of plant behaviour. The first is feature extraction, the second direct recognition of plant state, and the third identification of plant state in those cases where direct recognition is not possible.

There are a number of error types which are directly related to these kinds of activities. In the first area there are errors arising from low alertness or attention wandering (one might also include some kinds of distraction errors here). Another class of errors is absent-mindedness, for which some cue to action is recorded, but its significance not recognized.

There are at present no methods for predicting the probability of such errors, or even of describing a full range of error possibilities. But it is possible to derive some kind of design criteria which can prevent some such errors. For example, important cues should remain on display until acknowledged by the operator. If there is a response time limit for a cue, it should be accompanied by some kind of audible or flashing alarm. (The number of such alarms should be limited to prevent over-use and annoyance on the part of the operator - this means that the extent of reliance on rapid operator reaction should be limited).

It is also important that the number of cues arising at one time is not so large that extraction of a single one of importance is impossible. This is not the case for many process plants today - an extensive accident can well produce just "a mass of flashing red lights" on alarm panels.

Turning to the pattern-recognition function of Fig. 6.3 the most significant (known) error type is "familiar association", arising because the information presented to the operator (or that part he looks at) is ambiguous. He acts then on the familiar interpretation of presented information, and not on the true situation.

The Action Error Method

Action error analysis is one of the simplest methods for identifying possible operator errors. It starts with a listing of individual actions to be carried out on the process plant interface. These are drawn up as a sequence, and for each action, the consequences of the action on the plant are noted. They are added to the sequence, so that the diagram takes on the structure -

Action - effect on plant -- action - effect on plant.....

The "actions" here are defined in terms of the possibilities of operating on the plant, pushing buttons, opening valves etc. such as "starting a pump") are broken down into the individual steps at an elementary level. The effect of error on the plant can be evaluated, simply using the A procedure for consequence diagram construction.

The range of errors taken into account is given in Table 6.1.

It is possible to carry out such an analysis to include twofold or threefold combinations of action errors on plant of moderate complexity, provided that safe states can be identified fairly easily, so that not all sequences have to be followed. On a plant with twenty man-machine-activating devices, the possible number of sequences including threefold error combinations would be of the order 60*60*60 = 216,000 - an impossibly large number to analyse.

Fig. 6.5 gives a standard format for a cause consequence diagram for action errors for a single step in an operation procedure. In use, one such diagram should be filled out with details for every action of a procedure. Further sheets will generally be required to record possible causes or errors, consequences of errors if there is not room on the standard diagram.

There will generally be a very large number of possible "wrong actions" or "actions on wrong objects". If the limitation described above - that the error leads to a safe state - or the error prevents further progress in the procedure, or if the error is impossible to overlook and is easy to correct, then a complete analysis will be possible.

Cessation of a procedure	
Excessive delay in carrying out an action or omission of an action	
Premature execution of an action	- too early (timing error)
Premature execution of an action	- preconditions not fulfilled
Carry out action on wrong object of action (press button A instead of B)	
Single extraneous action	
In making a decision explicitly included in a procedure (good/bad, high/low, etc.), taking the wrong alternative.	
In making an adjustment or an instrument reading, making an error on the high or low side of tolerance limits	

Table 6.1 Action error modes.

Otherwise cut-off rules will need to be applied to limit the number of "wrong actions" and "actions on wrong object" considered. Actions on objects which are physically or psychologically near to the intended action are the one which presumably should be selected. These could include:

- action on objects of same type
- actions on buttons, knobs which look similar
- actions on objects which are situated similarly on different cabinets or consoles
- actions on objects with similar identifications.

In practice, it is generally found that only single errors can conveniently be taken into account with this method if operating procedures are long. In some cases it is possible to use heuristic rules to identify directions in the diagram for which a double failure development might be worthwhile, for example if an error leaves a potentially dangerous substance in an exposed place (remains in a tank which will be re-used etc.).

The method suffers from a qualitative limitation. It is difficult to take account of several errors arising in separate procedures at the same time. However, the method can be used to support pattern search and sneak path methods, overcoming these qualitative limitations to some extent.

The method is not at all suitable for quantitative (probability) analysis in the form shown, because there is too little information provided on the causes of errors.

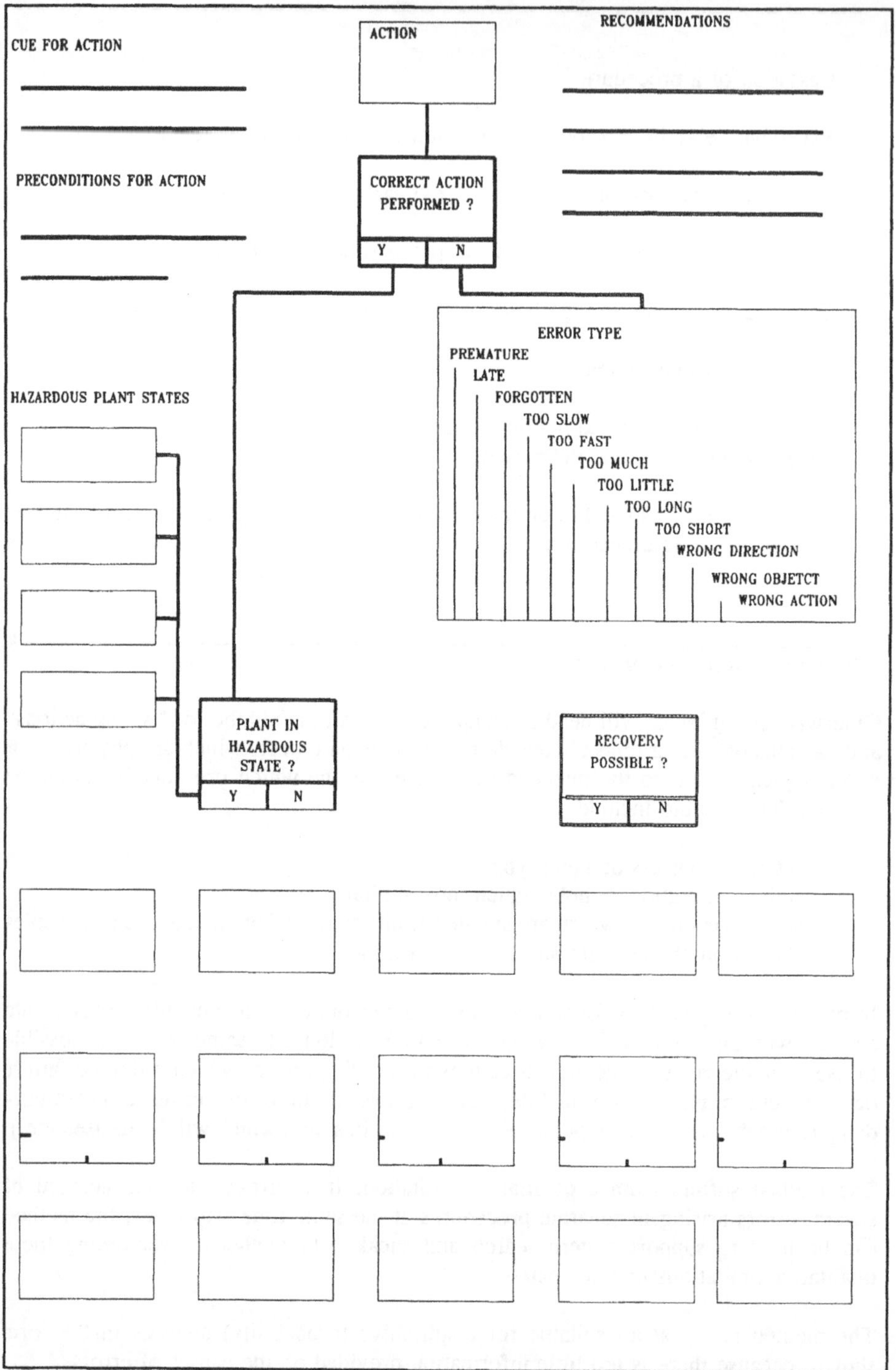

Fig. 6.5 Analysis sheet for action error analysis.

In general, the spread of error rates on individual action errors will be too wide to allow probability calculations to be meaningfully made with just action error descriptions along. In particular, it is important to know whether the action is cued - for example by a completion light from the previous procedure step, -whether there is immediate feedback to confirm the result of the action, and the type of procedure - whether it is freely planned by the operator, or a trained procedure. The method has been used as a basis for interlock design (Taylor 1975). Error rates can generally be estimated to within about an order of magnitude of uncertainty.

Through practical experience, some particular points have been found which can ease the use of the method. One question is - if a possibility has been found, which does not immediately give an accident, but only a potentially dangerous state, what should be done next. It has been found best to analyse any further errors, which can turn the potential into a real accident immediately. Trying to remember or record a complete list of potential dangerous states, which may or may not turn out to be relevant, is time-consuming and confusing.

The unusual plant states, which can interact with an error or normal action to give a hazard are entered onto the diagram in Fig. 6.5 in condition boxes on the left hand side. The number of such conditions can potentially be very large. In practice, the list should be limited. The parts of the plant directly affected by an action should be studied. Only the abnormal states in these parts of the plant need to be taken into account at the start of the analysis of a particular action. Other conditions can be accounted for as they arise naturally in the later parts of the consequence diagram.

Error Cause Analysis

The purpose of this method is to supplement action error mode analysis, by providing information on causes of errors and, if possible, to set probabilities on the errors found. The method is applied only when significant error modes have been found. The procedure is as follows:

1. For each significant error, the action associated with the error, and the error mode, is tabulated.

2. Potential error mechanisms and causes for each error mode are tabulated. The given check-lists are used to support the tabulation.

3. Performance shaping factors and situation factors, which can affect error probabilities are noted.

4. Possibilities for error detection and error recovery possibilities are noted.

The criterion for significance here is that a combination of failures has been found which can cause a serious safety problem. The levels of error causes are built around the model of operator error given earlier, and are arranged in levels of error mode, error mechanism, and error cause. Fig. 6.6 has been found useful in recording the possibilities for error cause.

Building a thorough list of possible error causes is in itself quite difficult. One of the most direct approaches is to take a model of the operator or maintenance technician, and apply a failure mode and effects analysis <u>to the model</u>. This gives a very thorough approach to error cause analysis. An initial stage in this approach is shown in fig. 6.6. When the approach is taken to its limit, the resulting list of potential error causes numbers several hundred. Use of computers provides the only practical way of applying the analysis method. A check list of error modes and error causes suitable for control room use is given in appendix A to this chapter.

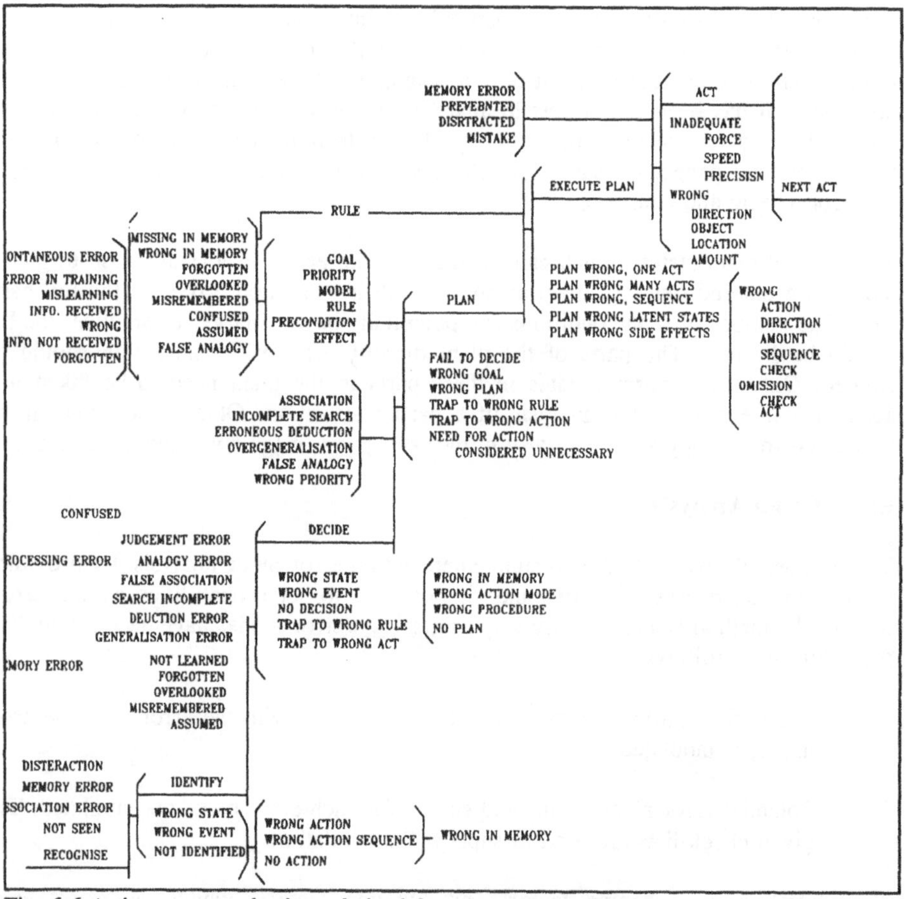

Fig. 6.6 Action error mechanisms derived from operator model

Special psychological and situational error mechanisms

Through study of case stories such as those just described it is possible to identify a number of characteristic error types. These may, in action error terms, seem quite complex, yet nevertheless they recur. They have been summarised here, and listed in table 6.2. Some of these mechanisms will be recognised from the earlier model of operator error. Most are however, quite specific. They occur again and again in accident reports.

- improvisation
- short cut (omission and better sequencing)
- reversion to stereotype action/standard action/earlier norm
- omission due to distraction
- procedure shift due to distraction
- neglect side effect of an action } forget, overlook, simply don't know
- neglect precondition for an action } forget, overlook, simply don't know
- "tunnel vision"
- information overload
- fixation
- shift change
  - omission of procedure
  - change of procedure
  - neglect a condition
- reinterpretation of indication as instrument error
- forget - remember syndrome
- interhange of objects of an action
- misinterpretation of ambiguous message or label
- dismantling active equipment
- ignore nuisance alarms, and also serious alarms
- switching off safety systems
- personalised emergency procedures
- ignore hidden warnings
- no allowance made for individual specialties
- extraneous acts
  - dropped tools
  - damage while cleaning
  - damage while testing or repairing

Table 6.2 Special situational and psychological error mechanisms

Improvisation

When something out of the ordinary occurs, there is often a perceived need to adapt equipment to meet the need. An example is the case of acid leaking from a road tanker. This was pumped over to an available tanker, to stop the leak, but as a result was contaminated. Subsequent purification attempts led to an explosion, because the extent of the contamination was unknown.

The incident at Flixborough in 1984 was caused by a similar improvisation - a damaged tank was bypassed, and the bypass was not correctly designed. As a result it buckled when the operators attempted to use it.

Typical improvisations arise when there is some substance in the wrong place, some standard equipment damaged or missing, or a lack of some necessary material. Use of non standard piping, non standard bolts, or "any pump which is to hand" are especial problems. Several accidents have been caused by improvised use of pumps capable of delivering much higher pressure than the equipment is designed to withstand. An error in regulation then leads to disastrous consequences.

Shortcuts

It is often tempting to omit or shorten steps in a procedure which do not contribute directly to the success of the procedure, but only to its reliability or safety. Waiting for

a batch of product to cool is an operation which might be shortened for example. Subsequent mixing with other reagents can lead to unwanted reactions.

There can be many reasons for the operator to take such shortcuts, from the desire to maximise production, to the need to complete a batch before a shift ends or the weekend begins, There are particular dangers if the operator does not know the reason for a procedure step. Operators will sometimes rail at "stupid design engineers" who insist on doing things the slow way. Such opinions are reinforced if the operator has experience of actual design errors, so that his estimation of design engineers is low.

Sequencing of operations can often be poorly planned, requiring the operator to walk back and forth between equipments. Such situations are an invitation for operators to optimise procedures, by regrouping steps. Of course not all operators will do this. The tendency depends very much on operator experience and awareness of danger. It depends very much, too, on the degree of emphasis management places on safety.

Reversion to stereotype

If a control is unusual, it may well be possible to train operators nevertheless to use it reliably. However, under stressed circumstances, the operator may well revert to stereotype. Examples are operators trying desperately to loosen a lefthanded threaded screw, by turning to the left.

Non stereotype designs involve lefthanded screws, instruments which show increases by moving left, down, or counter clockwise. Engineers desire for symmetrical control rooms have in the past been a cause of having two standards.

Designers should also be aware that stereotypes change from culture to culture.

While it is fairly easy to detect non stereotype equipment, the analyst should also be aware that the operator may have norms of action which have been learned earlier.

Particularly, he may have learned other operation sequences, or limits for action, on other equipment. These can resurface in times of stress, such as in emergencies.

Omission due to distraction

Omission of acts due to distraction, or interfernce of other tasks, is always is a potential problem. There is no real way of preventing distraction - virtually all operators have to be available on the telephone, have personal needs, and have to split their time between tasks.

Special distractions arise, especially during emergencies, if many other people enter control rooms. High level engineering and administrative staff can distract operators with questions, or just by their presence. error rates rise during such circumstances.

Procedure shift

It is possible for operators to lose the context of their work, particularly if the work is not very demanding and they are thinking of something else. If they have left the operating room, or left responsibility to someone else for some time, they may not appreciate the situation on return. As a result, they may take up some completely different activity than that in which they were previously engaged.

A special problem here is handover of responsibility between two people. A may not realise a change of situation has taken place. B will, unless he is directly aware of the problem, not realise that a is unaware of the shift. As a result, two operators work on two different sets of assumptions.

Neglecting side effects.

Operators may frequently oversee a side effect of some of their actions.

An example, which has occurred many times, is that an operator takes a sample or opens a drain, without using the proper procedure, for example removing pressure from the plant, or closing separate shut off valve. By careful manipulation of a valve he may be able to take a sample. But if there are crystals or crud in the sample flow, the valve may then be blocked open. He may be unable to reach the correct shut off valve, and the drain will continue to leak.

A sequence something like this occurred in the Feyzin disaster, where a drain valve on an LPG storage sphere blocked open. The result was a major release, a fire, and a subsequent BLEVE which killed several firemen.

In another incident, a very slow leak of liquid sulphur through a blocked open sampling valve caused a process room to be filled with sulphur to a depth of a few feet. Once stiffened, the sulphur had to be removed with the aid of pneumatic drills.

An even more direct "example of overlooking side effects occurs when a maintenance team receives instructions to overhaul or test a pump or motor etc. Before starting, they should check that the pump is not required for standby purposes. In several cases, pumps have been taken out of operation, and plants have as a result shut down, because the "second reserve" pump had failed.

In cases where side effects such as starting a motor or applying electricity could have wide ranging side effects, it is usual to have a careful system of permits and tagging of equipment before operation is allowed. Such permit systems are not always perfect, however, and do not always take into account the full range of side effects which can occur. Procedures such as action error analysis can be used to validate permit procedures.

Overlooking a precondition or latent state

Many actions in an operating procedure are conditional on a whole series of preconditions being fulfilled. The operation should not be carried out until the precondition is ensured. Checking of preconditions is often though, an "unnecessary" part of a procedure, in that in most instances, the procedure can be safely carried out without checking. Only in seldom circumstances with operation without checks be dangerous, such situations are especially error prone.

A classic example of this, which has been repeated in most process plants around the world at some time, is to begin pumping from a tank before ensuring that there is an adequate flow of air into the tank. Ventilation lines may be blocked because of valve closure, or because of ice, waste etc. For most storage tanks the result is that tank walls are "sucked in" and the tank collapses.

It is not even necessary, in such instances for a pump to be started. If the outflow line from the tank has a big enough fall, the vacuum generated by the falling liquid alone can be enough to damage a tank.

Another example arose when an operator wished to replace a regulating valve in a liquid gas system. He chose to make the replacement in a "rapid" way and closed block valves at either side of the regulating valve. He was aware that these might leak, and therefor closed the block valves very tightly. What he was unaware of was that a safety burst disc by passed one of the block valves, and served to prevent the regulating valve from overpressuring. The burst disc had in fact failed. On unbolting the regulating valve, this latent condition became apparent in the form of escaping liquid gas. Even then, the leak appeared only to be a bleed off of the contents of the regulating valve, until the valve was totally unbolted......

Too narrow basis, for decision - "Tunnel Vision".

"Operators faced with a mass of instrumentation, often select a few indicators as primary indications of the state of the plant and its subsystems. With sufficient experience, they come to base their decisions on these "reliable indicators" alone. This means that they may well neglect other indicators, which could give them useful information during disturbances.

More common than this is the tendency to look at single indications at a time, and not to relate indications to each other. During the Three Mile Island incident, there was sufficient information in the form of temperature and pressure indications, to reveal that water in the reactor cone was boiling. However, this information was not put together and conclusions drawn until very late in the course of the incident.

Fixation

If an disturbance has occurred, then one of the first natural reactions is to investigate possible causes. If a hypothesis has been proposed and accepted, then there is often a tendency to cling to this hypothesis, even after there is positive evidence that the hypothesis is wrong.

The incident at the Three Mile Island reactor provides an example. Once the operators had hypothesised that the reactor was "overfull" with water, they stopped emergency core cooling pumps. They retained their worry about overfilling the reactor, even after the water level in the reactor was excessively low.

If an assumption is made, on purely hypothetical grounds, by one person in a group, and the assumption fits some observed facts, others in the group may regard the assumption as proved. Such an effect can be self reinforcing, and can lead to diagnoses with no foundation in fact at all.

Shift change errors

Shift change is a situation during which errors often arise. At shift change, one operator must inform another about plant state, and about the stage reached in a production schedule. Anything forgotten during this orientation, or expressed ambiguously, or misunderstood, will provide a cause for later problems.

Reinterpretation as instrument error.

When an abnormal indication is shown on a display, the cause may be either a plant disturbance, or an instrument failure. The operator must decide between these alternatives. He will be especially prone to doing this, if his experience includes more instrument failures that plant disturbances. Such is the case for much instrumentation intended to monitor for emergencies. The problem can be so extreme, that when an emergency arises, the operator recalibrates the instrument.

The forget - remember syndrome.

If an operator forgets a particular step in a procedure, this may not have any effect. However, if he later remembers his omission, it is only natural for him to hasten to correct his mistake. For some plants, the correction can be disastrous.

A particular situation of this kind is that of applying heat to a batch reactor, prior to addition of a reagent. If heating is applied, the slow addition of reagent will be accompanied by production reactions which prevent the reagent accumulating. If heating is forgotten, the concentration of reagent may on the other hand, gradually build up. The late addition of heat can then lead to a runaway reaction.

Information overload

Information display systems work well while the plant is in normal operating state, or with minor disturbances. In large process plants, serious disturbances often generate display information at a rate which is far beyond any operator's ability to comprehend or to respond to. A single operator can respond to two or three alarms requiring his immediate reaction. He can interpret patterns with up to five or seven alarms or abnormal indications. Beyond this, he must use various serial search and priority strategies. Displays must be very carefully designs, and the operator must be trained in response to each form of emergency, if he is to respond quickly in the face of more information than this.

Dismantling active equipment

Almost every year, accidents occur because maintenance teams dismantle working equipment. The cause may be misinterpretation of instructions, wrong instructions, lack of coordination, or ambiguity in instructions. More often it arises because two similar production units are confused.

If flanges are partially unbolted, then equipment under pressure can, and sometimes does break the remaining bolts, and cause pipes and vessels to fly apart. Opening doors behind which pressure has accumulated can cause similar violent mechanical accidents.

Ignoring nuisance alarms

Many alarms arising in a plant are "nuisance alarms", which are the direct and expected result of start up or other routine operations, If such alarms come to be ignored when expected, they can also come to be ignored under unexpected circumstances.

Switching off safety systems

Safety systems such as cooling systems, pump away systems, dump and relief systems may be switched off by operators fear side effects of the safety system, or feel that the safety action is incorrect. This will especially be a problem in cases where there are conflicting safety goals. It may also be a problem if operators have experience of earlier design errors in safety systems.

Personalised emergency procedures

Operators with long experience on a plant may get to know its behavior better than the plant's designers. They may distrust safety procedures, and feel they can do better. This is a good thing, provided the revised procedures are discussed in safety committee meetings and formalised. sometimes, through such communication is difficult, and can lead to conflicting safety responses.

Ignoring hidden warnings

If safety indicators are alarms are covered by log books, coffee cups, hanging labels or safety tags, or even just another operator, they may well be ignored.

Individual specialties

The problems here concern particularly operators who are left handed, colour blind, or partially deaf. If allowance is not made for these, errors may readily result , through misunderstanding.

Sneak path analysis

One of the problems of the methods so far described for error analysis is that the potential for error in even a moderately sized process plant is enormous. Only a small number of such errors will lead to any kind of serious consequence. Particularly if "extraneous acts" and human activities going on in parallel are taken into account, the problems of making a good analysis are enormous.
The method described here is based on the theory that accidents in which human error is involved can be described by a "sneak path" (named by analogy with a sneak circuit). A number of "unfortunate situations" coming together set up a path leading to an accident. The objective is to find such unfortunate combinations of situations and actions as directly as possible.

The method for identification of such sneak paths starts by looking at potential sources and targets for hazards. Sources of hazards could for example be identified using the check-lists for hazard tree construction given earlier.

The next step in the procedure is to identify events and conditions which could bring source and target into contact, to establish an accident "route". This does not just involve identifying how a path might be opened up from a source to a target. There may be involved a breaking loose of a missile, an establishment of an ignition source etc. Establishing a route from an energy source to a person may involve the person moving to a point in an energy pathway.

At this point chains of events and conditions have been found which are necessary for an accident to occur. A number of possible accidents have been "designed". It is now necessary to find possible causes of events and conditions which can facilitate the accidents.

Five classes of events and conditions can now be identified, each a potential contributor to the accident.

- Normal operator actions, or operator actions which are unusual but not unreasonable or explicitly disallowed.

- Operator and maintenance errors.

- Technical failures on the "route" to the accident.

- Technical events, with causes within a more extended technical system.

- Extraneous acts or "blunders" while performing unrelated tasks on other equipment, or just moving through the plant.

Search for operator and maintenance errors now involves checking those operator procedures or actions which are "near" to the necessary errors,in the sense that the operations have something in common. Nearness might be temporal, spatial, or psychological. Examples of psychological nearness might be operating pump A. when pump B should be operated.

The final step is to check that a sequence of already identified events can actually occur - that a sneak path exists. Checks, for example, may be that a particular necessary operation can be performed in the presence of a necessary technical failure, or that two procedures which contribute to the accident can reasonably be performed at the same time.

In order to use the procedure, a check-list is required for types of psychological or topological nearness. A limited attempt has been made in Table 6.3 to give such a check-list, but since the method is relatively new, the list is incomplete from any point of view and should be regarded as an example rather than a final result.

<u>**Sneak Path Identification Procedure**</u>

1. Identify possible sources of hazard and targets for hazard.

2. Identify physically possible chains of events and conditions which can bring sources and targets into an accident relationship ("necessary elements of accident").

3. If a necessary element can be caused directly by a human action, seek for existence of the action in a normal, or an unusual but reasonable procedure, including actions by "strangers".

4. If a necessary element can be caused directly by a human error, seek for operations which are physically near or psychologically near (see check-list) to the necessary element to occur.

5. If a necessary element is an equipment failure, record this.

6. If a necessary element can be caused by a chain of technical events in equipment. Trace these chains backwards (sequential fault tree procedure) on finding events which can be caused by human actions, apply the reasoning used in steps 3. and 4.

7. When a chain of operations, errors, and failures has been established as a potential cause of an accident, check the chain for self-consistency, using consequence analysis.

TYPICAL HAZARDS TARGETS

Humans
Expensive equipment
Long-delivery items
Fuels
Potential reactants

TYPICAL HAZARD SOURCES

Energy
Poisons
Corrosives
Sharp edges
High-relative velocities
Oxidants
Ignition sources
Potential reactants
Moving objects

Table 6.3 Typical Sources and targets for sneak path analysis.

Example

An example can be given here for a chemical reactor.

The first accident to be designed is release of energy and material as a result of opening a charging lid. Such an action is "near" to

- the normal opening of the reactor
- opening for cleaning
- opening for maintenance
- opening in error, in mistake for another reactor.

An accident will occur, though only if these actions occur while there is an excess pressure in the reactor. This condition is part of the design of the accident.

The analysis can now proceed to check

- The reasons why there should be an over-pressure at normal opening of the lid.

- Reasons why the over-pressure should not be noticed

- Reasons why the normal procedure should be operated at an unusual time.

An analysis of technical failures prior to the normal opening, together with an analysis of why a previous overpressure check should have failed, completes the analysis for "normal opening". Similar steps are required for "cleaning", "maintenance" and "mistake".

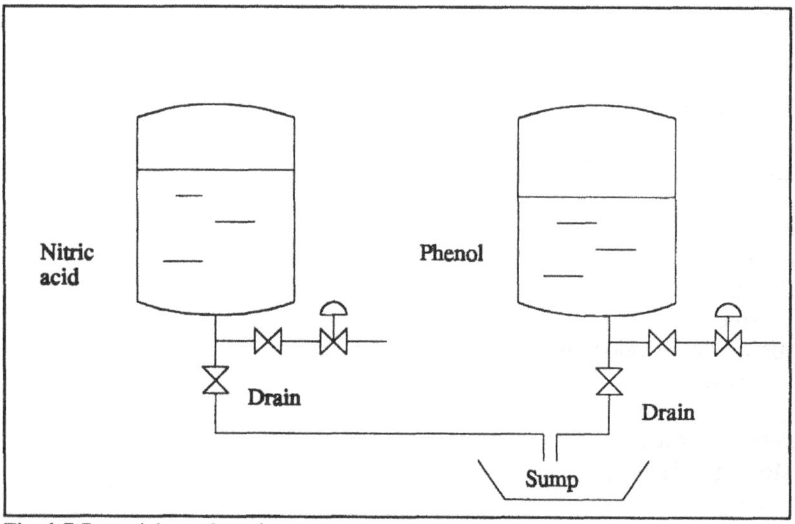

Fig 6.7 Potential sneak path

__Example__
As a second example, fig. 6.7 shows a system in which a tank contains phenol, another nitric acid. They drain into a common sump. If the nitric acid reaches the phenol, then a serious accident might occur. However, the sump is well protected,so that it is difficult for harm to occur to human beings.
Examining possible "routes" from hazard source to target:

The "source" is nitric acid. The target is phenol, and also a human being.

1. "Routes" consists of. Nitric acid flows to a sump through two valves, and so does phenol, man comes to a sump.

2. Nitric acid flows to sump, is sucked back to phenol, and man is at phenol tank.

3. Phenol flows to sump, is sucked back to nitric acid and man is near nitric acid.

Examining the second "route", this involves the nitric acid drain valve being opened, the phenol drain valve being opened, the phenol waste automatic valve opening, the phenol waste automatic valve opening, the phenol outlet line being closed, and the phenol cooling, and the phenol tank ventilation line being blocked, and a man at the tank.

These seven events constitute a design for an accident. Of these, the phenol drain valve opening is a normal state. The opening of the nitric acid drain valve with sufficient nitric acid in the tank to fill the sump is a simple error. This leaves events such as cooling of the phenol and the phenol tank valve line-up as remaining elements.

Here reducing heating flow and opening results in a normal state. The opening of the nitric acid drain valve with sufficient nitric acid in the tank to fill the sump is a simple error. This leaves events such as cooling of the phenol and the phenol tank valve line-up as remaining elements.

Here reducing heating flow and opening the outlet valve is a normal procedure. If there is a topological or psychological nearness between the drain and outlet valve, then failing to open the outlet valve, but opening the drain valve may result from the same error. Of the remaining contributions to the accident, blockage of the ventilation line is a fairly probable failure of a phenol line. Involving only cooling, and presence of a man at the tank is a normal condition.

Here an accident involving seven events and conditions, five of them abnormal, have been identified very quickly, and a sneak path established with a minimum of error analysis.

THERP

THERP (Technique for Human Error Rate Prediction) is the principal technique in use today for predicting process plant operation error rates. A better description of THERP can hardly be given than that of the author of the method himself (A. D Swain 1977). "THERP is a method to predict human error rates and to evaluate the degradation to a man-machine system likely to be caused by human errors in association with equipment function, operation and practices, and other system and human characteristics which influence system behaviour".

The steps in the method are given as:

1. Define system failure(s). These are the failure events for which the influence of human error is to be estimated.

2. List and analyse the related human operations. This step is "task analysis", and must take into account "performance shaping factors".

3. Relevant error rates are estimated.

4. The effect of human errors on the system failure events of interest are estimated. This step usually involves integration of human reliability analysis.

5. Changes which can improve reliability are proposed and new system failure modes are calculated.

Step 2 involves a "task analysis" of the kind described by R. D. Miller (1953). All actions prescribed for the operator are written down in the context of system operation, for example in a table. The results are then analysed in order to focus on "error likely situations".

The classes of error considered are:

1. Failure to perform a task or part of a task, e.g. a step (error of omission).
2. Performing a task or step incorrectly (an error of commission).
3. Introducing some task or step which should not have been performed (an extraneous act).
4. Performing some task or step out of sequence (A sequential error).
5. Failure to perform a task or error within the allotted time (a timing error).

In making the analysis, performance-shaping factors, that is the significant factors affecting error probability, are taken into account. A table of such factors is reproduced here as Table 6.4.

In making reliability estimates it will also be necessary to take into account, in addition to basic task reliability, the possibility of error correction, both by the operator and others.

Error probabilities are assigned from standard data base tables. The data for these is a series of studies performed in the USA, of which the primary source at present is the AIR data store. Transfer of such data to a particular plant involved very much engineering and psychological judgement. Techniques have been developed within THERP to allow systematic utilization of "expert judgement". Examples of some data estimates, based on engineering judgement (WASH 1400) are included in Table 6.4.

The technique for probability calculation used is essentially that of event trees. The correct operation sequence and all alternative sequences are drawn up, probabilities assigned, and the probability of following a particular route through the diagram is the product of the probabilities at each decision point along the path.

Swain has pointed out the importance of dependent or conditional probabilities, for example that the probability of a second error of type A can be very different from the probability of the first error occurring. For example in calibrating an instrument, the probability of a first error may be as low as 0.01, but the probability of repeating the error on a second instrument may be as high as (approximately) 1.0.

The method used by Swain for estimating the probabilities of dependent events is to give a probability for the first repetition of an error, given that a first error has occurred (a dependent probability) and then a probability for second and third repetitions. If the first error has a probability P, then the probability of two errors is typically given as 0.1 * P, i.e. The corresponding factor for a second or third repetition is given as 1.0.

This model of successive errors seems reasonably applicable to cases where an action may be repeated, where there is little feedback to indicate what kind of error has been made, or that an error has been committed. Swain has developed more precise models to treat core specific situations. It is probable that several such models will be needed to treat relevant problems in the general run of process plants.

It can be seen that the qualitative analysis in THERP is very similar to that of the action error analysis. The treatment of error types is almost identical (the action error analysis includes an additional class, premature action). The explicit use of performance-shaping factors in THERP, though, should allow good error probability estimates to be made for errors in trained procedures, provided that relevant data is available, and that the main objective is task reliability analysis.

Errors involving multiple actions appear in THERP in the form of dependent probabilities. The qualitative coverage of such multiple-action errors, arising from misidentification of plant state, or decision errors, depends very much on the task description taxonomy used. At present, the major multiple-error patterns which are treated seem to be repetition of similar errors. There is little information provided by the method to allow a search for systematic causes of errors. High-error probabilities, and especially multi-error probabilities, are not covered explicitly, but through engineering judgement.

As a quantitative <u>risk</u> analysis technique, THERP suffers from the problem described earlier, that it is difficult to obtain sufficient quantitative data from some events. Also the error causes treated are restricted largely to those associated with errors in execution of procedures. Error causes arising in plant state identification decision and planning are not explicitly identified. This is especially important insofar as it affects estimation of probabilities of multiple-action errors and the probability of otherwise unlikely "wrong actions". Nevertheless, where "separate-action" models of operator performance are applicable, and where error-rate data are available, THERP provides a fundamental method of analysis.

In order to reduce the difficulties in determining human error failure rates, Swain and Guttman have written an extensive handbook on human error analysis (1982), covering among other things the systematic allocation of human error rates. The error types covered are primarily simple errors, and recovery actions. This approach has been extended by Embrey (1988) and the author (1991), by the use of expert system software for allocation of error probabilities. The use of computer based aids of this kind allows a reasonably effective allocation of probabilities based on individual error causes, rather than just error types and performance influencing factors. It is possible to narrow the uncertainties and variations in error rate estimates in this way, giving considerable improvements in predictive accuracy as a result.

Equipment	**Physiological**
Speed, strength, accuracy required	Fatigue
Number of indicators	Boredom
Complexity of signals	Vibration
Speed of feedback	Temperature
	Freedom of movement
Task	Constriction
Complexity	Air freshness
Criticality	
Number of parallel tasks	**Individual**
Organisation between persons	Training
	Skill
	Motivation
	Knowledge
	Physical condition
	Attitude
	Interpersonal relationships

Table 6.4 Some performance influencing factors

*A systematic list of error causes is given in the appendix to this chapter. Most of these provide an altenative to consideration of performance influencing factors, or reducing the need for them. Performance influencing factors such a stress, temperature etc. are needed, where the influence on error rates is "parametric", that is changed by a factor. For discrete error "causes" on the other hand, the question is whether the cause is present or absent, and the probability of the cause must be added to the overall error probability.

Note that many of the error causes, such as distraction, are not a property of the operator, but of the situation. It is clear that a situation analysis is necessary to allow probabilities to be allocated.

The cause analysis which can be built up on the basis of the check list can be used as a basis for setting probabilities for error. It has been found that such detailed causal analyses allows the uncertainty in probability estimation to be reduced considerably. Several authors have provided "error taxonomies" similar to that in the appendix, with greater or lesser coverage. In principle, it does not matter too much which is used. In practice it is worthwhile to focus on just one such list, in order to allow data to be collected, and to provide experience in use. The list in the appendix has been used for about 20 full scale analyses, and has been used as a basis for data collection involving about 1000 hours of simulation, at the time of writing.

Hazardous Action	**Near to**
Simple action	Same action in normal procedure. Danger due to - latent failures in plant - lack of previous safety check on conditions or lack of previous safety action - normal procedure carried out under dangerous conditions - procedure error.
Simple action on object A	Same action on object B - near to A - situated in same position as A on a different panel - for system of same type as for A - same shape or form as A - achieves same objective as A (has some side effect, though).
Extraneous operation of component A	- by maintenance man testing system - by maintenance man testing wrong system - by interfering stranger - as a result of damage to component A
Damage to component A	- repair or maintenance nearby A. - repair or maintenance at other side of wall from A - transport by vehicle by A (trolley, forklift truck, car or lorry, truck)
Damage to component A (A can be damaged by falling tools)	- repair above A
Damage to A due to climbing on A to reach B	- A is a convenient stepping point
Operating A deliberately	- B achieves similar class of objectives as A (e.g.stop pump x - close valve leading to x)
Simple action Q	Another action in procedure which contains Q (sequence error) - immediately previous or following step in procedure - as a result of missing out a complete (goal-oriented) section of a procedure - as a result of omitting step previous to Q.

Table 6.5 Nearness principles for sneak path analysis

References

Aeronutronics, AD-431 611, Human Factors Aspects of Reliability, Newport Beach, California, January 1964.

Askren, W.B. and Regulinski, T.L., Mathematical Modelling of Human Performance Reliability. 1969

Davis, R., 1958, Human Errors and Transport Accidents.

Meister, D., 1971 Comparative Analysis of Human Reliability Models, Human factor dept., Bunker Ramo Corp., Westlake Village, California, AD-734 432.

Swain, A.d., 1973 Improving Human Performance in Production, Industrial and Commercial Techniques Ltd., 30-32 Fleet St., London EC4.

Rasmussen, J. Om ulykkers anatomi og bekæmpelse - en skitse. Risø-M-1696, 1974, 23 pp.

Rasmussen, J. Characteristics of Operator, Automatic Equipment and Designer in Plant Automation. Risø-M-808, 1968, 30 pp.

Rasmussen, J. The Human Data Processor as a System Component, Bits and Pieces of a Model. Risø-M-1722, 1974, 51 pp.

Rasmussen, J., and Taylor, J.R. Notes on Human Factors Problems in Process Plant Reliability and Safety Prediction. Risø-M-1894, 1976

Rasmussen, J. Man as a System Component. in "Man-Computer Research" H. Smith and T. Green (Eds.), Academic Press.

Swain, A.D. System and Task Analysis, A major Tool for Designing the Personnel Subsystem, SCR-457, Sandia Laboratories, Albuquerque, NM 1963.

Swain, A.D. and Guttmann, H.E. Handbook of Human Reliability Analysis, NUREG/CR 1278, USNRC 1983

Embrey, D.E et al. SLIM-MAUD: An approach to assessing human error probabilities using structured expert judgement. NUREG/CR 3158, USNRC 1984

Taylor, J.R. Using Cognitive Models to Make Plants Safer, experimental and practical studies, in *Tasks, Errors, and Mental Models*, ed Goodstein, Andersen, and Olsen, Taylor and Francis, 1989

Appendix to Ch. 6
Check lists for action error modes, mechanisms, and causes

Premature

- No details
- Observation error
- Identification error
- Diagnosis error
- Decision error
- Planning error
- Execution error
- Motoric error

Delay

- Dysfunction
- Observation error
- Identification error
- Diagnosis error
- Decision error
- Planning error
- Execution error
- Motoric error
- Hindrance

Omission

- Dysfunction
- Observation error
- Identification error
- Diagnosis error
- Decision error
- Planning error
- Execution error
- Motoric error
- Hindrance

Too much force

- Dysfunction
- Observation error
- Identification error
- Diagnosis error
- Decision error
- Planning error
- Execution error
- Motoric error

Too little force

- Dysfunction
- Observation error
- Identification error
- Diagnosis error
- Decision error
- Planning error
- Execution error
- Motoric error

Too long

- Observation error
- Identification error
- Diagnosis error
- Decision error
- Planning error
- Execution error
- Motoric error

Too short

- Dysfunction
- Observation error
- Identification error
- Diagnosis error
- Decision error
- Planning error
- Execution error
- Motoric error

Too fast

- Dysfunction
- Observation error
- Identification error
- Diagnosis error
- Decision error
- Planning error
- Execution error
- Motoric error

Too slow

- Dysfunction
- Observation error
- Identification error
- Diagnosis error
- Decision error
- Planning error
- Execution error
- Motoric error

Too far
- Observation error
- Identification error
- Diagnosis error
- Diagnosis error
- Decision error
- Planning error
- Execution error
- Motoric error

Wrong direction
- Observation error
- Identification error
- Diagnosis error
- Decision error
- Planning error
- Execution error
- Motoric error

Wrong path
- Observation error
- Diagnosis error
- Decision error
- Planning error
- Execution error

Wrong object
- Observation error
- Identification error
- Diagnosis error
- Decision error
- Planning error
- Execution error
- Motoric error
- Mislabelling
- Ambiguous label
- Missing label

Wrong implement
- Observation error
- Identification error
- Diagnosis error
- Decision error
- Planning error
- Execution error
- Motoric error

Wrong command
- Observation error
- Identification error
- Diagnosis error
- Decision error
- Planning error
- Execution error
- Motoric error

Wrong value
- Dysfunction
- Observation error
- Identification error
- Diagnosis error
- Decision error
- Planning error
- Execution error
- Motoric error

Wrong position
- Dysfunction
- Observation error
- Identification error
- Diagnosis error
- Decision error
- Planning error
- Execution error
- Motoric error

Dysfunction
- Attention failure
- Comfort call
- Distraction
- Action prevented
- Overload
- Priority error

Observation error
- Hidden cue
- Mistaken cue
- Cue overlooked
- Missing cue
- Hidden symptom
- Parallax

Identification error
- Thought process error
- Misremember
- Ambiguous symptoms
- Missing symptoms
- Symptom confusion
- Misleading symptoms
- Misleading labelling
- Misleading text
- Error in procedure
- Communication error
- Hidden symptom
- Information overload
- Change in system
- Location error
  - Mistake

Diagnosis error
- Thought process error
- Misremember
- Ambiguous symptoms
- Missing symptoms
- Misleading symptoms
- Symptom confusion
- Multiple disturbances
- Misleading labelling
- Misleading text
- Error in procedure
- Communication error
- Confusion
- Overlook symptom

Decision error
- Thought process error
- Misremember
- Misleading information
- Missing information
- Communication error
- Decision paralysis
- Diagnosis difficulty
- Decision paralysis

Planning error
- Thought process error
- Misremember
- Plan short cut
- Experimentation
- Priority error
- Too short planning horizon
- Error in goal

Execution error
- Misremember
- Forget
- Lack of training
- Judgement error
- Mislearning
- Procedure confusion
- Error in procedure
- Short cut
- Trapping error
- Communication error
- Timing error

Motoric error
- Stiffness
- Jamming
- Tiredness
- Overenthusiasm
- Lack of training
- Change in system response
- Too high force required
- Too high speed required
- Too high precision required

Distraction
- Other task
- Telephone
- Colleague
- Boss
- Disturbance
- Equipment failure

Thought process error
- Overlook side effect
- Overlook precondition
- Deduction error
- Overgeneralisation
- Erroneous analogy
- Error in mental model

Error in procedure
- Omission of step
- Confusing text
- Ambiguous text
- Error in text
- Mismatch to actual equipment

Communication error
- Ambiguous message
- Noise
- Incomplete message
- Erroneous message
- Mixed communications

Decision paralysis
- Stimulus overload
- Shock
- Fear
- Lack of knowledge
- Information overload

Overlook cue
- Distraction
- Cue hidden
- Cue non arrival
- Cue confusion
- Multiple cues
- Information overload

Overlook symptom
- Distraction
- Symptom hidden
- Symptom confusion
- Information overload

Hindrance
- Equipment failure
- Countermand
- Localisation problem
- No light
- Smoke
- Access difficulty
- Lack of air
- High temperature
- Hazardous material
- Radiation
- Flooding
- Zero gravity

Access difficulty
- Fire
- Distance
- Height
- Obstructing equipment
- Narrow access
- Locked door
- Jam

7 Release of Liquids and Gases

Most chemical plant accidents involve a release of liquid, gas or both. The size of the accident will generally depend on the length of the release period, and the rate of release. This in turn depends on the size of the hole, the form of the hole, its position, and pressure.

Fig. 7.1 illustrates some of the different kinds of releases. One example is a liquid phase release from an open tank. The rate of release is dependent on the height at which the leak occurs. In a practical situation, most leaks occur near the base of the tank, partly because of corrosion, partly because the pressure is higher here, and partly because there are more fittings, pumps, etc to disturb the flow.

A second example concerns a liquid release from a tank through a pipe. The situation is similar to that for the first case, with the exception that the flow rate is limited by resistance in the pipe, and any fittings such as valves, elbows, etc. in the pipe.

A third case concerns gas releases from a pressure vessel. Here, the substance may initially be a liquefied gas, or a compressed gas. If the gas is liquefied, the behaviour depends very much on the size of the hole. If the hole is large, the pressure in the vessel drops, and the liquid boils vigorously. The boiling can be so violent that liquid is thrown out of the tank together with the gas. If the hole is small, only gas is released.

A fourth case shows a release of liquified gas from below the liquid level in the tank. Here liquid is released initially, but part of the liquid flashes to gas very rapidly after the release. The amount of vapour flashing depends on the heat available in cooling the liquid to its boiling point, (or even below its boiling point).

The fifth case shows a full scale rupture of a pressure vessel containing liquified gas. When the rupture occurs, boiling is very violent, and liquid and gas are thrown out of the tank. This is described in more detail in the chapter on heavy gas dispersion.

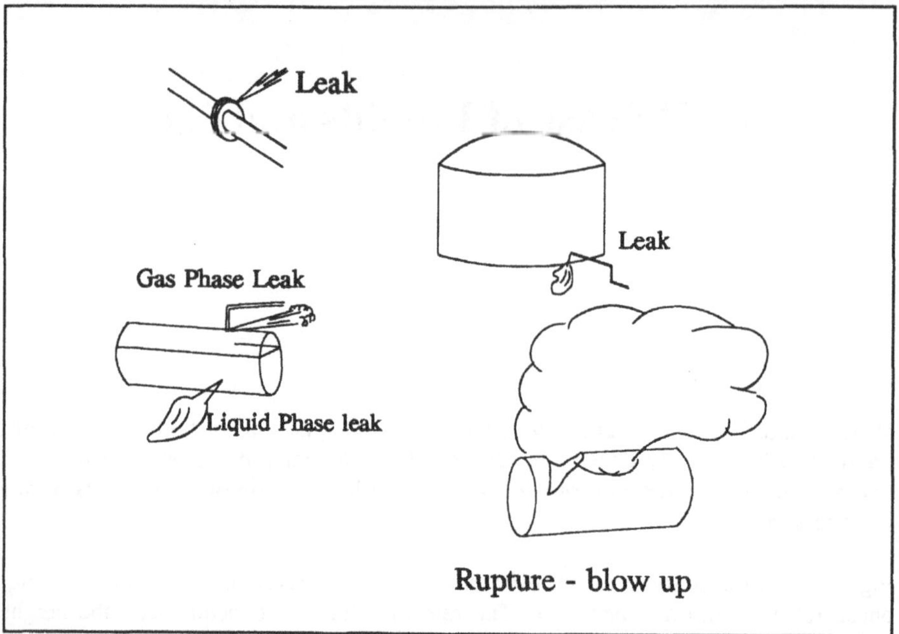

Fig. 7.1 Various cases of release of liquids and gases.

Release of liquids.
The rate of flow of liquid through a hole is given by solving Bernoulli's equation for the flow (see e.g. Levenspiel). The rate of liquid flow through a hole is given by :

$$\dot{M}=A\ C_d\rho\sqrt{2\ g\ h}$$

for a hole in an open tank

or

$$\dot{M}=A\ C_d\rho\sqrt{2[\Delta P/\rho+g\ h]}$$

where
$\dot{M}$ is mass flow rate
A is area of the hole
C_d is the coefficient of discharge
g is the gravitational constant
h is the head of liquid
ΔP is the pressure at the liquid surface

for all cases

The liquid head is defined as the height difference between the hole and the free liquid surface.

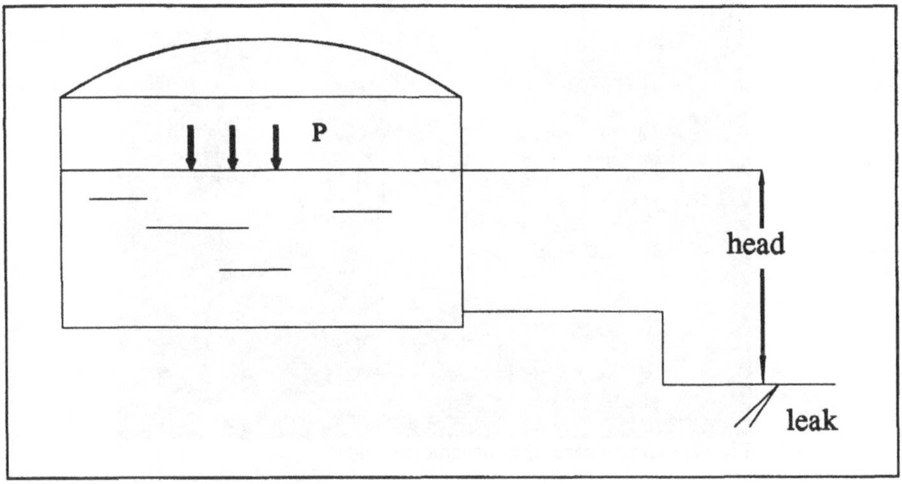

Fig 7.2 Definition of liquid head.

The discharge coefficient is an empirical constant, with a value of 0.61 for a simple sharp edged hole. For a rounded hole the value can come close to 1.0. For a short pipe section the value is approximately 0.81

The rate of flow from the tank depends on the head of liquid, and will therefore fall as the head falls. This is given by:

$$dh/dt = \frac{Mdot}{A_{tank}\ \rho}$$

where
A_{tank} *is the tank cross sectional area*

The distance travelled after leaving the tank by the liquid can be important. This is given by simple ballistic equations, which apply unless the flow is so rapid that the jet of liquid breaks up into a spray. The distance of flow is given by

$$x = u\ \cos(\theta)\ t$$
$$y = u\ \sin(\theta) t - g t^2$$

where
u is the jet velocity

$$u = \frac{\dot{M}}{\rho\ A}$$

ρ *is liquid density*
θ *is the* $\angle$ *of the jet to the horizontal*

Fig. 7.3 Jet release of hydrochloric acid

If y is set to zero, this gives:

$$u \sin(\theta) = gt$$
$$x = u^2 \sin(\theta)\cos(\theta)/g$$

x is then the distance which the liquid is thrown, as a jet, at ground level.

In actual practice , a leak under pressure tends to form a spray. Fig. 7.3 shows such a leak, of hydrochloric acid, from a leak in a valve body. The basis for determining drop size is the solution to flow instability equations, given by Rayleigh, for which drop size is $1.89{*}D_{jet}$. In practice, for high velocity jets, the droplets break up if the Weber number falls below about 20. The Weber number is defined as follows (Hinze 1953):

$$N_{We} = \frac{u^2 \rho_{gas} d_{droplet}}{\sigma g}$$

where
u is the velocity of the droplet relative to the air
ρ_{gas} is gas density
$d_{droplet}$ is droplet diameter
σ is surface tension
g is the acceleration due to gravity

Liquid releases from a long pipe.

If the liquid release takes place from a long pipe, it becomes necessary to take fittings valves bends etc. into account. For practical purposes in risk analysis, one can say that a pipe becomes "long" if more than 50% of the pressure drop takes place in the pipe itself. This pressure drop has about a 20% effect on flow rate.

The head loss in a pipeline can be represented by:

$$h = \sum_i K_i Q^2$$

where

K is a hydraulic resistance factor

Q is the volumetric flow rate.

This formula depends on head loss being proportional to the square of flow rate, which is good approximation for most practical cases in accidental situations, when flow is well away from the laminar flow regime.

For a length of pipe:

$$K = (1/2g)(4f\, l/d)(1/A_p^2)$$

where

l is the length of the pipe

D is pipe internal diameter

f is a friction factor

A is the pipe cross sectional area

The friction factor is usually expressed as a function of roughness and Reynolds number. Reynolds number is defined for pipe flows by

$$R_e = \frac{D\,Q}{\nu}$$

where

ν is the kinematic viscosity of the liquid

For water, ν varies between 17.8 m²/s at 0 °C and 2.8 m²/s at 100°C. Light hydrocarbons have values between 8 and 1.5 at temperatures between 0 and 100°C.

The friction factor is generally given by means of a diagram, the Moody diagram, or by means of empirical formulae. The most widely used of these is the Colebrook-White equation:

$$\frac{1}{\sqrt{\lambda}} = -2\log_{10}\left[\frac{2.51}{R_e\sqrt{\lambda}} + \frac{K}{3.7D}\right]$$

where

$$\lambda = 4f$$

A simpler equation, which takes into account variations in pipe roughness is:

$$f = 0.001375\left[1 + (20\ 000\frac{k}{d} + \frac{10^6}{R_e})\right]$$

where

$\frac{k}{D}$ *is the relative roughness*

Relative roughness is defined as the ratio of "roughness size" to pipe diameter. Typical values of roughness size are:

Coated cast iron	0.13 mm
Uncoated cast iron	0.20 mm
Uncoated steel	0.005 mm

Values vary widely however, depending on the level of corrosion. Variation usually involves an increase so that the above values may be regarded as a conservative lower limited accidental release cases.

For pipe fittings, valves etc, similar friction factors can be measured. examples are For an orifice

$$K = (1/2g)\ ((1/C_d)\ (A_{pipe}/A\ -1))^2\ (1/A_{pipe})^2$$

Where A_{pipe} is the pipe area
A is the hole area
C_d is the discharge coefficient for the hole

The value of C_d varies from about 0.61 small holes up to a value of 1 for large holes This formula is useful in determining the size of orifice plates to be used in limiting the maximum rate of flow in accidents.

For smooth pipe expansions and contractions there is little head loss.
For 90° bends

$$K = C_b/2gA^2$$

C_b is a constant which depends on the ratio of pipe diameter to pipe bend diameter. It varies from about 0.5 for rough pipes with a sharp bend (bend radius/pipe diameter = 1) to 0.2 for a bend radius to pipe diameter ratio of 7. For smooth pipes, the ratio varies between about 0.1 and 0.2.

The friction coefficient varies with the angle of bend. For 45° bends of sharp radius a value of 0.3 to 0.4 is appropriate.

The resistance of other fittings can be expressed in a similar way e.g.

	$K.\,2g/A^2$
Globe valve, fully open	6.4
Globe valve, half closed	9.4

Pipeline releases of liquids.

The situation for large releases from pipelines is complicated. For small releases, the simple orifice release equations given above may be used. For a large hole on rupture, in a liquid pipeline the situation depends very much on the initial flow rate, and on the slope of the pipe. For a complete rupture the following cases can be distinguished.

1. If the pipeline is on level ground, the flow continues at the pumping rate until the leak is detected, and pumping stopped.

2. For inclined pipes, the flow can be calculated as proceeding at the pumping rate, or at the rate of flow corresponding to the total head of the liquid from the highest point adjacent to the rupture. This will give an overestimate of flow rate, since bubbles may pass backwards along the pipe and siphon effects may reduce the effective head (while at the same time increasing the total mass available for release).

3. In the case of a level pipe, for which flow has been shut off, so that only the liquid between the shut off valve and the hole can flow, channel flow equations (outside the scope of this book) can be used, as the liquid flows out of the pipe.

For holes which have an intermediate size, in which the head and pumping rate are both significant, the calculations are very complex.

One of the typical kinds of liquid release calculations in risk analysis concerns release form a pipe attached to a tank. If the hole size chosen is a small one, the flow equation given in the first equation in this chapter will give a accurate result. The pressure valve to be set into the equation is the lead of liquid, between the liquid surface in the tank and the hole position, if there is no initial flow in the tank. If there is some flow from the tank through the pipe itself, then the pressure drop along the pipe may be taken into account, using the equations given on page 225.

For larger holes, the flow rate through the pipe must take into account any continuing flow downstream of the hole, and pressure loss along the pipe. Two equations can be set up, with the flow rate in the pipe and through the hole, as unknowns, and with upstream and downstream pressures as inputs.

Example

An oil tank, on a hill 30 m above sea level, delivers oil to a terminal through a 1.2 km pipe, 16 inches in diameter. A 3" hole is created by the impact of a truck. What is the release rate of oil? The hole is at 400 m from the tank, and at a height of 10 m above sea level.

In this case, the release rate depends on whether oil is being delivered to a ship or not. If it in not, then the oil pressure head is the height of oil in the tank, which will be assumed to be 8 m and the difference between the tank height and the hole height that is:

$$h_e = 8 + 30 - 10 = 28m$$

The discharge coefficient for the orifice and friction loss are given by the equations on page 225. Further loss is given by any valves, bends etc, in the pipe, but these will be ignored here.

If the pipe is delivering oil under gravity flow, at the time the release occurs, then there are two flows to take into account, upstream of the hole, and downstream.

First, the flow rate in the absence of the hole is calculated. This gives a starting value for flow, which allows an initial value of friction factor and K to be calculated. Then for the upstream and downstream sections.

$$Q=\sqrt{(h_e/K)}$$

Then an estimate of flow rate the downstream section, Q_2 is calculated on the basis of the head at the hole. The equation for flow through a hole, gives

$$Q_1 - Q_2 = A\ C_d \sqrt{(h_2)}$$

These equations can be solved algebraically. Generally however, it is easier to solve them iteratively, using a computer program, especially since calculation of new values of K_1 and K_2 should strictly be made, as improved values for Q are found.

A second typical case involves holes in pipes through which liquid is being pumped. The method of solving the equations is in this case very similar to that given in the example above. However, in this case it should be remembered that delivering pressure from a pump generally falls, when flow rate increases. It may be necessary to obtain flow/head characteristics for the pump. (Often though, the uncertainties in determining hole size make such accurate calculations rather superfluous).

Fig. 7.4 Pipeline leak

A very common case for calculation concerns releases from pipelines through which oil is being pumped. If the pipeline has a significant length, then a number of factors become important. The calculation methods described above may be used to calculate flow rate initially, until the pipeline is shut down.

Once the leak has been detected, then pumps will be shut down, or pipeline valves closed. For a small hole, detection may take a long time, depending largely on visual observation. For a large hole, the pressure drop at either end of the pipeline will provide an indication. It will, in the case of long pipelines take a few seconds for the pressure drop to reach the end of the pipe. It will take a further period, generally between 5 and 30 seconds to shut valves. During this time, the leakage will continue unchecked.

After the pipeline is shut down, oil will still continue to be released, due to the contraction of the pipe steel and expansion of the oil, as pressure is releases. These compressibility effects can be significant for a long pipeline section.

Oil may continue to flow too due to the variations in pipeline height. The pressure head of between the hole and the highest point in the line needs to be taken into account.

It is usual to use the flow equations to use the flow equations given above to make calculations of release after the pipeline has been shut down. In fact, results calculated in this way overestimate the oil release, since air must enter the pipeline in order to replace the oil, and this causes additional friction. Also, air entering the pipeline is likely at some stage to cause an air lock (bubble preventing flow) at the high points between the hole and the highest point on the pipe.

Risk analysis calculations of this kind often show that small holes in remote areas can give larger total releases than larger ones, due to the difficulty in detecting them, and the resulting long period over which releases takes place. To reduce this problem, it has become increasingly common to fit computerised leak detection systems, to pipelines which cross critical areas, such as water catchment areas and nature reserves.

Flange and seal leaks.

Releases through pump seals, valve seals, and flange seals, are generally more common than pipe leaks through holes. The flow path is annular or flat, and the flow may be laminar. The flow equations given above may be used to calculate flow rates, using a crack width of between 0.1 and 1 mm to calculate an equivalent area. This gives a very approximate rate of flow.

For pump seal leaks the flow is annular, and an equivalent hydraulic diameter should be calculated instead of pipe diameter. This is equal to the flow area divided by the sum of the inner and outer circumference of the annulus. For a crack, or flange leak, the effective hydraulic diameter is calculated as the flow area divided by the circumference of the hole.

A check of flow rate should be made to ensure that the Reynolds number is in the region giving turbulent flow.

A more accurate result is obtained from direct observation. Some flow rates measured by the author in actual failure situations have varied as follows:

Such leaks are themselves significant only if they occur indoors or in still weather if they involve highly toxic substances, or if they are ignited.

However, a fire at a flange will tend to heat the pipe on one side, and bend it, causing the flange to open further. This can lead to much larger holes forming.

Pipes containing heated oil (250 - 400^0 C) are very difficult to keep tight because of expansion and contraction of bolts and piping itself. There is almost always some seepage from these sources. These present a hazard if, for example the oil collects in a sump, or if insulation is wetted by the oil, since under these circumstances auto ignition is possible for some oils.

Steam pipes similarly, are subject to many leaks unless maintenance levels are very high. Steam traps, in particular are a source of frequent leaks.

Case	Flow rate
80 bar, gas and oil valve seal	10 cc per min.
Liquid Ammonia, valve seal loose	0.5 l per min.
Butane, valve seal loosened by vibration (8" valve)	4 l per min.
Isobutane, broken pump seal (carbon face seal), with no secondary seal). ($2H_p$ pump)	3 l per min.
Ammonia, broken pump seal, (no secondary seal) ($8H_p$ pump)	2 l per min.
Flange seal, ethylene (30 inch flange, leak between two bolts)	0.2 l gas per min.

Table 7.1 Release rate examples

The size of a hole

Most risk analysis calculations are made on the assumption of round holes with sharp edges. These assumptions are made for simplicity, and for lack of better information. In practice, the assumption is sometimes reasonable for total pipe ruptures, but seldom for simple holes. The form of flow through actual cracks has been investigated by Friedel and Westphal (e.g. 1989), and by Collier (1984). A study of naturally occurring cracks was made by Gardner and Tyrell.

For risk analysis purposes, it is desirable to know the form and the size of holes as a function of failure probability. Very little information has been collected concerning these aspects of failure statistics. The data in chapter some statistics for process plant piping, and in chapter 15, some information is given concerning pipeline failure.

Release of gases

For gases at low pressure, the flow through an orifice is governed initially by the equations of isentropic (constant entropy) flow, by solving the conservation of energy and mass equations. When the ratio of upstream pressure to downstream pressure is lower than the so called critical pressure, the mass flow through the orifice is given by:

$$\dot{M}=A\ Cd\ P_0\sqrt{\frac{2gM_w}{R\ T_0}\frac{\gamma}{\gamma-1}\left[\left(\frac{P_2}{P_0}\right)^{2/\gamma}-\left(\frac{P_2}{P_0}\right)^{\frac{\gamma+1}{\gamma}}\right]}$$

where
P_0 is the stagnation pressure upstream
P_2 is the downstream stagnation pressure
M_w is the molecular weight of the gas
R is the universal gas constant
T_0 is the upstream stagnation temperature
γ is the ratio of specific heats for the gas

The velocity of the gas is given by

$$u_2=2\ C_d\frac{\gamma}{\gamma-1}\frac{P_0}{\rho_0}\left[1-\left(\frac{P_2}{P_0}\right)^{\frac{\gamma-1}{\gamma}}\right]$$

where
ρ_0 is the initial density of the gas at P_0, T_0

These equations are useful for slow flows, as for example from a ventilator.

If the downstream pressure is continually reduced, the flowrate reaches a maximum when the critical pressure is reached:

$$\frac{P_{crit}}{P_0} = \left(\frac{2}{\gamma+1}\right)^{\frac{\gamma+1}{\gamma-1}}$$

At this pressure, the flow rate is just at the speed of sound in the throat of the orifice. The pressure ratio at which this is achieved is approximately .5, as follows.

γ	1.135	1.300	1.400	1.667
P_{crit}/P_0	0.577	0.546	0.528	0.487

For most risk analysis situation, the pressure at the throat of the orifice is the atmospheric pressure, or above this, so that for upstream pressures greater than about 2 bar absolute, flow will be critical, or supersonic. For these conditions:

$$\dot{M} = A\ C_d \sqrt{\gamma\ P_0\ \rho_0 \left(\frac{2}{\gamma+1}\right)^{\frac{\gamma+1}{\gamma-1}}}$$

$$u_{throat} = \sqrt{\gamma\ R\ T_{throat}}$$

$$\frac{T_{throat}}{T_0} = \frac{2}{\gamma+1}$$

Further reduction of pressure after the throat has no effect on the rate of flow of upstream flow conditions, and so flow rates, and momentum can be calculated from the equations given above. The downstream conditions change considerably, however.

If the pressure is above the critical pressure ratio, the gas continues to expand outside the orifice, with a corresponding cooling, and acceleration to supersonic velocity. Shock waves form in a complex structure, with both expansion and compression waves At first , with a moderately underexpanded jet there is a core of diagonal expansion shock cells. The core is roughly conical. When the ratio between the upstream and atmospheric pressure is greater than about 4, the fully developed jet appears, with a normal shock wave, the so called Mach disk, and an expanding fan of gas flowing at supersonic speed outside the central core. (Ramskill)

Theoretical descriptions of this process are very complex. Fortunately, for risk analysis purposes it is often only necessary to know the behaviour of the jet at large distances. The momentum of the gas at large distances can be found from equations . The temperature falls dramatically at the throat of the orifice, but as the jet of gas slows outside the orifice, kinetic energy is converted to heat. As a result, only relatively small changes in temperature are observed overall.

Gas flow in pipelines

Gas flow in pipelines can be calculated using the principles for determining flow resistance as described for liquid flow above, as long as the flow is steady. The flow equations are derived from an energy balance for the situations. The Mach number for the flow is defined as:

$$\text{Mach} = u/a$$

where

u is gas velocity
a is the speed of sound

The energy equation reduces to:

$$\frac{\gamma+1}{\gamma}\ln\frac{P_0T_0}{P_2T_2}+\frac{\gamma-1}{2}\gamma\left(\frac{P_0^2T_0^2-P_2^2T_2^2}{T2-T0}\right)\left(\frac{1}{P_0^2T_2}-\frac{1}{P_2^2T_0}\right)+\frac{4fl}{d}=0$$

This equation can be used to determine T_2. Then

$$\dot{M}=A\sqrt{2g\frac{M_w}{R}\frac{\gamma}{\gamma-1}\frac{T2-T1}{(T_0/P_0)^2-(T_2P_2)^2}}$$

These equations can be used as long as the flow remains subsonic. If the flow becomes sonic then the energy equation becomes:

$$\frac{\gamma+1}{2}\ln\left[\frac{2+(\gamma-1)Ma_0}{(\gamma+1)Ma_0^2}\right]-\left[\frac{1}{Ma_0^2}-1\right]+\gamma\left[\frac{4fl}{d}\right]=0$$

where Ma_1 is the upstream mach number, i.e ratio of the velocity to the velocity of sound.

$$\dot{M}=Ma_0P_0\sqrt{\frac{\gamma g M_w}{R T_0}}$$

In order to solve these equations it is necessary to assume a value for the friction factor f, based on an initial guess of flow velocity. The equations are then solved, and a new value for f derived so that a more accurate result is reached.

For a short pipe, attached to a large gas reservoir, the above equations give an accurate result, since a steady state is reached very quickly.

A common problem in risk analysis is the release caused by a hole or rupture in a gas pipeline. The flow rate for this case can hardly be regarded as steady, in fact it falls very rapidly after the initial release, as gas leaves the pipeline. The equations are based on the same principles as above, but it is necessary to take account of the reduction in pressure, and acceleration of gas in each section of the pipe, and the fact that the pressure wave in the pipe propagates at the speed of sound. In practice, this means that quite sophisticated computer programs are needed, using either the method of characteristics or finite element methods (see e.g. Fox 1977).

The actual flow pattern depends also on the possibility of closing the pipeline with shut off valves, and the response time for this close down. In determining shutdown time, it is necessary to take into account:

- the time for the pressure wave to travel to the nearest pressure sensor
- the time for the operator to respond, if shut down is not automatic
- the time for valve closure (usually 10 to 30 seconds, unless valves are of the slam shut type)

Fig 7.5 shows input data for such a program, and the dynamic variation in flow for a methane pipeline.

Two phase flow

Many of the most significant flow situations for risk analysis involve the escape of liquefied gas, stored at high pressure, and above its boiling point. As such gases are released, the gas boils, the heat energy due to the difference in temperature between the initial temperature and the boiling point being converted to latent heat of vapourisation of the gas. The actual process of vapourisation, or "flashing", takes time (a few milliseconds) so that if the escape is from a hole in a vessel, the equations for liquid release may be used. The degree of evaporation is expressed as the flash fraction, which is the ratio of gas to the total mass flow.

```
DATE:12-30-1990                                          PAPA V2.4
TIME:14:33:12                 BLOWPIPE gas release
                                            Analyst:JRT
                                            Case :RUPTURE AT END OF PIPE
                              Size of hole (inches):10
                Gas temperature in reservoir (deg C):25
                      Reservoir gas pressure (bara):54
                            Gamma value for gas:1.11
                        Molecular weight of gas:16.04
                        Viscosity of gas (Pa s):.000017
                            Pipe roughness (mm):.02
                Initial gas flow rate (Nm3/hr):36000
        Length of pipe up to break position (m):6000
                          Pipe diameter (inches):10
                        Distance to ESD valve (m):1000
                        Time to ESD closure (s):20
                            Timestep fraction:.1
```

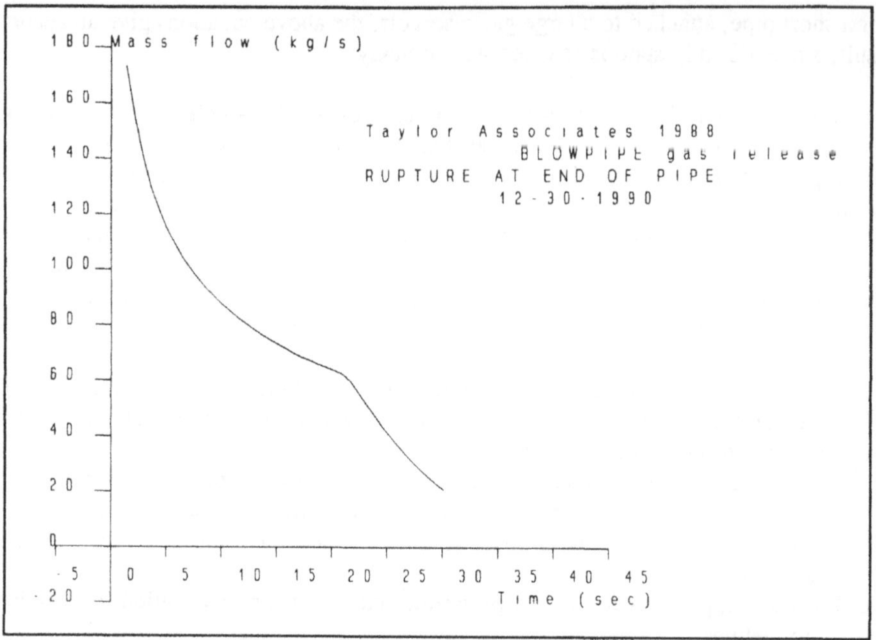

Fig 7.5 Flow from rupture in gas pipeline, showing effect of emergency shutdown

$$\chi = \frac{C_p \, \Delta t}{h_v}$$

where
χ is the flash fraction
C_p is the specific heat of the liquid
Δt is the difference between the initial temperature and the boiling point
h_v is the latent heat of vapourisation

For releases from pipes, the flow situation is much more complicated, with many different flow regimes. For the lowest degree of flashing, flow takes the form of slugs of liquid. As the pressure, and velocity increase, flow forms as droplets, as an annulus on the wall of the pipe, and finally, as a mist. For most of the cases of interest in risk analysis, mist flow is the most significant, but other flow regimes can be of interest, for example flows of hydrofluoric acid or butane at temperatures near to their boiling point, and mixtures of gas and oil in gas well blow outs.

Releases of liquefied gas have been studied especially by Fauske (see e.g. 1985) and by Fletcher (1984). Two decidedly different processes appear to be involved. Within the first meter or two, the rate of nucleation dominates the vapourisation process. For longer pipes, the vapourisation reaches equilibrium, and the degree of vapourisation is determined by pressure drop. Fauske's most recent models take these effects into account as follows:

$$\dot{M} = A\ C_d \sqrt{2(P_2 - P_v)\rho_l + F\ \frac{h_{fg}^2}{v_{fg}^2} \frac{1}{N\ T\ C_p}}$$

where
P_v is the liquid vapour pressure
P_2 is the external pressure (atmospheric)
h_{fg} is the heat of vapourisation
F is a friction correction factor
N is a correction factor for short pipes

The factor F is an empirical factor. Fauske has given value of F for pipes up to 200m. as a function of pipe length to diameter ratio. N is given by:

$$N = \frac{h_{fg}^2}{2(P_v - P_0)\rho_l\ C_d^2\ v_{fg}^2\ T\ C_p} + l/l_e$$

for $l < l_e$

$$N = 1$$

for $l > l_e$

$$l_e = 0.1m.$$

These equations apply in the steady state. In many cases, such as butane or ammonia pipelines, the flow rate is not steady, because the upstream flow is limited, for example by a pump or valve. Tam has derived empirical relations for such cases, based on experiments with 100 m. pipes and LNG and LPG releases.

$$\dot{M} = -\alpha M$$

where
M is the total mass in the pipeline

α is here an empirical constant, a function of pipe diameter, hole diameter, and excess temperature.

Much more sophisticated methods have been developed for the dynamics of two phase flow, but have found their primary use in examining loss of coolant accidents for nuclear reactors.

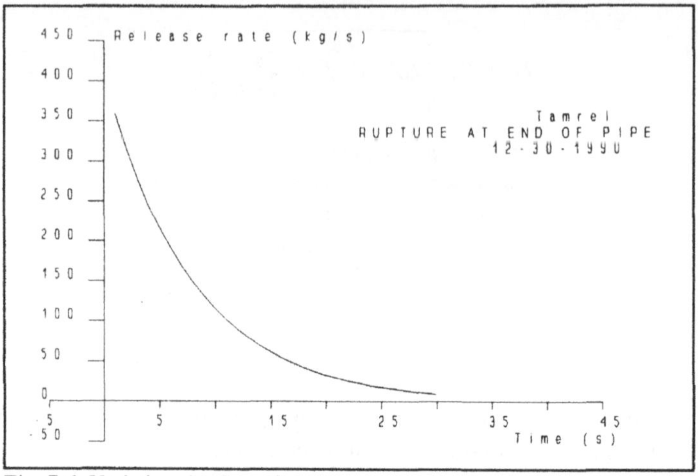

Fig 7.6 Variation in flow from a pipe containing LPG, rupture of a shutdown pipe.

References

Hinze 1953	Am. Inst. Chem. Eng. Jnl., 1, 289-295, 1953
Friedel,L. and Westphal,F	Leakage Through Cracks, IChemE Symposium Series No. 102
Collier, R.P.	Two Phase Flow through Intergranular Stress Corrosion Cracks, EPRI-NP-3540-LD (1984)
Ramskill, P.K	A Study of Underexpanded Gas Jets, SRD R 302, HMSO 1985
Fauske, H.K.	Flashing Flows, Some practical Guidelines for Emergency Releases, Plant/Operations Progress 4,132-134 (1985)
Fauske, H.K. and Epstein, M.	Source Term Considerations in Connection with Chemical Accidents and Vapour Cloud modelling, in *Int. Conf. on Vapour Cloud Modelling*, CCPS (1987)
Fletcher, B.	Flashing Flow Through Orifices and Pipes, AIChE Loss Prevention Symposium, Denver, 1983
Tam, V.H.Y.	A Simple Transient release rate Model for Releases of LPG From Pipelines, 6th Int Symp. "Loss Prevention and Safety Promotion in the Process Industries", Oslo, 1989
Fox,J.A.	Hydraulic Analysis of Unsteady Flow in Pipe Networks, Macmillan Press, 1977.

8 Evaporation of Liquids

When a liquid escapes from a pipe or a vessel, its behaviour depends on the liquid temperature in relation to its boiling point at atmospheric pressure. If the liquid temperature is above the boiling point, the liquid will boil and cool. If the initial temperature is high enough, all of the liquid will boil away to gas, and the rest of the liquid will be carried away as mist.

For liquids which have a boiling point close to the initial temperature, a pool of liquid will form. Three cases in particular need to be distinguished.

1. A liquefied gas at elevated temperatures (above the boiling point)
2. A liquefied gas at cryogenic temperatures.
3. A volatile liquid at ambient temperatures.

In the case of a cooled liquefied gas, the evaporation depends initially on the transfer of heat from the ground. The heat causes the liquid to boil. The rate of evaporation depends very strongly on the spreading of the pool, since heat conduction from the ground is primarily in a vertical direction. Once the upper layers of the ground have achieved a similar temperature to the liquid, the amount of heat obtained from this source is limited, and the atmosphere, (wind) becomes the most significant heat source.

For volatile liquids, for which the initial temperature is close to ambient temperature, and with boiling point above ambient temperature, evaporation due to boiling is not relevant. However, evaporation due to the vapour pressure of the liquid (vapour diffusion) and transport by the wind, can be significant.

Basic equations for flashing

When a liquefied gas is released at a temperature above its boiling point, a portion of the liquid will immediately (i.e. within milliseconds) be converted to gas. In doing so, heat is converted from sentient heat to heat of vapourisation. The temperature drops. The temperature which is reached depends on the difference between the initial temperature, and the boiling point of the liquid. If the difference is large, all of the liquid may evaporate. the final temperature is then above the boiling point. If the heat available is lower due to a lower temperature difference, some of the liquid will remain unevaporated. If the release is violent, the both the gas and liquid are cooled to less than the boiling point, a phenomenon known as supercooling. The temperature can be up to 10°C below the boiling point. For less violent releases, the liquid and gas are in equilibrium and the temperature is very close to the boiling point.

The proportion of liquid which evaporates is called the flash fraction. It is given by:

$$ff=\frac{sp(T_{init}-T_{bp})}{L}$$

where
ff *flash fraction*
sp *specific heat of liquid*
T_{init} *initial temperature*
T_{bp} *boiling point*
L *Latent heat of vapourisation*

The flash fraction can never be greater than 1.

For liquefied gases under pressure, the release is often very violent. The liquid breaks up into droplets during the release, producing a mixture of cold gas and mist. The liquid may gradually "rain out" from the cloud of gas, depending on the droplet size. Equations for rainout of droplets are given above, page 224.

Evaporation of a cooled liquid due to ground heat.

For a large pool of cooled liquid, the transfer of heat across the liquid interface with the ground is very rapid. For shallow pools, the transfer of heat into the liquid, and the boiling, is also very rapid. The factor limiting the rate of evaporation is then conduction from the ground. The following analysis is based on the work of Opschoor.

Assuming uniform heat flow properties and a semi infinite "ground", the temperature at time is given by Fourier's equation:

$$\frac{\delta T}{\delta t}=a_s\frac{\delta^2 T}{\delta z^2}$$

with boundary conditions

$T = T_s$ for t=0 and z=0
$T = B_p$ for z=0
$T = T_s$ for $z=\infty$

where

B_p is the boiling point of the liquid
T_s is the ground temperature initially
z is the depth into the ground
a_s is the thermal diffusivity of the ground

The solution of this equation is

$$T(z,t)=B_p+(T_s-B_p)erf\frac{z}{2\ (a_s\ t)^{1/2}}$$

The heat flux is given by

$$q=l\ dT/dz$$

Substituting, this gives

$$q=\frac{l\ (T_s-B_p)}{\sqrt{\pi\ a_s\ t}}$$

The mass rate of evaporation is then given by

$$\dot{M}=A\frac{q}{h_v}$$

Where h_v is the latent heat of evaporation of the liquid.

Note that the evaporation here falls in proportion to the inverse of the square root of time i.e. with a slowing rate of fall. At some stage however, the rate of evaporation due to boiling will fall to less than that due to wind evaporation.

The value lambda used here is the coefficient of heat conduction of the ground, and a_s is the thermal diffusivity of the ground. The values of the constants in these equations depend on the porosity of the ground. Substituting standard values for the constant produces values for evaporation rate which are significantly lower than those observed in practice. (Drake and Reid 1975). Opschoor, interpreting this data, uses a correction factor of 8 for dry soil.

The values for a_s and l_s are given in table 8.1 (Carlslaw and Jaeger 1959).

Substance	l_s	a_s
Average ground	0.9	$4.3*10^{-7}$
Sand (dry)	0.3	2.3
Sandy ground (dry)	0.3	2
Sandy ground (damp)	0.6	3.3
Gravel	2.5	11
Carbon steel	45	127
Concrete	.42	$13*10^{-7}$

Table 8.1 Values for soil heat condutivity and thermal diffusivity

Note that soil conductivity about doubles with an increase on water content from 2% to 10%.

A spreading pool of boiling liquid.

In many cases, the size of a liquid pool is determined by the physical limitations, such as bund sizes. In some cases though, the liquid will be free to spread as a pool. The initial advance of the edge of the pool is often very fast, because the boiling of the liquid supports, large droplets, which move over the ground on a cushion of gas. A velocity of between 5 and 20 m/s is typical for small pools, but there is a limiting minimum depth to such a pool, (about 1/2 to 1 cm droplet depth) which limits the size of the pool.

For larger releases, the rate of advance of the pool is given by

$$dr/dt = (C\ g\ (h - h_{min}))^{1/2}$$

using standard hydrodynamic theory for thin liquid films.

C is here a constant, about 2

h is pool depth

h_{min} is minimum pool depth

h_{min} is determined by pool roughness, and surface tension and wetting tendency of the liquid.

The volume of the pool is given by

$$Q^e = \pi\ r^2\ h$$

For a continuing release, dQ_e/dt will often be constant, or only slowly changing.

This gives

$$dr/dt = (Cg\ (Q_{e(t)}/\pi r^2 - h_{min}))^{1/2}$$

The rate of evaporation depends not only on the size of the pool, but on the rate of spreading of the pool, since evaporation at the pool edge is more rapid than at the centre. For this reason it is necessary to treat each ring of the pool separately, and to integrate the evaporation equations for the pool, for each instant in time.

The equations can be solved analytically for constant Q_e or for Q_e rising at a constant rate. In practice, there will often be an initial high rate of release, or instantaneous release, followed by a release which falls with time. Computer programs therefore uses a numerical integration procedure to calculate pool diameter and evaporation rate. These values are plotted as curves against time.

In order to be able to transfer data to some of the later calculations, for example simple dispersion calculations, a single value of evaporation rate must be chosen, unless dynamic dispersion equations are used. The values chosen may be the peak evaporation rate, or some representative value, if the peak lasts too short a time to give a good impression of the risk. For toxic substances for example, the peak evaporation rate may have no significance because it lasts too short a time to give a significant toxic load.

The actual form taken by the release rate as a function of time varies widely, according to whether the release is limited by bund walls, curbs etc. and whether there is an initial release. Fig 8.1 shows some of this variation. The sharp drop in evaporation rate for the spreading pool corresponds to the complete evaporation of the outermost area of the pool shortly after outflow stops. In reality, the process here is very complicated, and it is possible for the outer edge of the pool to oscillate back and forth. The simple curves in fig 8.1 represent therefore only an approximation.

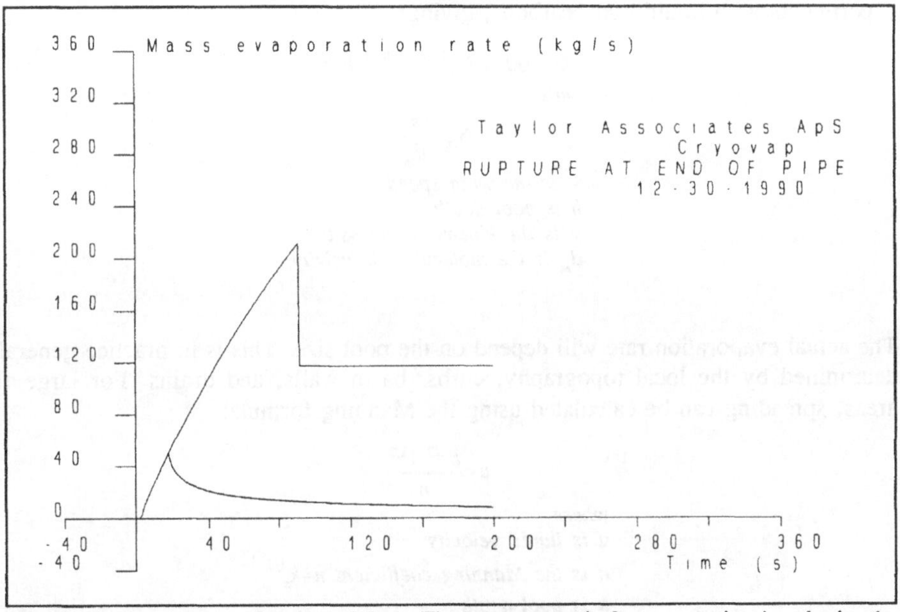

Fig 8.1 Evaporation as a function of time. The lower curve is for evaporation in a basin, the upper curve for the same case, but with unrestricted spreading.

Evaporation of volatile liquids

Liquids such as gasoline, hydrogen fluoride in cold climates, solvents, acrolein, etc. are not cryogenic. Heat from the ground generally does not contribute significantly to their evaporation. Evaporation is due to heat convected by the wind, and to heat radiation from the sun and clouds.

The evaporation rate is given by (Fleischer 1980):

$$\dot{M}=\frac{k\,A\,P_v\,M_w}{R\,T_a}$$

where
$\dot{M}$ is the evaporation rate
k is the mass transfer coefficient
A is the pool area
P_v is the liquid vapour pressure
M_w is the liquid molecular weight
R is the universal gas constant
T_a is the ambient temperature

Various correlations have been given for the mass transfer coefficient k, e.g. by Perry and Green, by Sutton. Mackay and Matsugu (1973) based their correlation on experiments with liquid hydrocarbons, giving:

$$k=0.0048\;N_{sc}^{-0.67}\;w^{0.78}\;h^{-0.11}$$

with

$$N_{sc}=\frac{\nu}{d_m}$$

w is the wind speed
h is pool depth
ν is the kinematic viscosity
d_m is the molecular diffusivity

The actual evaporation rate will depend on the pool size. This is in practice generally determined by the local topography, curbs, basin walls, and drains. For large flat areas, spreading can be calculated using the Manning formula:

$$u=\frac{h^{2/3}\;i^{1/2}}{n}$$

where
u is liquid velocity
n is the Manning coefficient $n=k_s^{1/6}$
h is pool depth
i is the slope
k_s is the surface roughness depth

The equation applies only for fully rough flow. k_s has values varying from .03 m. for rough concrete to .06 for rough terrain.

References

Opschoor, G. — Evaporation, in *Methods for the Calculation of the Effects of the Escape of Dangerous Materials*, Dutch Ministry of Labour, 1978

Drake, E.M. and Reid, R.C. — How LNG Boils on Soil, Hydrocarbon Processing, Vol. 54, No. 5 (1975)

Carlslaw, H.S. and Jaeger, J.C. — Conduction of Heat in Solids, Clarendon Press, (1959)

Fleischer, M.T. — SPILLS, an evaporation/air dispersion model for chemical spills on land. Shell Development Centre, Houston, 1980

Sutton, O.G. — Micrometeorology, McG-H, 1953

Makay, D. and Matsugu, R.S. — Evaporation Rates of Hydrocarbon Spills on Land and Water, Canadian Jnl. Chem. Eng. Vol. 51, August 1973

9 Dispersion of Gases

When a release of gas, or of gas and mist has occurred it disperses into the atmosphere. Dispersion is initially due to violent mixing, in a turbulent jet, if the release is through a hole, or in a "blow up" of gas (violent mixing). These phenomena often dilute the release by a factor of 10 or more, within a short distance from the origin of the release. After this point, further dispersion depends on whether the mixture of gas and air is heavy or not. If it is heavy it will disperse initially as a layer, close to the ground. If it is lighter than air the dispersion will take the form of a rising plume.

Jet Mixing

When gas is released from an opening, it spreads in a cone, and entrains air. The amount of air entrained is important, in order to determine the density of the gas as a basic parameter for subsequent dispersion calculations, and also as a basis for jet flame calculations.

Calculations for such jets have been described by Cude (1974) based on experimental work by turbulent jets, which will be the case for most accidental gas escapes of significance. The earlier Ricou and Spalding theory assumes a uniform cross section across the jet.

Other models have been developed, for example by Ooms et al.(1974) which assure a gaussian distribution across the jet. However, the differences in results from the models do not justify using the more complex models unless there are other advantages. One such advantage is that the transition to plume models is made easier. This is the case for example for the TECJET model (Emerson 1986). The practical differences between theses models have been shown to be not very significant in themselves for risk analysis (Lee 1989) but can make significant differences in the way that the jet is continued.

For the uniform distribution the volume discharge rate is

$$Q=\frac{\pi}{4}\,d^2\,u_o$$

where
d is the orifice diameter
u_0 is the initial velocity

The momentum conservation equation is

$$\pi\, r_0^2\, \rho_{gas}=\pi\, r^2\, \rho_{mix}$$

where R is the radius of the jet at distance x
u is the average jet velocity at distance x

From observations, we know that jet radius grows linearly with distance x

$$r=\alpha\, x$$

We can then solve for Q, the volumetric flow rate at distance x

$$\begin{aligned} Q &= \pi\, r^2\, u \\ &= \alpha\, x\, \pi\, r\, u \\ &= \alpha\, \pi\, \frac{d}{2}\, x\, u_0 \left(\frac{\rho_{gas}}{\rho_{mix}}\right)^{1/2} \end{aligned}$$

α is observed to be about 0.16, corresponding to a jet angle of 17°.

For high degrees of mixing,the gas mixture density can be approximated by the air density. For intermediate values, the mixture density is given by:

$$\rho_{mix}=\frac{Q_0+(Q-Q_0)\rho_{air}}{Q}$$

Rather than solving the resulting equations explicitly, it is easier to get an approximate value Q/Q_0 using the density of air as an approximation to mixture density and then to use the resulting value to get a better approximation for the mixture density.

It is often desired to find the point at which concentration falls to a non-flammable or non-toxic value. The relative concentration is given by:

$$\frac{\chi}{\chi_0}=\frac{Q_0}{Q}$$

For example, lower flammability limits are often given by the volumetric percentage concentration of hydrocarbon, LFL. The distance to the lower flammability limit is then given by:

$$x=\frac{d}{2\alpha}\left(\frac{\rho_{air}}{\rho_{gas}}\right)^{1/2}\frac{1}{LFL}$$

Note that no allowance has been made here for wind or atmospheric turbulence. Each of these factors will give a higher degree of dilution than in the above equations.

Experiments carried out by the author on impinging jets indicate that, for a flat surface, there is no significant additional dispersion. This accords with the fundamentals of the momentum jet, since mixing at a point on the jet shear surface where it meets the still air is proportional to velocity of the jet. Both velocity and jet surface area are conserved, to a good approximation, when the impinged surface is flat. For concave surfaces, rooms, etc. mixing is reduced, in the limit to such an extent that no jet dispersion takes place.

It is customary to regard the jet mixing to be complete at the point where jet velocity becomes equal to wind velocity (at least for the usual levels of wind velocity. For very low wind speeds, the jet is judged to be terminated when movement due to momentum is negligible when competed with movement due to buoyancy or atmospheric turbulence). The length of jet is then given by:

$$x_{max}=\frac{u_0}{w}\frac{d}{2\alpha}\left(\frac{\rho_{gas}}{\rho_{mix}}\right)^{1/2}$$

Two Phase Jet Dispersion

For flashing gases, the above analysis still applies, provided that the liquid is atomised, so that the liquid momentum can be regarded as part of the jet momentum. However the initial area of the jet is obtained by considering the volume of the initially expanded vapour. The length of the jet should in principle be measured from the point of the virtual source, obtained by projecting the jet backwards. The initial volume flow rate is (approximately) the flow rate of the flashed portion of the liquid.

For liquids with a high degree of superheat at atmospheric pressure, any liquid which does not flash will tend to form a mist of small particles. The term for density in the momentum equation should be adjusted to account for this additional momentum.

For a flashing liquid too, there will be an additional evaporation, beyond the immediate flashing, due to the heating effect of the entrained air. The degree of additional evaporation is given by:

$$\Delta M = (Q - Q_0)\frac{\rho_{air}\, C_{p_{air}}\, T_a}{h_v}$$

This will give an additional contribution to the volume flow rate, increasing Q, but this effect is usually negligible in practical calculations.

Two phase releases of gas and oil, from pipes with a pressure of several bar up to several tens of bar, produce violently turbulent jets. Two phase non flashing jets have apparently not been studied experimentally, and non of the standard models are directly relevant. The model given here,is based on observations of a two phase flashing jet of water and steam with an initial pressure of 10 bar. These observations allow us to evaluate the effect of rain out of the liquid phase from the jet.

The immediate result of a jet of highly compressed gas is an expansion, within the pipe from which the release issues. This expansion tends to break up the liquid phase into droplets, due to the accelerating forces behind the liquid, and frictional forces over the surface of the liquid.

Outside the throat of the pipe there is a sideways expansion of the jet, down to atmospheric pressure, (actually slightly below due to the velocity of the jet). The jet continues forward under its own momentum, in the form of a simple cone.

The radius of the cone grows linearly, with an angle from 17° to 20° as a result of air entrainment.

In a flashing jet with an appreciable degree of superheat, the liquid phase breaks up not only because of the acceleration, but also because of bubble formation. In ref.1, break up of the liquid jet can be seen to be delayed in some cases, by the amount of time required for bubble growth.

As the jet proceeds forward, air is drawn in, and initially causes intense shearing. This effect too seems capable of breaking up droplets. From observation of a 10 bar steam and water jet, droplets of sizes from 1 mm down to 10 microns are formed.

As the jet proceeds, some of the liquid phase "rainout" due to gravity, and is removed from the jet. Initially, the rainout droplets have an appreciable velocity in the direction of the jet, but after a few meters, the direction of rainout is more or less vertical.

After several tens of meters, the jet has mixed with air to such an extent, that its velocity has fallen to a normal wind velocity, and the release ceases to have the character of a jet. It has more the character of a plume.

Applying an analysis similar to that above we get:

$$Q=\alpha\ x[(1-\beta)u_{liquid}+\beta\ u_{gas}]^{0.5}\left(\frac{M_0}{\rho_{mix}}\right)^{0.5}$$

where β is the flash fraction

$$\rho_{mix}=\frac{\beta\ M_0+(1-\beta)rf\ M_0+Q-Q_0\ \rho_{air}}{Q}$$

rf is here the rainout fraction, that is, the part of the liquid which falls to the ground. The equations assume vertical falling, which accords with observation for horizontally or upward directed releases.

These equations can be applied in an iterative scheme, as for single phase jets.

Rainout

For violent two phase jets, that is for any two phase release with more than 1 bar overpressure, the liquid in the jet will be rapidly atomised. The size of droplets was described in chapter 8, page 214. If the release is not too violent, the liquid may "rain out" of the jet. If it does so it may subsequently evaporate, so that the net release is the same. Concentrations will be lower however, and the dispersion more prolonged.

The rate at which droplets fall is given by:

$$u=\frac{F}{6\pi\ r\ \nu}$$

where
F is the drag force
r is doplet radius
ν is the kinematc viscosity of air

$$F=\frac{12}{R_e\ \pi\ r^2\ u^2\ \rho_{air}}$$

for non turbulent flow of air around the droplet.

In practice it is necessary to decide how far the droplet must fall, in order to determine how long rainout takes. In practice this is taken as the depth of the jet, plus it height of the release point. If the time is long, i.e if the droplet speed is much lower than wind speed, rainout can be ignored.

Dispersion of light toxic gases and toxic smoke.

The dispersion of light toxic gases is important for many types of chemical plant accident, f.ex.

- release of a light gas from a leak in a pipe or vessel

- release of a cloud of light gas from a rupture in a pressure vessel.

- evaporation of toxic vapours from a pool of liquid.

- production of toxic smoke from a fire.

There are several physical effects which are important in this context. For releases of gas which are pressurised, the gas in release forms a jet. The jet is cone shaped, growing wider along its length, as air is drawn into the jet. The jet is regarded as terminating when its velocity falls to that of the wind, or when the concentration of toxic substance in the jet falls below the threshold limit of interest.

Once the jet has terminated, the gas disperses as a plume. The dispersion results from turbulence in the atmosphere, and also from turbulence in the cloud itself due to its buoyant motion if the gas is very light. Turbulent eddies of air move into the plume, and also push the plume sideways. Turbulent eddies of gas move out of the plume.

At large distances, (over 1 km.) the large scale turbulence in the atmosphere causes the plume to wander from side to side,(meandering).This can be important for toxic releases,because it means that a larger population may be exposed to toxic effects, but for a shorter time, as the different loops of the meander wander over the landscape.

Toxic smoke is formed when toxic combustion products are formed by burning or pyrolysis, or when toxic vapour fails to be burned in a fire. Air is drawn into the rising plume of smoke, in the same way as a jet of gas, as a result of the momentum of the smoke column.

The height to which the plume rises is very important for calculating toxic effects to the population. If the plume rises high in the air, then there is little chance that persons on the ground will be poisoned. The height to which the plume travels depends on its buoyancy (density relative to air) which in turn depends on the density of the toxic gas, and any other gases in which the toxin is mixed. It also depends on the temperature of the gas.

The height reached by the plume of light gas depends also on atmospheric conditions, and particularly on the way in which air density varies with height. If the atmosphere is neutrally stable or unstable, then the plume will continue to rise. If the atmosphere has a density inversion, with a rapid fall in density at a given elevation, then the plume will tend to rise to the level of rapid change, and to stop there. Only if the plume is lighter than the air above the inversion layer will it "punch through" the layer and disperse at a higher level.

Neutral Plume dispersion.

The study of dispersion of gases and smoke in the atmosphere date back to work at Porton Down in the 1920,s and to theoretical studies by O.G. Sutton, in the 1930,s through to 1950,s (Micrometeorology, Mc G Hill 1953). The field can be studied in many different ways,for example experimentally, in terms of dimensionless numbers, and in terms of partial differential equations. The differential equation approach poses many difficulties, because analytic solutions are generally not available, and because numeric solutions require an ever increasing degree of detail in order to be able to compute turbulent flows. As a result it is generally necessary to combine differential equation approaches with experimental or observational results, so that turbulence effects at the finest level are treated empirically. In what follows, the approach and notation follows F, Pasquill and F.B. Smith (Atmospheric Diffusion, 3rd edition, 1983.)

Neutral dispersion (empirical results)

Dispersion of gas which has the same density as the ambient air is called neutral dispersion. The pattern of dispersion is usually modelled semi empirically for practical purposes using the Gaussian plume equation (Pasquill and Smith , P. 320)

Pasquill and Smith write -

" Observations of passive plumes have confirmed that the Gaussian form is a satisfactory description for cross wind distribution in an ensemble average sense."
"For vertical distribution it has been found that the shape is not generally Gaussian (except when the plume is elevated----) but this deviation has tended to be disregarded as the practical consequence is not very significant"

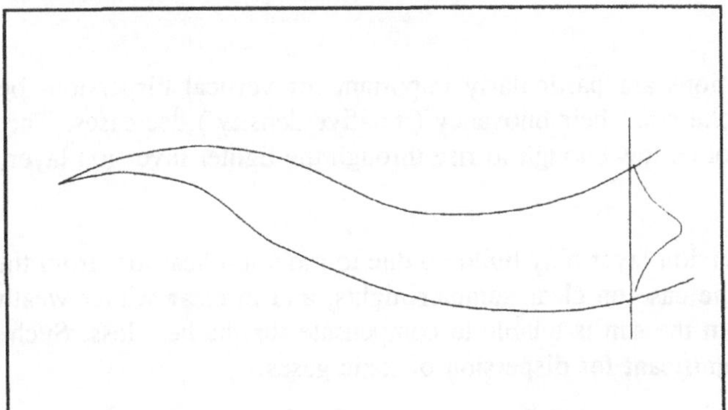

Fig 9.1 Concentration profiles in plume.

For an elevated source with plume rise the dispersion is modelled in a similar way (Pasquill and Smith, P.333)

$$\chi(x,y,z)=\frac{Q}{2\bar{u}\sigma_x\sigma_y}.\exp\left(-\frac{y^2}{2\sigma_y^2}\right).\left(\exp\left(\frac{-(h_s-z)^2}{2\sigma_z^2}\right)+\exp\left(\frac{-(h_s+z)^2}{2\sigma_z^2}\right)\right)$$

The values of σ_y and σ_z are derived empirically. Various tabulations have been given, as a function of stability category and surface roughness length (F.B. Smith),as a function of stability type (M.E. Smith), and as a function of stability category and terrain (open country or urban) (Briggs)

For calculations here, the values due to M.E. Smith are used for elevated releases and releases with a rapid and significant plume rise, and F.B. Smiths values for surface releases. There is currently a move towards deriving σ values from fundamental measurements of turbulence, velocities, and Monin Obukov length, rather than the qualitative stability categories. However wind rose probability data in these terms are (as yet) seldom available.

Effect of special meteorological conditions.

The rate at which air cools due to adiabatic expansion is called the dry adiabatic lapse rate Γ. It is 5.4 degrees F per 1000 ft. or 9.84 degrees C per 1000 m. If the air is well mixed vertically, then the temperature profile with height is linear and follows this rate. The stability of the atmosphere is said to be neutral. If a parcel of air is displaced vertically, it will neither begin to rise further, nor sink back.

If the lapse rate is less than Γ, that is the air at higher level is warmer and lighter than in the neutral case, the atmosphere is said to be stable. As a result, air displaced from a lower level air mass is warm and the lapse rate is greater than Γ, and there will be a tendency for convection currents to arise, and for a good deal of vertical mixing to take place.

Stable conditions are particularly important for vertical dispersion, because as light gases or smoke rise, their buoyancy (relative density) decreases. The density of the gases may not be low enough to rise through the lighter inversion layer, and the gases are trapped.

A stable inversion layer may build up due to radiation heat loss from the ground. This is typically the case on clear summer nights, and in clear winter weather, when heat radiation from the sun is unable to compensate for the heat loss. Such conditions are especially significant for dispersion of toxic gases.

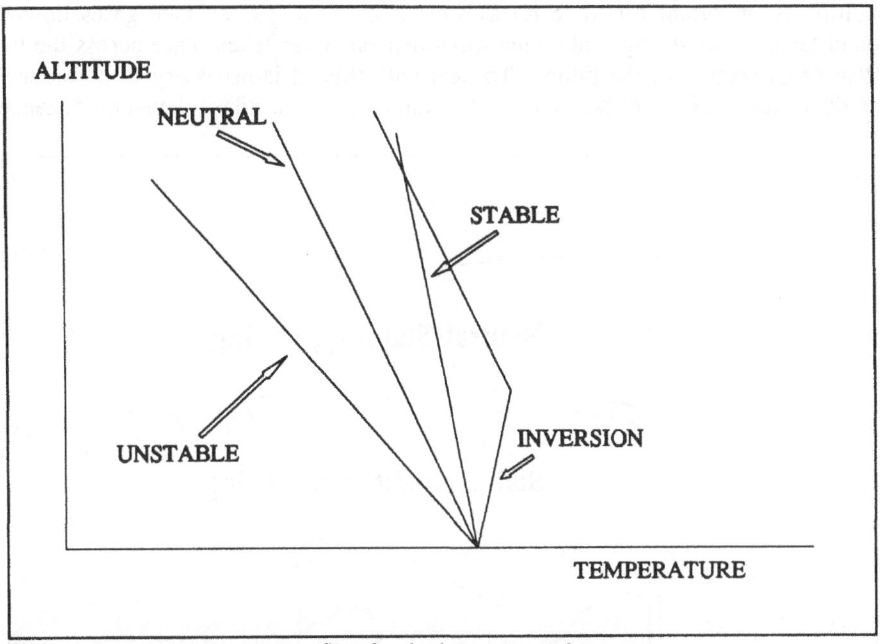

Fig 9.2 Vertical temperature profiles for the lower atmosphere.

Pasquill and Smith give a simple mathematical treatment of the condition with a finite mixing depth for the plume. They give (page 328):

$$X(x,y) = \frac{Q}{u.h_c(2\pi)^2\sigma_y}\exp\left(-\frac{y^2}{2\sigma_y^2}\right)$$

when the vertical distribution has become uniform.

u is here an average velocity for the mixed layer, and h_c the depth of the mixed layer. They give, as a practical procedure for calculation of dispersion in this case -

- use of σ_z following usual practice, until a value of 0.8 is reached

- use of $\sigma_z = 0.8h_c$ and the above equation thereafter.

The actual pattern of turbulence gives differences in the pattern of dispersion. In neutral conditions, gases disperse in a roughly symmetrical cone, and the condition is called coning. In stable conditions vertical dispersion is suppressed and the plume spreads as a horizontal fan. The condition is called fanning.

In unstable conditions, vortices in the wind flow are often of the same dimensions as the plume diameter (tens of meters) and as a result successive sections of the plume may be carried upwards and downwards. This condition is called looping. This condition is important for toxic releases,because the loops can take gases down to ground level, even at stages at which little dispersion has taken place across the time instant cross section of the plume. To deal with this, it is necessary to consider the time dependency of the dispersion and the values of the standard deviation distances.

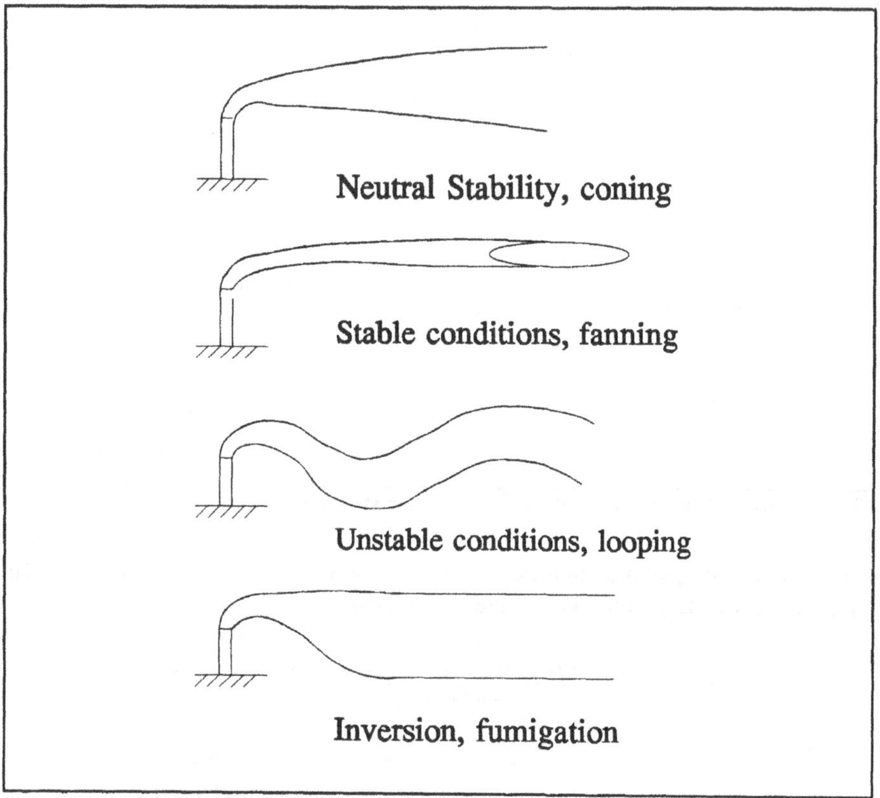

Fig. 9.3 Various plume profiles, depending on atmospheric conditions.

Effect of measurement time on dispersion.

The width of the dispersing plume depends on the period of time over which the width is measured. As the plume widens, but often the dense part of the plume will be much narrower, and the concentrations higher, if the width is measured at a particular instant.

Turner treats this problem in his work book in terms of average concentration ratios. For exposures greater than a few minutes, he gives:

$$X_s = X_k \left(\frac{t_k}{t_s} \right)^P$$

where t_s is the desired exposure time

t_k is the standard exposure time (generally 10 min)

P is an empirically derived constant, 0.17

These calculations do not take into account the effect of looping close to the source, and Turner explicitly states that his calculations cannot be used within a distance of three release heights.

M.E. Smith treats this problem and gives

$$\frac{X_{peak}}{X_{mean}} = \left(\frac{3600}{t}\right)^r$$

which assumes that the mean concentrations have been calculated on a per hour basis.

The values of given are:

Conditions	r
Very unstable	0.65
Unstable	0.52
Neutral	0.35

These calculations do not take into account the effect of looping close to the source.

The authors personal observations of looping behaviour at one particular site over several years loops which are approximately semicircular in the of vortices. The plume reaches the ground if the vortices are of the same dimension as the plume height, which is sometimes the case, it is the plume centre line concentration which applies at ground level,for a short period. The length of the period can be calculated from the wind speed and vortex dimension, assumed to be equal to plume height for cases for relevance. This gives:

$$t = h/u$$

where h is plume height, u is wind speed. Typically with h about 20 m. and u about 5m/s the high concentration loops will pass over a point at ground level in 4 s. The short exposures may be repeated regularly if the looping behaviour is strong.

Plume rise.

A plume consisting of a light or hot gas will rise due to its buoyancy. (the difference in density between the plume and the surrounding air). As the plume rises, it will disperse, and the plume include more air. As a result, the buoyancy and rate of rise

will decrease. It is easily seen that under neutral or unstable conditions, the plume will rise indefinitely. In practice, the plume rise is always stopped at an inversion layer. For very large fires though, the rise may be as much as 10.000 ft
(In such cases, local toxic concentrations at ground level are completely negligible).

A simple approach to the theory of plume rise has been given by Scorer. Using Newtons law:

$$d/dt(\rho w \pi r^2) = \pi r^2 \rho \frac{\Delta T}{T}$$

Assuming no loss of heat from the plume

$$d/dz(\pi r^2 \rho C_p \Delta T) = 0$$

Assuming a linear growth of plume radius with height

$$r = a.z$$

Solving these equations gives

$$w = bz^{-1/2}$$

$$\Delta T = cz^{-2}$$

where

$$b = \left(\frac{2gQ_h}{3\pi a^2 \bar{u} T}\right)^{1/2}$$

$$c = \left(\frac{Q_H}{\bar{u} z a^2}\right)$$

This gives

$$dz/dt=\left(\frac{Q_H}{\overline{u}z}\right)^{1/2}$$

$$z=\frac{Q_H^{1/2}x^{2/3}}{\overline{u}}$$

This simple calculation illustrates the form of the buoyancy equations, and applies equally to buoyancy arising because the gas dispersing is light or hot. The assumptions of linear growth of plume radius is supported by observations for the early stages of plume rise.The constant in the plume rise equation must be determined empirically. Also, the calculation applies only when departures from neutral atmospheric stability can be ignored.

The prime sources of data on plume rise are then the papers by Briggs (1965,1969) although there have been many variants on Briggs work by other authors.

Briggs defines the following

- the stack parameter

$$F=\frac{gQ_H}{\pi\rho C_p T_0}$$

- the stability parameter

$$s=\frac{g}{T}d\Theta/dz$$

(Θ is the potential temperature of the atmosphere which is the temperature that air would achieve if compressed to standard pressure)

- the wind speed

$$\overline{u}$$

For neutral, windy conditions, Briggs then gives, for plume rise

$$\Delta h = 1.6F^{1/3}u^{-1}x^{2/3}$$

This applies up to a distance

$$x^{*} = 0.52F^{5/2}h_s^{3/5}$$

ft/sec units

In stable stratification,Briggs gives

$$\Delta h = 2.9F^{1/3}u^{-1/3}s^{-1/3}$$

(feet)

If the wind is so light that the plume rises vertically, Briggs gives

$$\Delta h = 5.0F^{1/4}s^{-3/8}$$

(feet)

For a buoyant plume to penetrate a ground inversion, Briggs gives the following condition:

$$Z_{inversion} < 4.0F^{0.4}b_i^{-0.6}$$

(calm)

$$Z_{inversion} < 2.0F^{0.5}u^{-1/2}b_i^{0.5}$$

(wind)

(feet)

Where the inversion parameter

$$b_i = g\Delta T_i/T$$

$$T$$

is average ambient temperature

$$\Delta T_i$$

is the temperature difference between top and bottom temperatures of the inversion.

Toxic substances in smoke

When substances burn, the smoke produced is generally to some extent toxic. Three relevant cases are distinguished here. Firstly the direct and normal products of burning may be toxic. This is true for most chlorinated hydrocarbon, which produce hydrogen chloride on burning. It is also true of most sulphur containing compounds.

The second case is that of burning with limited ventilation. The result of burning of organic compounds may then be a partial pyrolysis of the substance, cracking, and generation of gases such as carbon monoxide and hydrogen cyanide.

The third case is that of incomplete burning. This may leave existing toxic compounds undestroyed, or create new toxic compounds.

The second case, of production of toxic substances due to incomplete combustion, is very complex, and for the most part must be treated experimentally. It is in any case most relevant for indoor fires. A good deal of empirical data has been gathered by Hidalgo, for plastics, and the burning of urethane foam, which generates HCN, has been widely studied.

The production of toxins in fire is very complex also, and has only in a few cases been studied, the most important being for production of dioxins and difuranes. Some data has been collected on the production of phosgene in fires in which there are chlorinated organic compounds,in the presence of metal catalysts.

The case of incomplete destruction of organic compounds is an important one for pesticides. Very little experimental work has been carried out, and combustion properties vary widely from compound to compound. Some compounds are thermally unstable, and almost completely destroyed, while others are hardly changed by attainable fire temperatures.

Risk levels are generally found to be greatest with small fires, because the plume rise is least with these, and the temperatures in the fire are often lower. For substances which burn easily, the largest risk levels arise when the pool is small and only partially burning.In this case radiation from the fire can cause evaporation in the non burning part of the pool. Only part of the evaporated vapour will be drawn into the flames.

A rough estimate of the amount of toxic vapour can be obtained as follows -

The radiation levels near a pool fire are between 1/8 and 1/5 of the surface emissive power of the fire, up to a distance of 4 pool diameters. For Kerosine fires, this gives a radiation level of about 5 KW/m2. After a short time, most of this heat will be absorbed by evaporation of liquid.

The latent heat of vapourisation of organic pesticide lies typically in the range 80 cal/g or 20.000 joules /kg. Radiation from a pool fire will therefor be able to evaporate

$$\frac{5.000}{20.000} = 0.25 \quad \text{kg/s m}^2$$

Over a 10 m2 area, this will give 2.5 kg/s. At least one third of this will be drawn into the fire plume, namely that from the upwind side of the pool fire, but observations of actual pool fires in wind show that the vapour can disperse at ground level, with very little being drawn upwards into the plume. This is especially so for small fires, which do not create such a strong updraft and surrounding wind field as large fires. A good deal more needs to be learned here.

Dispersion of Heavy Gases

Many of the most serious releases of material in process plant accidents involve heavy gases. Gases such as chlorine, sulphur dioxide, and hydrogen sulphide provide examples with considerable toxic risk. Gases such as propane, butane, ethylene and propylene provide other examples where there is a risk of cloud fires.

In modelling such releases, there are two cases of significance - continuous releases over an extended period of time, and short of "puff" releases over an extended period of time, and short "puff" of the kind which might be generated by a "blow up" release, as in the previous section.

Modelling of heavy gas dispersion

There are two main groups of theories which have been applied to modelling heavy gas releases (see the review by Havens 1982). The first group is the so-called "Top Hat" models. The concentration profile across the plume is considered to be uniform. Here, the initial spreading is modelled as a gravity flow. After some distance, turbulence in the cloud and at the spreading front of the cloud causes entrainment, dilution, and a growth in height of the cloud, while gravity-spreading continues. In the far field, passive dispersion due to atmospheric turbulence causes further dispersion.

The second group, "K theory models" assumes a relation between the flux of gas through the cloud and into surrounding air due to turbulence, and the velocity, temperature and concentration in the cloud. They then establish partial differential equations for mass, momentum, and turbulent energy flow.

A third group overlaps these two categories by assuming a fixed form for the concentration profile in the cloud (similar to the top hat profiles, or more typically, a gaussian distribution). This is then combined with a K-theory-representation for turbulent mass transfer.

There are a number of physical phenomena to be taken into account in the models. Firstly, there is the physical spreading due to gravity of the heavy gas. Secondly, there is the entrainment of air at the front of the spreading cloud due to its spread velocity. Thirdly, there is the entrainment of air into the top of the cloud due to atmospheric turbulence and turbulence due to heating of the cloud by the ground.

Further effects are important in the far field - additional entrainment due to convective turbulence as the cloud is heated by the ground. For continuous releases, dispersion effects due to atmospheric instability, causing the plume to "wander", become important.

K theory models are essentially computer based, and require large amount of computer time to apply.They essentially solve gas flow equations usually regarding the gas as incompressible since the pressure differences are very small, and solving the turbulent mass transfer as a diffusion. K is then the diffusion coefficient. The reader is referred to the references for further details.

Similar profile models which use a K-theory-representation for turbulent diffusion have been described by Riele (1977), Flothmann and Nikodem (1980), Ooms (1974), and Colenbrander (1980). These models require less computer time but are still relatively complex.

One of the first "top hat" approaches to modelling was that of Van Ulden (1974). Gravitational spread is here described, on dimensional grounds, by an equation of the form:

$$U_f = C\sqrt{g_d\, h}$$

Where u_f is the rate of horizontal spreading of the cloud
g_d is the relative acceleration due to gravity of the cloud
h is the height of the cloud
C is a constant. Experiments indicate a value for C of approximately 1.

$$g_d = g\frac{\rho_{gas} - \rho_{air}}{\rho_{gas}}$$

Mixing at the front of the cloud during the gravity slumping phase is described by:

$$\frac{dV}{dt} = 2\alpha\ \pi\ r\ h\ \frac{dr}{dt}$$

where V is the cloud volume
r is the cloud radius
α is a constant

This assumes that entrainment is proportional to a speed of spreading.

Initially, gravity slumping is the dominant physical phenomenon. When the available turbulent energy becomes comparable with the potential energy difference between the cloud and the surrounding air, entrainment due to turbulent mixing becomes significant. The turbulent energy is characterised by a friction velocity U*.

The relative potential energy of the cloud is given by the transformation of potential energy of the cloud height the kinetic energy. The turbulent energy is given by

$$2\rho_{air}U_*^2$$

This indicates that turbulent mixing will be significant for $U_* > U_f$. In Van Uldens model there is a transition to turbulent dispersion at this point.

The models of Fryer and Kaiser 1979, which are used in the DENZ program, take into account not just frontal mixing, but also mixing through the top of the cloud. This uses a classical formula for the entrainment:

$$\frac{dV}{dt}=r^2\ u_v$$

where V is the cloud volume,
r is the cloud radius
u_e is the vertical entrainment speed given by

$$u_e=\beta\ \frac{u_*}{R_i}$$

here ß is a constant and
R_i is a Richardson number, the ratio of kinetic to buoyancy forces

$$R_i=g_d\frac{\rho_{gas}\ l}{u_*^2}$$

Eidsvik 1979 points out that the turbulence level in a cold heavy cloud is high, due to convective effects. As a result, the cloud is relatively homogeneous, with only low concentration gradients. By contrast, there is a large gradient at the top of the cloud, but relatively little transport.

This means firstly, that "top hat", models, with uniform concentrations within the cloud appear at first sight. Secondly, it means that it is very difficult to apply numerical turbulent diffusion theory methods to the spread of heavy clouds.

The dispersion equations resulting from the above cannot be solved analytically. However, assumptions can be made which allow a piecewise analytic solution (Jensen 1981). These are that during an initial gravity slumping phase, entrainment is insignificant; that turbulence within the cloud is large compared with that in the surrounding air; and that top entrainment does not interfere with cloud dynamics except through its effect on T and p. These assumptions are the basis for the following models.

The initial volume of the cloud V, assumed to be in the form of a cylinder. With no initial entrainment of air, we have for the first phase of the process:

As above

$$V = \pi r^2 h$$

$$u_f = C g_d h$$

Putting u_f equal to dr/dt and integrating

$$r^2(t) = r_0^2 + 2C (g_d V t / \pi)^{0.5}$$

The cloud moves with the wind, so that t=x/w where w is the windspeed and x is the distance from the source.

The cloud passage time, which is important later in calculation of exposure and dose rates, is calculated from $\tau = 2\ r(t)/w$

When the turbulent energy becomes comparable with the kinetic energy of spreading, the height of the cloud becomes dominated by the turbulence effects.

This calculation give an initial estimate of the way in which a heavy gas cloud spreads, and may be used close to the point of release.

A model for heavy gas cloud release

In this section some simple "top hat" models for heavy gas clouds are described. When a heavy vapour such as hexane or butane, or a heavy gas such as chlorine, are released the gases tend to fall to ground level, and are then dispersed by wind turbulence. The degree of mixing is strongly affected by the difference in density between air and the gases. The gas generally falls to form a "pancake cloud".

If the gas is very dense, the normal gas dispersion processes (neutral density dispersion) are to a large extent suppressed. A turbulent eddy of air may move down towards the cloud, and by its momentum penetrate the cloud. However the density difference will reduce the penetration, the eddy "floating" on the dense cloud. Similarly, an eddy of dense gas will tend to fall back into the cloud.

The actual form of dispersion depends very much on the way in which the gas is released. Five different cases have been distinguished here:

1. A release of gas as a jet from an opening.
2. A release of liquefied gas as a flashing jet from an opening.
3. A release of subcooled liquefied gas as a stream, forming a pool, which evaporates.
4. A release of liquid with a low vapour pressure, which then evaporates.
5. A release of liquefied gas under pressure due to a vessel rupture ("blowup").

These cases differ in two primary factors. Jet releases and blowup releases lead to mixing with air, and dilution of the gas, whereas pool releases generate more concentrated gas. Liquefied gas releases are cooled due to the latent heat of evaporation of the gas, and there may be an excess of liquid mist which gives a continuing cooling effect. By contrast, releases of gas lead only to expansion cooling, and the effect of cooling of air which is mixed into the gas is much less.

Blow up releases of liquefied gas

When a vessel containing liquefied gas under pressure ruptures, the result is a violent ejection of liquid, a so called "blowup". In the very short time after the rupture occurs, the pressure in the vessel is reduced below the boiling point of the liquid, and the liquid boils violently. If the hole is below the surface of the liquid, then all of the liquid will be thrown out. If the hole is above the surface of the liquid, then the amount thrown out depends on the degree of subcooling, that is, the difference in temperature between the initial temperature and the boiling point of the liquid at atmospheric pressure.

This phenomenon has been investigated by Fletcher (B.Fletcher,The Sudden Discharge of Superheated Fluids to Atmosphere, in The Assessment of Major Hazards, I. Chem. E. Symposium series, 1982). He found a remaining amount for a large hole of between 10 and 50%, when the degree of superheat was greater than 10°K.

When the liquefied gas is thrown out of the vessel, part of the liquid evaporates to form vapour, and the remaining part is broken up into fine droplets (Resplandy,A., Experiments with Ammonia to determine the properties that condition environmental safety measures around industrial storages. Chim. Ind. Genie Chim., 102, 691-702,1969). For ammonia these have a diameter of about 10 to 30 microns, so that settling velocity of the droplets is very low. It is observed that rain out of droplets from sulphur dioxide, chlorine, or ammonia stored at ambient temperatures is minimal.

The released gas usually has considerable momentum. As a result, there is considerable turbulence and mixing with air, and the initial gas cloud is expanded by a large ratio. For the Potchefsroom incident (Lonsdale,H. The Potchefstroom incident, private communication) this ratio was 10, and these values have been confirmed with experiments by Maurer on propylene (Maurer, B., Hess,K., and Giesbrecht,H., and Leukel, W., Modelling of Vapour Cloud Dispersion and Deflagration of Liquefied gas after Bursting of Tanks Filled with Liquefied Gas. 2nd International Symposium series on Loss Prevention and Safety Promotion in the Process Industries, Heidelberg, 1977). Note that the cloud can cool to far below its boiling point (subcooling).

Calculation of the resulting concentration, temperature, and density of the gas are all important for the ensuing dispersion.

The initial mass of gas evaporated is given by

$$Mg = M_0 \ Cp_{gas} \ (T_0 - T_b)/L_g$$

Where Mg is the initial mass of gas,
M_0 is the initial mass of liquefied gas
T_0 is the initial temperature of the liquefied gas

T_b is the boiling temperature of the gas at atmospheric pressure (assuming no subcooling)
Cp_{gas} is the specific heat of the gas (average for the T_0 to T_b temperature range)
L_g is the latent heat of vapourisation of the gas.

The initial volume of the gas is then given by

$$Qg = Mg/Rhg$$

Where Rhg is the density of the gas at its boiling point.

The amount of air mixed into the gas is then given approximately by

$$Qa_0 = 10\ Qg$$

The mass of air in the cloud is given by

$$Ma_0 = Qa_0\ Rha\ T_0/T_b$$

Where Rha is the density of air at the ambient temperature.

The additional amount of gas which evaporates is given by

$$Mg_+ = Ma_0\ \ Cpa\ (T_0\text{-}T_b)\ /\ Lg$$

Initial spreading of a heavy gas cloud

A blow up release of liquified gas gives a cloud consisting of air, liquid droplets and gas. Most such releases will be non uniform, and irregularly shaped. However if the release is a simple one, in which a hole appears in the top of the vessel, or if the vessel explodes into small fragments, the cloud will spread sideways and upwards, forming a cylindrical column.

The column of air and gas is initially roughly cylindrical, and is generally assumed to have about the same height as breadth. The uncertainties in these assumptions are very high, in that the actual initial conditions of dispersion will vary very widely according to the form of the hole from which the gas is released. The Hecate program, described later, has been used to investigate the dependence of dispersion on the height/diameter ratio of the gas cloud, and shows that the effect on dispersion at a distance over 100 m for a 10 t release is not large.

Once the initial, momentum driven, expansion is over, the column of gas and air begins to sink under the effect of gravity. If the difference in density between the gas/air mixture and the surrounding gas is large enough, the cloud will develop considerable momentum. As a result, there often forms a doughnut shaped turbulent ring of gas and air. This effect is important for the later fate of the cloud, since it represents a turbulent mixing of the cloud with air.

Nearly all of the standard gas dispersion models treat this "slumping" phase of dispersion by considering a balance of potential energy and kinetic energy as follows.

Volume of cloud (assuming no entrainment)

$$\pi r^2 h = Q_{T1}$$

so that $dh/dr = -2 Q_{T1}/(\pi r^3) = -2h/r$

Defining δR as $(R_{M1} - R_A)$, we get, for a change in height dh.

Potential energy $d = \delta R (\pi r^2 dh) g h$
Where h is the cloud height
r is the cloud radius
R is the cloud density
g is the acceleration due to gravity.

The kinetic energy is

$$1/2 R_{M1} (2 \pi r h dr) v^2$$

where
v is the velocity of the cloud edge.
This gives:

$$\delta R (\text{pi } r^2 dh>) gh = 1/2R\, 2\pi r d r v^2$$

therefore

$$v^2 = \delta R/r \; dh/dr \; g$$
$$= 2 g \delta R/R \; h/r$$

so

$$r = (2 gd h)_1^{/2}$$

where gd is defined as

$$gd=g (R_{M1} - R_A)/R_{M1}$$

This can be taken further, by writing

$$dr/dt = (2 gd h)_{1/2}$$
$$= (2 gd Q_{T1}/ \pi r^2)^{1/2}$$
$$= (2 gd Q_{T1}/ \pi)^{1/2} 1/r$$

Integrating this gives

$$r^2 (t) = r_0^2 + 2 c \sqrt{\ } gd Q_{T1}/\pi t$$

c in this equation is given by $\sqrt{2}$, on the assumption that potential energy is perfectly converted to kinetic energy. An experimental value is found which is closer to 1.

This simple model is clearly an approximation, since it does not take into account entrainment of air during the slumping phase, does not take ground friction into account, and uses only a single parameter c to take into account the transformation of coordinated kinetic energy into energy of turbulence and to heat. More detailed studies have been made, notably by G.Schnatz (Investigation of energy fluxes in heavy gas dispersion, in Heavy gas and risk assessment II, ed S Hartwig, Reidel 1983). The above model, with the value of c set to 1.0, fits the data given by Picknett (Field experiments on the behaviour of dense gas clouds, Porton, 1978, report PTN/il 1154/78/1), and accords with the World Bank model collection (Technica 1985)

Dispersion of a heavy gas cloud

Initially, a heavy gas cloud of the kind described above will be relatively dense, if for no other reason than that it is cold. The density of the cloud hinders turbulent eddies in the cloud from rising into the overlying air, or atmospheric turbulent eddies entering the cloud.

The resistance to this transfer of mass across the density boundary can be measured in terms of the Richardson number, R_i which is a ratio of gravitational forces to kinetic forces.

$$gd\ l/u_1^2$$

where gd is the effective acceleration due to gravity of the gas mixture subject to buoyant force
u_1 is a turbulence velocity
l is a turbulence length

The turbulence length is determined empirically (Taylor, RJ, 1970) to be $l = h_r (h/h_r)^{0.48}$
where h_r is a reference height, 30.2m.

For the top surface mixing, Cox and Roe (1977) and later Cox and Carpenter (1980) use a top entrainment velocity given by :

$$U_t = \alpha\ U_i\ (R_i)^{-1}$$

where U_t is the top entrainment velocity
α is an empirical constant
U_1 is the turbulence velocity
R_i is the Richardson number, as above.

This applies, in Cox and Carpenters (1978) model, for high values of R_i If Ri is under 1.0 the dispersion is described by

$$U_e = \beta\ U_1$$

The value of the constant α is determined of the basis of experiment to be 0.1

To determine the value of U_1 an approach is taken which is simplified when compared with standard methods in the literature (see eg Pasquill and Smith 1984) The approach taken is to estimate the effects of atmospheric stability separately.

Firstly, the value of U_1/U_* is determined on the basis of the work of Monin (1962) giving

U_1/U_*	Stability class
3.0	very unstable
2.4	neutral
1.6	stable

The value of U_*/U where U is wind velocity, is given by Pasquill (1974) and a value of 0.1 for open terrain is used.

The model used for edge mixing is based on early work by Van Ulden and by Picknett with

$$U_e = ß\ dr/dt$$

where

U_e is the edge mixing velocity
r is the cloud radius
ß is a constant.

In order to fit experimental results, a value of gm of 0.6 is used.
Inside the cloud, the temperature is very low if a liquid gas is concerned. Heat is added to the cloud from the air, and from the ground. In the first instance, this heat will be used to evaporate any mist or droplets present. Once the cloud has been heated sufficiently to consist of gas alone, the heat serves to warm the cloud, and thereby to reduce the difference in density compared with the ambient air. In any case, the heating from the ground causes turbulence within the cloud, and a considerable degree of mixing.
The equations for dispersion are then

For a cloud

$$d\ M_T/dt = ß\ 2\ \pi\ r\ h\ dr/dt + \alpha\ U_1\ \pi * r^2/R_i$$

For continuous release

$$d\ M_T/dt = ß\ 2\ h\ U\ dy/dt + \alpha\ U_1\ 2\ y\ u/R_i$$

where

M_T is the total mass of gas and air in the cloud or mass per unit time for a continuous release plume.
r is the cloud radius
h is the cloud height, or plume height
y is the plume semi width.

For a jet or blow up, the initial mixing is such that the most of the cloud mass is air, and so the initial velocity, as a result of conservation of momentum, is close to that of

the wind. For evaporation from a pool, the momentum will be lower, and wind may "blow over" the cloud.
Bradley et al. have extended the Cox - Carpenter model to take account of this case.

More recently, large scale experiments have been performed, e.g. the Thorney Island experiments (McQuaid), and have allowed more precise adjustment of constants. The "generalised Picknett" model of Webber (1983) was fitted to the Thorney Island data by Prince, Wheatley, and Brighton, giving values for α of 0.6 and ß of 0.14. The model also makes a more detailed approach to determining cloud velocity.

Passive mixing

As the heavy gas cloud disperses and heats up, the difference in density between it and the ambient air is reduced. As a result, the dispersion mode changes to that of passive dispersion, following the equations given for light gas.

In the Cox - Carpenter model, this is considered to occur when the rate of spread of the cloud due to gravity, dy/dt, becomes equal to the rate of change of passive dispersion.

$$dy/dt = d\, \sigma_y/dt$$

Where sigy is the usual dispersion coefficient for the gaussian concentration profile for the cloud or plume.

This is a simple a convenient criterion computationally, but not well motivated as far as vertical dispersion is concerned.

The equation for passive mixing of the cloud is then

$$x\,(x,y,z) = 2/(2\pi)^{1/2}\ u\ \sigma_x\ \sigma_y\ \sigma_z$$
$$\exp\,(-\,1/2\,(x^2/\sigma_x{}^2 + y^2/\sigma_y 2 + z^2/\sigma_z 2)\,)$$

More recent models make the change to passive dispersion when the Richardson number falls to a low value, generally 1.

In order to continue the dispersion calculation in the passive phase it is necessary to match the two calculations. The source strength for the two must be identical, so for a point source there is only one parameter, the distance of the source, which can be adjusted. For this reason, an area source model is used, in which the diameter of the source and the upstream distance allow the centre line concentration and the plume width to be fitted.

Continuous release

The equations for slumping of a continuous release follow those given above for an instantaneous release, with the natural modifications arising from the change in dispersion pattern, with dispersion along the length of the plume being neglected. (In the steady state this dispersion will be negligible).

In the slumping phase the width of the plume is given by

$$dy/dt = (K\ H\ g_d)^{1/2}$$

following energy balance principles as described above. The mass conservation equations for the dispersion are given by

$$dM/dt = \beta\ 2\ h\ u\ \ dy/dt + \alpha\ u_1\ 2yu\ /R_i$$

as described above.

Transition to passive dispersion is considered to follow the same pattern as above, transition occurring when

$$dy/dt = d\ \sigma y/dt$$

passive dispersion is in this case described by

$$x\ (x,y,z) = Q/\pi\ u\ \sigma_y\ \sigma_z \quad \exp\left(- 1/2\ (y^2/\sigma_y 2 + z^2/\sigma_z 2\ )\right)$$

The Degadis Model

The Degadis models(Havens and Spicer 1988), based on the prior program HEDAGAS (Colenbrander), has achieved something of a special status, due to wide spread use, and the fact that the coding has been published. This has allowed the programs to be studied in detail and makes the assumptions involved open to inspection.

The Degadis model is a similarity model. The current program includes Ooms model for jet release, models for buoyancy dominated spread, for stably stratified, and for passive dispersion.

The model divides the dispersion into regimes according to Richardson number

$R_i > 30$	Negative buoyancy
$30 > R_i > 1$	Stably stratified
$R_i < 1$	Passive dispersion
$v/u > 16/(19+R_i)$	Jet dispersion

The model thus combines jet and plume dispersion directly.

The Ooms model treats a jet plume, taken up in the wind, by calculating the concentration as a gaussian distribution, across the plume.

$$R_i = \frac{g(\rho_{gas} - \rho_{air})h}{\rho_{air} u_*^2}$$

where
ρ is density
h is layer depth
u_* is friction velocity
v is the jet velocity
w is the wind velocity

The equations for the dispersion are similar to those given above, but are treated in a different way than for the top hat models.

For instantaneous or short term releases, both gravity and drag forces are taken into account to determine the size of a "gas blanket" which determines a source term for take up of gas. Calculation involves solving force/momentum equations.

The similarity model divides a long period release into several small releases, and follows the fate of each, integrating the effects of each subrelease. The overall result is that the method can take account of releases which vary with time.

Effects of Buildings

Buildings and other obstacles cause additional mixing by causing additional turbulence. The order of magnitude of this effect can be estimated by assuming effective entrainment speed equal to the wind speed

$$u_e = u/R_i$$

This gives a relative increase of $(u/u_*)^3$ in the entrainment speed, a factor which can be several orders of magnitude. The effect may still be an increase of a factor of 2-4, even after the cloud has reached a height greater than that of the buildings.

A simpler, rule of thumb approach is to regard a building as an obstacle which completely mixes the plume, in a local concentration behind the building. This concentration can then be regarded as a new source, of cross section similar to that of the building, and with concentration equal to the release rate of gas divided by the flow rate of air across an area equal to the building cross section.

Effect of sloping Terrain

If there is a slope at the place where spreading occurs, the cloud of plume will acquire an additional velocity in the direction of the slope. Except in the case of extreme slopes, the gravitational slumping will not be significant by affected (the effect is proportional to where is the slope angle). Considering the equations of motion we get

$$u_{slope} = \sqrt{\frac{h(\rho_{gas} - \rho_{air})\sin(\theta)}{C_f}}$$

where Θ is the slope angle. This give an increased turbulence, and more rapid dispersion.This factor is typically from 2 to 4 for steep slopes, giving a significant additional dispersion effect locally.

References

B.Fletcher, The Sudden Discharge of Superheated Fluids to Atmosphere, in The Assessment of Major Hazards, I. Chem. E. Symposium series, 1982

Resplandy,A. Experiments with Ammonia to determine the properties that condition environmental safety measures around industrial storages. Chim. Ind. Genie Chim., 102, 691-702,1969

Maurer, B., Hess,K. Giesbrecht,H., Leuckel, W. Modelling of vapour cloud Dispersion after bursting of tanks Filled With Liquefied Gas, Int. Loss Prevention Symp. Heidelberg, 1977

Leukel, W. Modelling of Vapour Cloud Dispersion and Deflagration of Liquefied gas after Bursting of Tanks Filled with Liquefied Gas. 2nd International Symposium series on Loss Prevention and Safety Promotion in the Process Industries, Heidelberg, 1977

G.Schnatz Investigation of energy fluxes in heavy gas dispersion, in Heavy gas and risk assessment II, ed S Hartwig, Reidel 1983)

Picknett, Field experiments on the behaviour of dense gas clouds, Porton, 1978, report PTN/il 1154/78/1)

World Bank model collection (Technica 1985)

Taylor, RJ Quarterly Journal Royal Meteorological Society 96 p750 (1970)

Cox, R.A.
Roe, D.R. A Model of the Dispersion of Dense Vapour Clouds, Proc 2nd Int Symp. Loss Prevention, Heidelberg, 1977

Cox, R.A.
Carpenter R.J. Further development of a dense vapour cloud dispersion model, in *Heavy Gas and Risk Assessment* ed. Hartwig, 1980

Pasquill Atmospheric Diffusion, 2nd edition, Ellis Horwood 1974

Van Ulden A.P On the Spreading of Heavy Gas Released Near the Ground, 1st Int Conf. Loss Prevention, The Hague/Delft 1974

Picknett, R.G. Field Experiments on the Behaviour of Dense Clouds Ptn/1154/78/1, Porton Down, 1978

Havens, J. A Review of Mathematical Models for Prediction of Heavy Gas Atmospheric Dispersion, in The Assessment of Major Hazards EFCE Publication Series No 25. 1982

Havens, J Dispersion Model for Elevated Dense Gas Jet Chemical Releases, EPA-450/4-88-006a, U.S.Environmental Protection Agency 1988

Colenbrander,G.W. A Mathematical Model for the Behaviour of Dense Vapour Clouds, 3rd Int Symp. Loss Prevention and Safety Promotion in the Process Industries., Basel 1980

Cude, I.L. Dispersion of Gases Vented to Atmosphere from Relief Valves, The Chemical Engineer, October 1974

Ooms,G., Mahieu,
and Zelis, F. The Plume Path of Gases Heavier than Air, First Int Symp. Loss Prevention and Safety Promotion in the Process Industries, Delft, 1974

Emerson, M.C. A New Unbounded Jet Model, 5th Int Symp. Loss Prevention and Safety Promotion in the Process Industries, Cannes, 1986

Lee, G.K. Evaluation of Jet Dispersion Models with Data, 6th Int Symp. Loss Prevention and Safety Promotion in the Process Industries, Oslo, 1989

Briggs, G.A. Plume rise, USAEC, 1969

Flothman, D. and Nikodem, H.J.	A Heavy Gas Dispersion Model, in *Heavy Gas and risk Assessment,* Reidel 1980
McQuaid, J.	Symposium on Heavy Gas Dispersion Trials at Thorney Island, Sheffield, 1984. J. Hazardous Materials 1985.
Eidsvik, K.J.	A model for Heavy Gas Dispersion in the Environment, Atmospheric Environment, 14, 1980
Ricou, F.P. and Spalding, D.B.	Measurement of Entrainment by Axisymmetrical Turbulent Jets, Jnl. Fluid Mechanics, 11, 1961.

10 Effects of Toxic Gases

Toxicology is an extremely complex subject and this chapter can hardly even begin to do justice to this complexity. Just the bare minimum of information needed for risk analysis is presented here. The reader is recommended in each specific case, to obtain as much information as possible concerning toxicity of the specific compounds which may be released from his/her plant.

Major hazards analysis involves study primarily of the acute toxic effects from the release of gases. This is (fortunately) an area where little direct experience exists. What information is available is obtained from animal experiments, and with very poor control, from accidents in which large releases of gases have occurred.

Toxic response

The effect of a toxic chemical depends to a very large extent on the amount or dose of the substance which is transferred into an organism. for gases, this is in turn dependent of the concentration of the gas, and the exposure time.

The effect of a chemical is characterised by a response, for example the degree to which respiration is reduced, the reduction of blood oxygen concentration, or the percentage of the population which dies. Individuals vary very widely in the degree of response to a given dose. This variation is generally assumed to follow a gaussian distribution. The information in this distribution is generally shown in a cumulative distribution curve (fig. 10.1) which shows the percentage of a given population giving a specific level of response.

This form of expression is useful in risk analysis, because it is possible to read from the curve the percentage of a population affected and to relate this to the concentration level. By calculating the area covered by a given concentration the specific percentage of the population affected in a given area can be obtained.

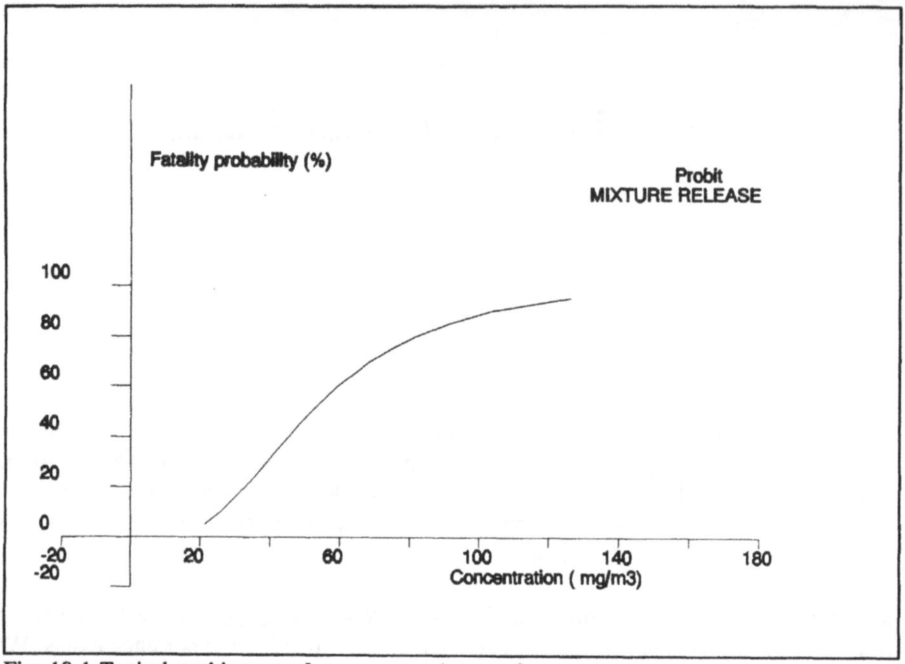

Fig. 10.1 Typical probit curve for a gas or mixture of gases

The reasons for the large variations in toxicity in a population, are manyfold. They include:

- the actual rate of ventilation (breathing), and metabolism.
- the excess of functional capacity, (such as breathing, oxygen uptake, muscle strength in the case of paralysing agents etc.)
- the extent of any secondary reactions, such as shock
- the general state of health of the individual
- for some chemicals, the rate of transfer of the chemical into the blood.
- for some chemicals, the rate of metabolism, which can in turn depend on the rate and efficiency enzymatic activity.

Generally speaking, infants, old persons, and sick persons are more susceptible, to toxic effects, which means in practice that more care, and improved safety criteria are necessary in the area of institutions such as hospitals.

The degree of variation in dose/response can be expressed in terms of the standard deviation of the gaussian response curve. It is usual to use a special terminology in toxicology (and risk analysis), that of probits. A probit of 5 corresponds to a 50% response. A probit of 6 to a 50% response plus one standard deviation i. e. 84% and a probity of 4 corresponds to a 16% response. The conversions are essentially an arithmetic convenience.

Definition of "dose" in the case of a solid or liquid is fairly straight forward. For gases and aerosols this is much more difficult, varying with rate of respiration, concentration, and exposure time.

To complicate problems even further, response to some chemicals depends very strongly on concentration. Examples are chlorine and ammonia. To allow for this, it is usual to define an exposure factor E

$$E=C^n t$$

Where

C is the concentration of toxic
n is a dimensionless exponent
t is the exposure time

Exposure limits

Risk analysis depends very much on collected tabulations of toxicity data. (See e.g. OSHA, EPA, Sax, Wunderlich, RTECS, A.I.I.H). These collections gives data in terms of limits. The simplest limits are lethal concentrations defined as follows.

Many of these data concern chronic exposure, fewer concern acute toxic effects. For most gases and vapours however lethal concentration data exist, either inferred from accident reports, or from animal experiments.

Lethal concentrations are defined as follows:

LC_{50} Lethal concentration 50% the concentration which, with a given exposure time, gives 50% fatalities.

LC_{10} The lethal concentration 10% the concentration which with a given exposure time, gives a 10% lethality.

LC_{lo} The lowest concentration which has been recorded as giving a lethal effect.

These values are by far the most useful when making actual risk analyses. In determining these values for humans, it is usual to apply safety factors in order to take uncertainties into account. This is especially so when values for humans are based on animal experiments. As a result of this, risk calculations often overestimate hazards from toxic releases. The problem is more than an academic one, since the most accurate knowledge available is necessary when making contingency plans. Overestimation can lead to errors in deploying possibly scarce resources.

The American conference of industrial Hygeinstists publishes threshold limit values, TLV. For risk analysis purposes, these are in themselves of little use, since they are intended to give a no effect limit for 8 hour 40 day per week exposure. The short term exposure limits, STEL, however, are intended to give a no effect level for 15 minute exposure. (Time weighted average concentration). As a result, these levels can be used as a guide to the limit of a completely safe zone.

The American Industrial Hygienist Association has published a series of "Emergency Response Planning Guides". These contain a review of available toxicological data and recommendations concerning emergency planning levels. ERPG-3 is defined as the maximum airborne concentration below which it is believed that nearly all individuals could be exposed for up to one hour without experiencing or developing life-threatening health effects. ERPG-2 is the maximum concentration below which it is believed that nearly all individuals could be exposed for up to one hour without experiencing or developing irreversible or other serious health effects or symptoms which could impair an individual's ability to take protective action. ERPG-1 is the maximum airborne concentration below which it is believed that nearly all individuals could be exposed for up to one hour without experiencing other than mild transient adverse health effects or without perceiving a clearly defined objectionable odour.

These ERPG reviews represent at present the most valuable collected and reviewed data on acute toxicology, since they include human accidental data as well as animal data. In all cases however, a review of the original literature is needed in order to be able to judge relevance of the data and avoid pitfalls.

Exposure outdoors

Outdoor, public exposure depends not only on concentration and duration of a release, but also on peoples response to these concentrations. For some releases, it is possible to walk or run out of the release. conditions which make this possible are:

- the release must be visible or detectable by smell.
- the release must have a sufficiently low concentration as not to be immediately lethal.
- it must be possible to see (ammonia, for example, is very damaging to eyesight).
- the release must take the form of a narrow plume, since it is necessary to be able to see the edge in order to be able to walk out of the plume.
- young persons and infirm persons, cannot be expected to rescue themselves in this way.

The conditions for this form of "self rescue" are somewhat relaxed, if instead of walking out of a cloud, persons can go indoors. As will be seen in the next section, going indoors provides an effective way of avoiding serious consequences, in the case of short timed releases.

What this means in practice for risk analysis is that exposure times used in determining hazard concentration levels may be reduced for some 90% of the population, for cases in which the above conditions are fulfilled.

Exposure indoors

The concentration of toxic gas inside a house will initially be zero, when a plume of cloud begins to pass. The concentration rises as gas passes into the house. The rate at which concentration rises is initially dependent on the ventilation rate in the building. As concentration inside the house increases, an exchange begins, in which some gas enters and some leaves.

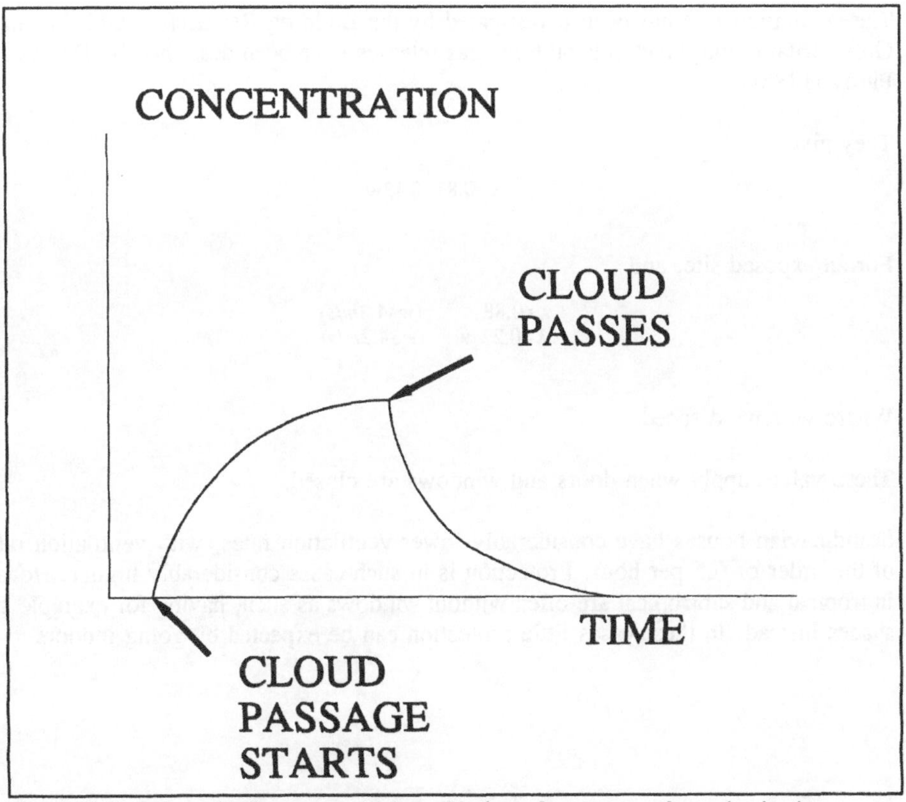

Fig. 10.2 Variation of indoor concentration with time, for passage of a toxic cloud.

The result is given by the equations:

$$\frac{dC_i}{dt} = \lambda (C_0 - C_i)$$

where
C_i is indoor concentration
C_o is outdoor concentration
λ is rate of ventilation in air changes per unit time

This gives, for constant outside concentration:

$$C_i = C_o (1 - e^{-\lambda t})$$

Following the passage of the plume, there is a corresponding fall in concentration over time. (Fig. 10.2)

The ventilation rate has been investigated by the Building Research Establishment in Great Britain, and the effects on toxic gas releases have been described by Davies and Purdy (1986)

They give:

$$\lambda = 0.87 + 0.13w$$

For an exposed site, and

$$\lambda = 0.88 \qquad (w < 4.2 m/s)$$
$$\lambda = 0.22\,w \qquad (w \geq 4.2 m/s)$$

Where w is wind speed

These values apply when doors and windows are closed.

Scandinavian houses have considerably lower ventilation rates, with ventilation rates of the order of 0.5 per hour. Protection is in such cases considerably higher. Houses in tropical and subtropical are often without windows as such, having for example sun shades instead. In these cases little protection can be expected by going indoors.

References

Sax,I. and Lewis, R.J.	Dangerous Properties of Industrial Materials, 7th edition, 1984, Van Nostrand Reinhold.
AIHA	Emergency Response Planning Guidelines Series 1988- AIHA Emergency Response Planning Guidelines Committee 475 Wolf Ledges Parkway Akron OH 44311-1087
Eisenberg et al	USCG Vulnerability Model, Report CG-D-53-77 Feb 1977.
Wunderlich	Major Accidents of Industrial activities, Information on Dangerous Chemicals European Community Contract x1-B6614-2-24-88
Davies. PC Purdy, G.	Toxic Gas Risk Assessments in Refinements of estimates of the consequences of heavy toxic Vapour Releases. Ins. Chem. Eng. North Western Branch Papers No.1 1986.
OSHA	Occupational Safety and Health Administration, US Dept. Labour. Occupational Safety and Health Standards, Subpart z Toxic and Hazardous Substances 29CFR1910, 1000 Washington DC.
RTECS	Registry of Toxic Effects of Chemical Substances. National institute for Occupational Safety and Health. DHHS (NIOSH) Publication No. 83-107 (1983) DHHS Centre for disease control (1985-).

11 Fires

Fires are hazardous both because of their direct heating effect, by convection within the fire itself, and because of the radiation from the fire. For humans, the effects are on skin and on the lungs in the case of fire engulfment. Indoors, smoke rather than the fire itself, is the most frequent cause of death. Fire can constitute a hazard for process plant by destroying instrumentation, cutting off power, causing overpressuring, and weakening of structural and pressure bearing steel.

In order to determine the range of hazards from fire, it is generally necessary to determine the rate of burning, the flame form, and the heat radiation pattern.

The main cases which must be treated for risk analysis of process plant are:

- pool fires, in which a liquid is collected on the ground or on a platform, and burns as a roughly conical flame
- running front fires, similar to pool fires, but in which the flame front spreads due to the flow of the liquid fuel
- confined, or compartment fires, in buildings
- flash fires, in gas clouds
- jet fires, from escaping gas or liquid
- fires from boiling liquid expanding vapour explosions (BLEVES)

Pool fires

In pool fires, liquid at the surface of the pool is heated by radiation from the fire. As it evaporates, it becomes available for combustion. For gases such as LNG and LPG, the rate of evaporation may be initially boosted by ground heat.

Theoretical investigation of pool burning cannot at present predict burning rates or flame sizes and empirical results, bases on experiments, must be used.

The influx of radiation to a pool from a fire will increase as the size of the fire increases, as a result of an increase in view factor. This has been observed in experiments and is described by Wesson Brown and Welker (Brown 1975) by:

$$m = m \ (1 - e^{-BD})$$

where m is mass burning rate
m is a maximum mass burning
B is a constant
D is the pool diameter

For irregular pools, they give

$$D_{equivalent} = \text{Pool area /Pool circumference}$$

The given values of 2,54 x 10^{-4} m/s for the burning rate in pools of LNG. This may be compared with recent measurements in large pools by Mizner and Eyer (Mizner 1984).

LNG	2.37 x 10^{-4} m/s
LPG	2.17 x 10^{-4} m/s
Kerosine	0.76 x 10^{-4} m/s

The value given for B by Brown et al is 0.65/m.

Flame dimensions have been found to be best described in terms of a tilted cylinder (in fact, the flame shape is very irregular). For calculation of flame lengths, correlations based on conical flame shape are preferred by Moorhouse as given better estimates of risk.(Moorhouse 1984)

Thomas (1963) gives a correlation for ratio of flame length to pool diameters of

$$L/D = 42 \left(\frac{\dot{M}}{\rho_{air}} gD\right)^{0.61}$$

where
L is the flame length
D is the pool diameter
$\dot{M}$ is the fuel burning rate
ρ_{air} is the density of air

Mizner and Eyre found this correlation "adequate" to describe their experimental results, underestimating flame length by 12.6% for LNG, 9.4% for LPG, and between 13 and 36% for kerosine.

Under wind conditions a more complex correlation is given (Thomas 1965)

$$L/D = 55\left(\frac{\dot{M}}{\rho_{air}} g\ D\right)^{0.67} w_*^{0.21}$$

$$w_* = w/u_c$$

$$u_c = (g\ Mdot\ D/\rho_a)^{1/3}$$

where w is wind speed

The AGA experiments (AGA 1973) which are the basis for the work of Brown et al, and the work of Moorhouse (1982)give different values for the exponents in the above equation:

	A	B	C
Thomas	55	0.67	0.21
AGA	1	-0.19	0.06
Moorhouse (cylinder model)	6.2	0.254	-0.044
Moorhouse (cone model)	4.7	0.21	-0.114

These values give some idea of the uncertainties involved, an impression which is reinforced by the scatter plots given by Moorhouse. As a quick rule of thumb, L/D values of 2 are often used.

In a wind the cone of flame from a pool fire "tilts" in the wind direction. Flame tilt is given by Atallah and Raj (1974)

$$\cos \Theta = 1 \qquad w_* < 1$$

$$\cos \Theta = 1/\sqrt{w_*} \qquad w_* > 1$$

Flame tilt is given alternatively by Welker and Sliepevich (1966) as

$$\tan \Theta / \cos \Theta = 3.2\ (U_w D/v_a)^{0.07}\ (Uw^2/gD)^{0.7}\ (Rh_v/Rh_a)^{-0.6}$$

Rh_a is the kinematic viscosity of air.

The experiments by Mizner and Eyre confirm both these correlations, there being a discrepancy of at most 1.5° between the two correlations and the observation. The experiments by Moorhouse indicate that a better fit can be obtained by using other values for the exponents.

Moorhouse points out the importance of "flame drag" in determining radiation levels. Flame drag can have two effects on radiation. One is to increase the area over which radiation takes place. The other is to bring the flame near a target. Moorhouse gives the correlation

$$D^1/D = 1.5\ (w^2/gD)^{0.069}$$

D^1 here is the extended flame base diameter due to drag.

Fig 11.1 Experimental pool fire.

Given flame dimensions and flame radiation levels it is possible to calculate radiation flux at a target.

One way of calculating radiation levels is based on the proportion of generated as heat which is radiated. This is given by Burgess and Zabetakis (1962) as

Substance	Pool diameter cm	% of Heat radiated
Hydrogen	33	0.25
Butane	76	0.27
Benzene	122	0.36
Methanol	122	0.17
LNG	76	0.23

A more direct approach is to take direct observations of radiation levels. A value of 170 KW/m^2 has been given by several authors (Craven 1976). Nash gives values from 30-189 KN/m^2 (1974).

Welker predicted a limiting flux for large LNG fires of 142 KW/m^2. Atallah and Raj found a limiting flux of 100 KW/m^2.

Mizner and Eyre have discussed their measured values. They found an average flux from LNG fires of 153 KW/m^2 and a peak flux of 200KW/m^2 in a 20 m pool. For LPG, values of 48KW/m^2 were found and values less than 35 KW/m^2 for kerosine. They noted a dependence on pool size. Generally, high emissive power relies on the presence of glowing soot particles. But as more soot is formed it can mask the fire, so that emissive power is reduced. The maximum emissive power is suggested to occur at the pointy at which black soot begins to appear in the flame. Increase in pool diameter will then reduce air flow and increase black soot formation, causing reduction from kerosine fires of 130 KW/m^2 from a 2 m.pool, reducing to 60 KW/m^2 at 10 m.

Flame temperatures are important in the discussion of formation of poisonous byproducts in flames.
The survey by Crawley gives:

Table 11.1 Flame temperatures

Pool Fire Type	Temperature °C	Source
Pentane	1020	(Hazogawa 1977)
LNG	1000	(AGA)
Methane, ethane, propane	870-984	
Medium size pool fire	630-720	
Large LPG	915-1223	(Hardee 1978)
Torch flames	1387-1440	
Torch flames	1300	(Crawley 1982)
Medium flames	1050	
Large smoking fires	800	

Jet Fires and High Pressure Flares.

When a flammable gas or a spray of oil is ejected from a pipe or vessel, there is a possibility that it will ignite. The resulting jet flame can be extremely intense, and can impose high heat loads on nearby equipment.

With low pressure in the escaping gas, the flame is laminar. In this state the flame length is proportional to the pressure of the escaping gas. At quite low pressures though, the flame becomes turbulent, and at this point the length of the flame becomes almost independent of pressure. The flame length is dependent primarily on the gas and on the diameter of the hole from which the gas escapes.

For jet flames, the classical empirical formulae developed by Hawthorne are used for flame length. These give

$$L = (5.3\, d / C_t)((Tf/\alpha T_n)\ (C_t + (1-C_t)\ Ms/Ma)^{1/2}$$

where

- L is flame length
- d is jet diameter
- C_t is molar concentration in stoichiometric fuel air mixture _
- M_n is molecular weight
- M_a is air molecular weight
- α is the ratio of unreacted and reacted mols in the stoichimetric mixture
- d here is the diameter of the expanded jet.
- T_f is the flame temperature
- T_n is ambient temperature

For fdth this gives:

$$Y = 0.29 \ (\ln (L/X)^{1/2}$$

where

$$Y = \text{flame width}$$

The models were based on experiments at relatively low pressures and with low hole diameters.

At higher pressures, there is a tendency for the flame to blow out. The high gas velocity, higher than the flame velocity at the release hole, causes the flame to lift. If the lift distance is too great, the jet velocity falls below the flame velocity. This effect though depends on hole diameter. With larger diameter release holes, the pressure required for flame blow off is higher.

The equations given above are based on hole diameter. However, for high pressures above a few atmospheres, this calculation is inaccurate. It is usual to base the calculation on the expanded jet size, where the pressure in the jet first falls to approximately atmospheric pressure.The effective diameter for the jet is given, using standard supersonic flow theory, as in Ch. 7.

When gas at high pressure is released from an opening such as a leak, the immediate effect is a large expansion, followed by a conical jet.The values for flame length used are those given by Hawthorn et al. as described above. This expression has not been validated for very high pressures, but Gugan comments that there is no reason to indicate that the formula is incorrect at these high pressures.

One thing is certain, however, that is that the diameter entering into this equation should not be the diameter of the hole from which the jet issues. This can be seen by considering that events in the supersonic region just outside the throat will be indistinguishable from events leading to a jet at the point where expansion is complete and at a subsonic velocity. The temperature of the jet which also affects flame length, is also strongly determined by adiabatic cooling.

While calculation of the state at the throat of the orifice is relatively easy, the state at the "mach plate" where the jet becomes subsonic is not easily calculated. It depends on the precise shock pattern at the jet outlet, and this in turn depends on the precise form of the jet orifice.

An approximate model is, therefore, used here, in which the initial diameter for the flame is determined by assuming:

a. That the pressure just after the mach plate is approximately 1 atm.
b. That the velocity just after the mach plate is just subsonic.
c. That the temperature of the gas at the mach plate is given by formulae for isentropic gas expansion.
d. That the diameter of the jet can, therefore, be determined initially by knowing the pressure, the temperature, and therefore the density of the gas, the velocity and the mass flow rate, and therefore the area.

This is an approximation because firstly, the gas expansion is not isentropic, and the actual gas temperature will be higher than predicted. Secondly, the pressure within the jet at the mach plate will be lower than atmospheric (the static head will be atmospheric pressure) and there fore the density will be overestimated.

Both of these factors tend to overestimate gas density and therefore jet diameter and flame length. On the other hand, the low temperature enters into the jet length calculation tending to overestimate flame length. The approximation works in both directions, but cannot be expected to cancel precisely.

The area of the throat of the jet (vena contracta) is given by

$$\underline{A \approx 0.8A.}$$

The pressure at the throat is the critical pressure, given by

$$P_c/P_1 = (2/k+1)^{k/k+1}$$

Mass flow can be found by substituting into the equations for gas flow through an orifice (see Ch 7). The equations for turbulent jet flow can then be used to give the mass flow in the jet, and assumptions of complete combustion can be used to calculate the expansion of the gas in the jet, and the jet temperature.

This is the procedure taken in the vesta4 model developed by the author. This calculates flame form, temperature, and radiation levels by taking the release and the equations as above, and the turbulent momentum jet equations from Ch. 11 to describe the flow of air into the flame. The combustion is then described assuming that all air entering the flame is immediately burned. This gives the flame form and length.

In the model, the development of the flame is described in terms of flame elements, in practice 20 sections along the flame. This allows the effect of buoyancy to be taken into account, so that horizontal and tilted flames can be treated. It also means that the radiation pattern around the flame can be calculated in detail.

Chamberlain (1987) developed an empirical correlation model for high pressure flares, based on experiments with jets with 0.15 to 1.07 m. and with velocities 2.5 to 554 m/s The model gives:

$$L_B = L_{BO}\,(0.15\, e^{-0.4\, w} + 0.49)[1 - 6.07 * 10^{-3}\,(\Theta - 90)]$$

Where w is wind speed and Θ os the angle between the initial jet axis and the wind vector.

Kalghatis work is used as the basis for vertical flame length:

$$\Psi = \left(\frac{D\ \beta}{L\ W}\right)^{2/3} = \Psi(\epsilon)$$

$$\Psi(\epsilon) = 0.2 + 0.024\ \epsilon$$

$$\epsilon = \left(\frac{g}{D^2 u^2}\right)^{1/3} L$$

Ψ is a non dimensional flame paremeter
D is the effective source diameter
W is the mass fraction of fuel in a stoichiometric mix
ε is a Richardson number for the flame
u is the jet exit velocity
β is a constant (Becker and Liang)

Chamberlain also give values for flame tilt and for flame width. These do not agree with the values given in the models above. Since Chamberlains model is empirical, with a wide experimental base, and good agreement with experiment,this needs explanation. Close observation of large flares shows that there is a central core of gas in a jet, but that the wind interacts with the outer surface of the jet, carrying gas away from the core. As a result, both flame tilt and flame width in a wind are increased.

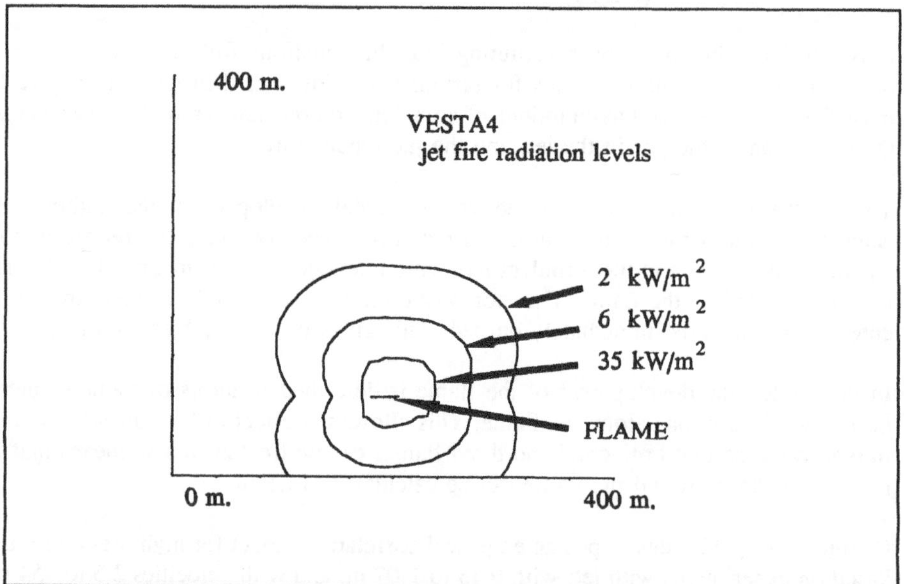

Fig. 11.2 Radiation pattern from a jet flame. Vertical projection

A jet flame may, for low exit speeds, begin burning at the jet exit. As velocity is raised, the flame "lifts" from the exit. the flame burns at a distance from the jet exit. As the velocity is increased further, the jet may become unstable and blow out.

The pressure at which blow out occurs has been evaluated both theoretically and experimental by Khalghati (Combustion Science and Technology 1981 Vol 26 p 233-239). He gives stability curves in terms of U_e, the velocity at the effective burner exit, and the hole diameter. There is a critical diameter, above which flames are stable at any pressure.

Liquid jets

For flashing liquids a formula based on experiments has been given by Werthenbach

$$q = 0.0006785\ L^{2.5}$$

where
L is the length of the flame
q is the mass flow rate of fuel.

Jet flame temperatures

Temperatures in jet flames have been measured by the author at 1200-1250°C for propane/butane. Heat transfer rates from flames impinging on a steel plate have been measured by Anderson and Stresino (Jul. heat transfer, Feb. 1963, pp 49-54) as, at maximum:

3.6×20^5 BTU/hr ft^2
= 1.13 MW/m^2
for methane in air.

Jet flames can also give significant levels of heat radiation. In order to calculate the radiation level, it is usual to distribute the heat release (ca. 18 MJ/kg fuel for paraffine) over the area of the flame. Not all the heat produced, however, is emitted as radiation. The emissivity factor E is given by Lees (Loss prevention in the Process Industries) as

luminous flames	0.8 - 0.9
non-luminous flames	0.1 - 0.3
hydrogen	0.1
ammonia	0.9

Heat radiation

Fires and jet flames produce radiation which can produce considerable heat loads at neighbouring vessels, and may cause harm to person.

Heat radiation is described by Stefan Boltzmann's law

$$E_s = \Sigma\sigma T^4$$

where

E is the radiated energy per unit area.
Σ is an emissivity factor and
T is absolute temperature, and
σ is Stefan Boltzmann's constant.

The heat transferred from a hot surface to a target is then

$$E_t = F\,\alpha\,\tau\,\Sigma\,\sigma\,T^4\,A$$

where

E_t	is the heat flow per unit area at the target
F	is a view factor
α	is the absorptivity of the target
τ	is the atmospheric transmitivity
A	is the radiating area.

The view factor depends on the distance of the target from the heat source, and the angle to the radiating surface.

For a simple source which can be regarded as a point, we have

$$F = \cos\Theta\ /(4\,\pi\,r^2)$$

For more complex surfaces, overall view factors are found by integration.

In complex cases, numerical integration is necessary. For example, the radiation pattern for a gas jet flame shown was derived by dividing the flame into 100 surfaces and calculating radiation levels from end surfaces.

The radiation levels for various types of damage are given below.

Absorptivity depends on distance from the flame, and on the amount of water vapour in the atmosphere. The degree of absorption rises from 0 to typically about 20% at distances of a few hundred metres, with little additional absorption after this.(NFPA 1989)

Table 11.2 Radiation effects levels

KW/m^2	Effect
6	Pain after 8 secs.
6	Blistering after 20 seconds.
6	Burns on rats after 20 secs.
23	Burns on rats after 6 secs.
66	Ignition of cellulose after 4 secs.
34	Ignition of cellulose basic material after 20 secs.
38	Blackened by fire, ignition after 1 min.
13	Minimum for wood ignition
66	1% lethality after 4 secs. exposure
20	1% lethality after 20 secs. exposure
117	50% lethality after 4 secs. exposure
35	50% lethality after 20 secs. exposure

Ordinary clothing gives good protection from radiant heat on the covered parts of the body. The model of human response to radiation given in API 521 is worth noting, that is: 3 seconds for a reaction, 10 to 15 seconds to seek cover. The cornea of the eye is particularly sensitive to heat radiation.

Flash Fires

Flash fire is the term applied to the case in which a large volume of flammable gas is ignited.

As can readily be seen from films of accidents, from experimental fires, and from direct observation, there are two kinds of effect involved.1 If the cloud is ignited before it has had chance to mix significantly with air, there is a region of fitful burning in the region in which the concentration of flammable material reaches the limit of flammability.

If there is a significant degree of mixing of air with the cloud prior to ignition, then a much larger volume of the cloud can ignite at one time. Eisenberg et al (1975) has given a model for this case. However, very little experimental work has been reported. Most research has concentrated on trying to predict overpressures and possibility of explosions.

If the volume of gas which burns is large enough, it can become sufficiently buoyant to rise as a fireball. The fireball is in fact a highly turbulent vortex ring, which draws in the air from below.

The size of a fireball was given by High (1968). High's formulae are given in terms of rocket propellant weight, but adapted to fuel weight by the author, they are:

$$D = 57\ M^{0.32}$$
$$T = 3.9\ M^{0.32}$$

where D is the fireball diameter in meters, M is the mass in tonnes, and T is the duration in seconds.

Fig. 11.3 Flash fire from gas leak.

Crawley reviewed the effects of flash fires as observed in actual incidents. His results are adapted here and extended using information from T.A. Kletz (private communication) and from a study by Kampmann and the author.

From the calculation of fireball size, based on the above equations, it can be seen that the results from rocket propellants underestimate the effects of fireballs in a gook number of cases. In some cases, the underestimate is as much as 100% of the calculated value. This can partly be explained by the difficulties in observation, of actual accidents, the different criteria for judging fireball size of the initial flash fire, rather than from resulting fireball.

Cooling Process Equipment/BLEVES

When a pressure vessel is submerged in fire, it is heated both by radiation and convection. Both the vessel and its contents are heated.

Heating has several effects which are important for safety.

- If the contents of the vessel are liquid, near the boiling point, then the liquids begin to boil.
- If the contents are gaseous, the gas expands.
- In both cases, the pressure will rise, and safety valves may open.
- The pressure vessel is heated. In the unwetted parts, not cooled by liquids, the vessel wall temperature may become so high (250 - 300°C for normal carbon steel) that it cannot resist the internal pressure.

When the pressure has risen and the strength fallen, so that the vessel ruptures, the result is a boiling liquid expanding vapour explosion (BLEVE). This usually takes some time, because of the thermal inertia of the vessel and its contents.

A BLEVE can be a quite violent explosion, throwing fuel and missiles into the air and causing a blast wave. By far the most important effect, though, is the rapid combustion of the fuel, creating intense radiation.

BLEVES are among the most destructive of accident phenomena, having destroyed several refineries. A BLEVE in Feyzin in 1966 destroyed three storage spheres. At Texas City in 1978 BLEVES in spheres and cylinders caused damage over a 200 m radius, with 7 deaths and 10 injuries. A BLEVE at San Juan Ixhuateapec, Mexico City in 1984 caused severe fires at up to 350 m. radius, and over 500 deaths.

One of the features of BLEVEs is the creation of explosion fragments and missiles. In the case of Mexico City these were thrown up to 1.2 km. (Pieterson 1986,Davenport 1986)

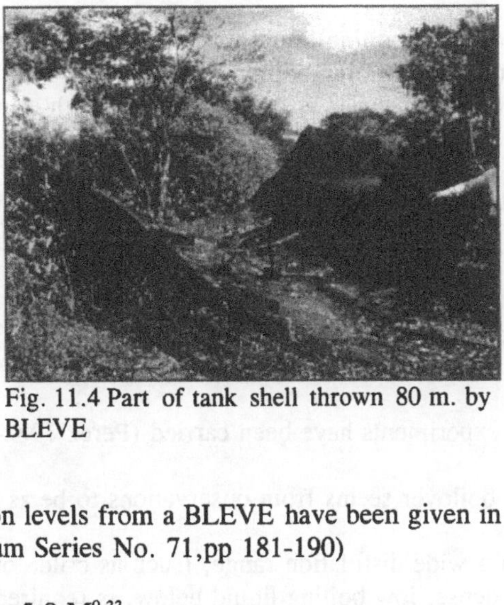

Fig. 11.4 Part of tank shell thrown 80 m. by BLEVE

The size and radiation levels from a BLEVE have been given in a review by Roberts (I Chem E Symposium Series No. 71,pp 181-190)
In free air

$$D_{max} = 5.8\ M^{0.33}$$

Where D_{max} is the maximum diameter in m.
M is the amount of fuel in kg.

Radiation from the fireball is estimated as between 20 and 30% of the total combustion energy in the fuel.

Boilover

A boilover can occur when a hot layer of high boiling point liquid overlays a cold low boiling point liquid, in a fire. If these become mixed, the low boiling fraction vapourises rapidly. As a result, the liquids are thrown out very rapidly, and very large fireballs can result.

Related phenomenal are:

- Roll over, where layering of low boiling, dense liquids, with an overlying layer of light, high boiling liquids occurs. When mixing occurs due to disturbance, the light boiling liquid, heated by contact with the low boiling liquid, boils violently.

- Steam or vapour explosion in which a low boiling liquid is pumped into a hot liquid e.g. water into hot oil, gasoline into a heated slops drum, or propane into a butane tank.

- Froth over, in which boiling liquid froths and froth spills over tank side. This may involve mixtures of high boiling and low boiling liquid or absorbed gases. It is particularly dangerous because burning froth can flow very quickly, propelled by the boiling.

Boilover effects are known from a large number of less serious accidents, with moderate effects very large accidents:

- Tacoa Venezuela 1984
- Milford Haven Wales 1978
- Thessaloniki Greece 1985

Some medium scale experiments have been carried (Perez 1982-3).

The mechanism in a boilover seems from observations to be as follows:

- A liquid with a wide distillation range, (such as crude oil) or a layered liquid body, with a dense, low boiling liquid below, is required.

- a fire starts, and heats the upper layer of liquid

- heat travels downward through the liquid. Any low boiling components in the heated layer are removed by diffusion, convection or low energy boiling. The heat wave travels at low speed, typically many hours for a full tank, which indicates that heat travels primarily by conduction, rather than by establishing convection cells.

- finally, the heat wave reaches a layer with a high density and low boiling point with a temperature sufficient to cause boiling. Boiling starts.

- the boiling itself may course burning oil to be thrown out of the tank.

- the boiling causes mixing of perhaps water, perhaps the low boiling oil fraction, with the hot oil at the top of the tank. Boiling becomes violent, and burning liquid is thrown into the air.

- the burning liquid forms a fire ball.

Note that some layering, either water or a low boiling layer at the base of the tank, appears to be required. This is shown by the fact that in many cases, the boilover only occurs when the heat wave reaches the tank base. The distillation of light fractions from the top of the tank does not appear, (at least in the cases observed) to be sufficient to cause sufficient layering. This is in itself surprising, since it might be expected that a fire might cause a hot, high boiling fraction at the tank top, and a cool, low boiling fraction at the tank base. Apparently, if the initial liquid charge is homogenous, then the heat conduction and distillation remain in sufficient equilibrium to prevent boiling.

On the other hand, a range of boiling points, allowing layering due to distillation of lighter fractions during burning definitely seems to be necessary - without this, in the experiments, no boilover occurred. With a small amount of low boiling fraction (20% kerosene in fuel oil), a slipover occurred. A wide boiling range, with significant amounts of low boiling fractions were necessary in order for full boilover to occur.

The mechanism therefore seems to be:

- boiling of a pre existing low boiling layer

- explosion due to the low boiling fraction of the oil, mixing with the high boiling fraction and vapourising rapidly.

The size of fireballs from this mechanism varies widely, with relatively low energy "slop overs", to large explosions with fire ball diameters up to 20 times the tank diameter. From observation of video films, the most typical appears to be 3 to 5 times tank diameter, but safety distances should be much larger than this if a boilover is expected during fire fighting.

References

Wesson, H.R., Brown, L.E. and Welker, J.R. — Predicting LNG Fire Radiation, Fire Prevention science and Technology, and Hydrocarbon Processing, 1974

Thomas, P.H. — The Size of Flames from Natural fires, Ninth Symp. Combustion, Academic Press, 1963

Hägglund,B. and Persson, L.B.	The Heat Radiation from Petroleum Fires FOA C20126-D6 1976
Mizner, G.A. and Eyre, J.A.	Large scale LNG and LPG Pool Fires, I.Chem. E. Symposium Series No. 71, 1982
Moorhouse, J. and Pritchard, M.J.	Thermal Radiation Hazards From Large Pool Fires and Fireballs, I.Chem E. Symposium Series No.71, 1982
Moorhouse, J.	Scaling Criteria for Pool Fires Derived From Experiment, ibid.
AGA	LNG Safety program, interim report, American Gas Association, 1974
Welker and Sliepevich	Fire Technology, 1966
Hazegawa, K. and Soto, K.	Study of the Fireball Following Steam Explosion of n-Pentane, 2nd Int Symp Loss Prevention and Safety Promotion in the Process industries, Heidelberg, 1977
Hardee, H.C., Lee, D.O. and Benedick, W.B.	Thermal Hazards from LNG Fireballs, Combustion Science and Technology, Vol. 17, 1978
Crawley, F.K.	The Effects of the Ignition of a Major Fuel Spillage, I. Chem. E. Symposium Series 71, 1982
Kalghati	Blow out stability of gaseous jet diffusion flames. Combustion Science and Technology 1981 Vol 26 p 233-239
Chamberlain, G.A.	Developments in Design Methods for Predicting Thermal Radiation from Flares. Chem. Eng. Research and Development, Vol 65, July 1987
Hawthorne, W.R., Weddell, D.S. and Hottel, H.C.	Mixing and Combustion in Turbulent Gas Jets, Third Symp. on Combustion, Flame, and Explosion Phenomena

Anderson and Stresino	Heat Transfer from Flames Impinging on Flat and Cylindrical Surfaces, J. Heat Transfer, Feb 1963
Pietersen, C.M.	Analysis of the LPG Disaster in Mexico city 5th Int. Symp. Loss Prevention and Safety Promotion in the Process industries, Cannes 1986
Davenport, J.A	Hazards and protection of Pressure Storage of Liquefied Petroleum Gases, ibid.
von Werthenbach, H.G.	Ausbereitung von Flammen an zylindrisch Behaltern, gwf-gas/erdgas 112 p.383-386 (1971)
Roberts, A.F.	The Effect of Conditions Prior to Loss of Containment on Fireball Behaviour, I.Chem. E. Symposium Series 71, 1982
Perez, P.	Lecture, Caracas 1990

12 Explosions

The Theory of Vapour Cloud Explosions.

Fig 12.1 Results of unconfined vapour cloud explosion, about 4 T of Hexane. Copenhagen, 1980

Vapour cloud explosions represent one of the most violent kinds of major accidents, and have been responsible for the destruction of many plants. The theory of vapour cloud and gas cloud explosions is not well understood, though many important experiments have been carried out in recent years.

When a cloud is ignited, one of the most important factors is whether the gas or vapour is already mixed with air. If the cloud is not mixed with air, then it will burn at its edge as a flash fire. If there is enough gas, and it burns for long enough, the buoyancy in the gas cloud will cause it to lift, and form a fire ball. In this case, the overpressure will generally be negligible and the term explosion will not be applicable.

If the cloud is in part premixed with air, then the part between the upper and lower flammability limits may take part in an immediate explosion. Gas mixture above the UEL(upper explosion limit) will subsequently burn or explode after mixing with air due to turbulence.

An explosion is characterized by a pressure wave, shown in an idealized form in Fig. 12.2. The pressure is caused by expansion of gas and air.

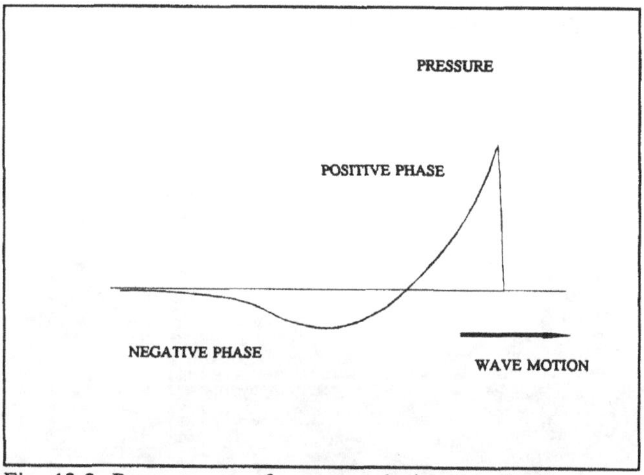

Fig. 12.2 Pressure wave from an explosion (idealised)

The pressure build up from the gas explosion is not instantaneous, because of the time taken for the gas to burn, and reflections and disturbances in the burning cause noise or extra peaks on the pressure wave.

The damage which can be caused by a pressure wave depends on the peak overpressure, but also on the negative phase pressure, because walls can be "sucked out", and on the duration of the wave, because this determines how far a massive, yielding, structure will be deformed.

Fig 12.3 Ibid, Explosion damage at 70 m. from explosion centre.

The overpressure in an explosion of this kind is caused by expansion of gas as it burns. If the gas was completely enclosed, then within limits the overpressure would be given directly by the energy released in burning and by the pressure/temperature relationship of the gas. However, since the gas here is unconfined it can expand, and accelerate away from its original position. If the burning is slow enough, the gas will expand without causing significant overpressure. The overpressure therefore depends on the speed of burning.

A number of theoretical models have been made, particularly for spherical clouds ignited centrally. A review of these is given in Strehlow, Luckritz, Adamczyk and Shimpi, "The Blast Wave generated by Spherical Flames", Combustion and Flame 1979, and an investigation is made of the relation between flame speed and overpressure. They show that for the usual flame speeds encountered in laboratory measurements, overpressure of the kind observed in accidents cannot occur.
Typical flame velocities are given by Lind and Whitson "Explosion Hazards Associated with Spills of Large Quantities of Hazardous Materials". Report CG-D-85-77, 1977. Flame speeds vary from about 3.6 to 15 m/s

An overpressure of 0.01 atmosphere would be expected from a normally burning gas cloud, using the Strehlow et al.'s model. In practice overpressure damage corresponding to .1 to .3 atm is in some cases observed in accidents.

Pressures of the order of 0.01 atm. are in fact observed in experimental explosions. However, this can hardly explain the enormous amount of damage which can be caused by unconfined and partly confined vapour cloud explosions. The mismatch between theory and observation is even worse, because the theoretical models overestimate the pressure in edge-ignited cloud.

The explanation of the much higher pressures obtained in some explosions is that the flame speed is raised from its normal value to a much higher value. Several explanations of this have been suggested.

One explanation is that turbulence can cause mixing of the gas with air, and mixing of the flame with unburned gas. Turbulent eddies can travel at high speeds and spread the flame through the cloud, increasing the speed of burning. It is certainly the case in experiments when placing piping etc. in the cloud causes higher pressure.

Another possible cause which has been suggested is the SWACER mechanism (Knystautas et al "Direct Initiation of Spherical Detonation by a Hot, Turbulent Jet". Seventeenth International Symposium of Combustion, 1979). In this mechanism, a small explosion in a confinement produces a jet of flame,and a high degree of turbulence, and pockets of gas surrounded by flame. The increase in flame surface results in an effective increase in flame velocity.

Experiments have been made to test these mechanisms (Van Wijngarten 1989, Harrison 1987, 1989). These show that while obstacles such as piping do cause flame acceleration, the acceleration is insufficient in the experimental scale to achieve the pressures observed in accidents. On the other hand, explosions in small enclosures can create sufficient turbulence to start a percussive explosion.

Clearly, pressures will be higher if the cloud is enclosed, or even just partially enclosed, because the expanding gas will have less opportunity to escape.

The dependence of explosion pressure on flame speed means that either flame speeds or overpressures must be determined empirically. A large amount of evidence exists from TNT explosions experiments, but also the results of "working backward" from known damage results.

One empirical result which has been known since 1915 is the Hopkinson Cranz Law

$$p = F (E\ 1/3\ /\ R)$$

This law describes the way in which pressure falls away from the explosion, and gives a way of escalating from one explosion size to another. This law and its extensions are the basis for plotting empirical data from experiments.

Some general observations about unconfined vapour cloud explosions can be made. Firstly, many releases occur without ignition. The ignition probability for small releases is about 0.03, for large releases 0.1 to 0.3, on historical records (Kletz). Only about 1/5 of the releases which are ignited and that could in principle lead to explosions do in fact explode.

Those gas clouds which do explode seem to generate overpressures between 0.1 and 0.3 bar. The normal flame velocities for a gas cloud are of the order of a few meters per second, or less. In order to generate an overpressure of 0.3 bar, a flame speed of about 200 m/s is needed. Except in the most reactive gases, true detonation (where flame speed reaches the speed of sound) has not been observed to occur. Detonation produces pressures at about 2000 m/s.

The amount of material entering into an explosion is not 100 % of that released. Most often, there is a considerable amount of material which is outside the flammable range. Of the material which does enter into the explosion, only a proportion of the energy is transformed into blast energy. The proposition has been calculated variously by different researchers, mostly working from backward extrapolation from accidental explosion effects. The quoted as being between 1% and 20%, with a consensus which seems to be forming for a value of 10 % of the material in the flammable part of the cloud. (Lewis 1984)

Explosions have not been observed in small quantities of released gas and vapour. The smallest well documented release giving an explosion involved kg 200, on this case of highly reactive gas. The lower limit for explosion in saturated hydrocarbons has been quoted by various authors as 1 ton, 5 tons, or 10 tons. The author prefers to use a limit of 200 kg of material in the flammable range, based on the more extreme case stories (see e.g. D.G.Lewis) for alkenes, 1 ton for alkanes.

The TNO yellow book (1988) model provides a simple way of calculating pressures as a function of distance. It is a correlation based on many experiments and practical experience from cases. Sample calculations show that the pattern of peak overpressure maintenance, which is characteristic of gas cloud explosions, is reflected in the models.

The model requires that the volume of flammable gas is calculated, and from this an energy-scaled distance can be worked out ($r/ ^{3E}$).The calculation then requires a judgement to be made concerning reactivity. Peak overpressures and positive phase pressure duration can then be read directly from plotted curves.

The TNO model is based on generally accepted shock wave models, on a detailed model for flame expansion (Wiekema, Journal of Hazardous Materials 3 No. 3) for deflagration. For the reactive gases, the model is based on experimental results for gas detonation obtained by Kogarko et.al. (Int. Chem. Eng. 6, 1966) and confirmed theoretically by use of the theory by Brinkley and Kirkwood (Determination of Explosion Yields, Loss Prevention 3, 1969).

Houwelling has reviewed the available theoretical and practical studies and provided a parameter-fitting summary of these. Turbulence-generating obstacles are represented by a parameter in this model.

Houwellings correlation gives

$$P_{so} = 0.139 \quad \psi^{0.235} \ (R/L)^{-0.965} \exp(\psi^{0.54})$$

$$I/L = 68.23 \exp(-(2*10^{-5})\psi) \ (R/L) \exp(-0.991\exp(-(2.6*10^{-6})\psi))$$

Here

P is the scaled pressure rise = $P-P_o/P_o$
ψ is a combined energy density and flame speed parameter.
$\psi = 0.5*10^{-16}\ TCR^{0.667}\ pn\ N/M\ H^2_c$
R is the distance from the centre of the explosion
L is a characteristic distance
$L = (E/P_o)^{0.333}$
I is impulse
E is the energy released and P_o is the ambient pressure.
I is the impulse in the positive phase.
H_c is the heat of combustion.
M is the molecular weight of the gas.
N is n + r + m-k-21/4
for a compound $C_n\ H_m\ O_e\ F_k\ S_r\ N_x$ (F is any halogen)

n is 0.5 for methane
1.0 for saturates
1.5 for unsaturated hydrocarbons
2.0 for alcohols, ethers, ketones, and halogenides.
3.0 for oxides.

P is the density of the gas
C is a concentration factor, = 0.1 at flammability limits.
T is a turbulence factor, 1 for a free field, 2 for rough terrain, 10 for pipe type obstructions, 30 for process equipment obstructions.

This model has been used by the author in post analysis of a number of accidents, with results that indicate agreement to within a factor 2, at worst, of observed damage effects (assessed as overpressure).

The models given above are based on an idealization of a gas cloud, as a hemisphere ignited at its centre. In practice, flammable clouds are usually either jet shaped or have a flat form, spreading with the wind. They are usually ignited at the edge.

This means that the standard models for blast pressure greatly overestimate explosion pressures. In an attempt to reduce this problem, Geiger (Loss Prevention and Safety Promotion in the Process Industries, Swiss Society of Chemical Industries 1980) and Auton and Pickles (CERL Laboratory Note RD/1/Nm210/78, 1978) have developed models for "pancake" clouds. The models are still based on central ignition, and so overestimate explosion pressures.

Blast Wave Propagation and damage

The effects felt by a building in the path of a blast wave depend on the static wave pressure p and on the impulse. It depends also on the so called "dynamic pressure" which is the wind pressure from the gases rushing from the explosion. This pressure is given by:

$$P_d = C_d \, 1/2 \, \rho \, u^2$$

Where

C_d is a coefficient to describe the degree of air resistance.
u is the velocity of the displaced gas.

Usually, this pressure is low compared with the static pressure for chemical industry explosions.

The pressure effects on a building depend also on the dynamics of pressure wave reflection. Roughly speaking, the front of a building which experiences the direct impact of blast wave moving normally to the wall, will experience twice the static overpressures. Sidewalls will experience just the static pressure. The rear surface will also experience an overpressure.

A more precise formula for "front on" overpressure is given by:

$$P_f = 2P + \frac{(\gamma+1)\,P}{(\gamma+1)P + 2\gamma}$$

where
P is the side on pressure
γ is the ratio of specific heats

The pressure exerted on a side wall is typically equal to the blast wave overpressure. However, corners and reflections can increase the effect of pressure waves. Behind the building there will also be a significant overpressure effect.

In some cases, surprising levels of damage may occur at long distances from an explosion. The reason is that blast waves may be refracted and focused by air layers at different densities. A full theoretical treatment for this kind of effect is not yet possible but experimental results have been obtained.

When an explosion occurs, the explosion energy is in part transferred to a pressure wave, in part to a ground wave. The ground wave takes only a small proportion of the energy from cloud explosions but can be responsible for failure of process equipment with weak supports.

There are two main approaches to estimating the extent of explosion damage:

- using empirical data from explosions records
- using theoretical results on propagation of blast waves, and on the impact of blast waves on persons.

The theoretical results are more difficult to apply because of the problems of reflection, the degree of dynamic coupling between the pressure wave and the affected buildings and because of the problems of diffraction.

Brasie and Simpson developed a table of damage extent by reviewing three chemical process explosion incidents.

Clancey has prepared more extensive tables based on a review of literature and Ministry of Home Security data.

Table 12.1 Some observed damage due to explosion overpressures

Overpressure bara	Effect
.01-.03	Window breakage
.03	Minor structural damage to buildings
.07	Major damage to houses
.07	Cone roof collapse
.08	Industrial buildings partially demolished
.12	Wired glass breakage
.2	Some broken piping, deluge and sprinkler piping broken
.2	Steel frame buildings distorted, walls damaged
.2	Vessels overturned
.2	Empty part of oil tank collapses
.3	Much broken piping
.4	Building steel cladding ruptured
.4	Damage to distillation columns
.5	Loaded wagons and trucks overturned
.5	Brick panels in steel or concrete frame ruptured
.7	Total destruction of buildings
1	Large filled vessels overturned
.1	Damage to human organs
.03	Damage to hearing

Baker et al ("Explosion Hazards and Evaluation", Elsevier 1983) provide a more sophisticated approach to empirical analysis, in which both peak pressure and impulse are taken into account. They warn that Brasie and Simpson's approach can give serious overestimate of damage from small explosions, with relatively low impulse values.

Baker et al have also given thorough "workbook" approach to explosion, in terms of peak pressure and positive phase impulse. This allows damage to be calculated more or less straightforwardly from the released amount of material.

Glass damage is one of the more important aspects of explosions. Glass breakage can be a significant part of the cost of an explosion in a built up area. The flying fragments of glass are additionally one of the most serious causes of injury from an explosion. Harris, Marshall, and Moppet described a series of experiments using explosions generated by gas balloons inside a bunker.

Scilly and High (1986) have summarised these approaches. In the work reported by Clancey, the following quantities were defined:

$$R = \frac{5.6M^{1/3}}{\left[1 + \left(\frac{3175}{M}\right)^2\right]^{1/6}}$$

Where R is the radius in metres for class B damage,(houses beyond repair), and M is the mass equivalent of explosive in kg TNT. Damage for values of other values of R are given as follows:

A	Houses demolished	r = 0.675 R
B	Houses uninhabitable	r = R
Cb	Houses needing extensive repair	r = 1.74R
Ca	House in need of some repair to be habitable	r = 3R
D	Houses habitable but needing repair	r = 6R

Scilly and High point out that the above equation of a special case of a more general equation:

$$R = \frac{a\,M^{1/3}}{\left[1 + \frac{b^2}{M} + \frac{c^6}{M^2}\right]^{1/6}}$$

The different terms in the equation correspond to strong and weak targets, the overpressure at greater distances being reduced, and the pressure wave broader.

Baker et al. give damage in terms of peak pressure and impulse. Some structures are strong and rigid, deforming only a little. These structures will fail primarily if peak pressure exceeds a given limit. Other structures are less strong, and so deform, but are resilient. These structures fail if sufficient energy is absorbed, and hence sufficient displacement achieved. This will depend primarily on the impulse. Fig 12.3 shows this the form of the resulting P-I limit curve for a structure. The curves are specific for each building type, and must be calculated independently for each structure, using structural dynamics programs. In practice, this is done for only the most critical cases in risk analysis.

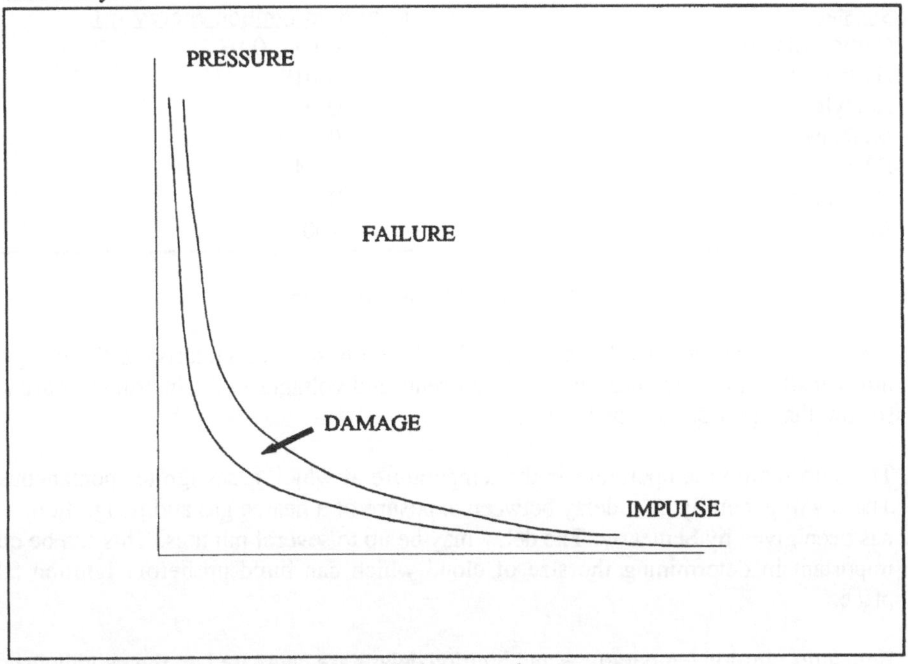

Fig.12.4 General form of P-I damage limit curves for structures.

Ignition of Gases.

For a cloud of gas to be ignited, the first requirement is that the concentration of flammable material is in the flammable range. Clouds will fail to ignite either if the concentration is too dilute or too rich. It has been known for clouds of over-rich gas to pass over and through a furnace but fail to ignite. The only effect was to cause thick black smoke to be emitted from a chimney.

The concentration limits which determine whether a mixture of gas and air is flammable are the flammability limits. The limits widen with increasing temperature and pressure. Mixtures with oxygen have also a wider flammable range.

The possibility for a liquid is also determined by the flash point of a substance which is the minimum temperature at which the substance gives sufficient vapour to be ignited (sufficiently high vapour pressure). This may be measured in several different ways, but the main distinctions are between open and closed cup tests which refers to the type of cup used to heat the liquid when making tests. In risk analysis, these values are important when studying ignition of heated pols.

In order to ignite a gas a certain amount of energy is required. This depends both on temperature and on concentration of gas. Table 12.2 shows minimum values for a few gases. Note the low values for carbon disulphide, hydrogen, and acetylene.

Substance	Minimum ignition energy mJ
Carbon disulphide	0.01 - 0.02
Hydrogen	0.019
Acetylene	0.02
Methane	0.24
Ethane	0.24
N. Hexane	0.25
Ammonia	100

Table 12.2 Ignition Energies.

Limitation of energy available for ignition is the basic idea underlying the design of intrinsically safe instrumentation - the currents and voltages in the instruments are kept so low that ignition cannot occur.

The auto-ignition temperature is the temperature at which gases ignite spontaneously. There will generally be a delay between exposure of a heated gas and its ignition. This has been given by Semenov. The delay may be up to several minutes. This can be quite important in determining the size of cloud which can build up before ignition takes place.

Both auto-ignition temperatures and ignition delays are decreased on the neighbourhood of heated solid surfaces. The reduction in ignition temperature can be as much as 200°C .

Sources of ignition include:

- lightning
- furnaces
- heated surfaces
- sparks from motors and switch gear
- sparks from tools, boot studs, etc. especially on aluminium.
- hot dust or soot particles
- other fires
- overheated pumps and motors
- pyrophoric substances
- vehicles
- electrostatic sparks and microwave sparks
- welding
- cigarettes
- sparks from broken or damaged cables.

In areas where there is often, or always, a flammable atmosphere, equipment which is flame proof or intrinsically safe is generally used. This is a requirement for some zones in a plant, according to the fire regulations in most countries. This means that motors, switch gear, and instruments in such areas should only constitute an ignition source if damaged.

Pyrophoric and self-heating substances include finely divided nickel and similar metals, iron sulphide scale (often present where sour gas or H_2S is treated) and asphalt. Fluorocarbons and aluminium can react to give ignition. Asphalt and heavy oil layers in piping and tanks can auto-ignite, as can oil-soaked rags or lagging. Lagging is particularly dangerous because oil is present dispersed over a wide surface, often where there are heated pipes and vessels.

Organic material, especially powders, if stored damp in heaps or in silos, can ignite as a result of heat generated by bacterial action.

Static electricity is generated by almost any movement of solids (including vehicles transport belts and persons) and liquids. It accumulates if there is an insulated surface so that the electricity cannot be conducted away.

Probabilities of ignition are very difficult to estimate and can only really be judged from actual accident experience. They depend very much on actual concentrations at the point where ignition takes place and on the distance a gas cloud must travel to reach the ignition source. Note that calculations of concentration using dispersion formulae will be misleading especially in the near field. Clouds and plumes are not homogeneous while some parts of the cloud may be too dilute to ignite, other may be sufficiently rich, beyond the distance at which concentration is calculated to be below the flammability limit.

For most gases, ignition probabilities for clouds which reach a naked flame with flammable concentrations seem to be of the order of 0.5. This does not apply to hydrogen, carbon disulphide, or acetylene (nearer 1) nor ammonia (nearer 0.1).

For substances released with a temperature near the autoignition temperature, the ignition probability is best regarded as being near 1.0, though the problem of ignition delay should be remembered (it can lead to explosion).

Probabilities of ignition for individual items of equipment are given in table 12.3. They assume that a cloud or plume reaches the equipment with a concentration within 20% of the flammable limit.The data have been adapted from work by HED Phillips, and modified on the basis of Danish and other accident statistics.

Source	Ignition probability
Motors, switch gear, pumps (not flame-proof)	$1 - 10*10^{-3}$
Switch gear, unprotected which is switched off during exposure	0.5
Electrical sub-station	$10-10*10^{-3}$
Analysers	$1-10*10^{-3}$
Control rooms, offices laboratories (not overpressured)	$50*10^{-3}$
Fans	$50*10^{-3}$
Operators	$1*10^{-3}$
Lights	$0.3*10^{-3}$
Instruments	$0.3*10^{-3}$
Operational maintenance Hot working during maintenance	$5*10^{-3}$
Each person during maintenance	$2*10^{-3}$
Background items, static	$0.1*10^{-3}$
High static operations, drying etc.	$1*10^{-3}$
Transport, visitors etc.(assuming normal process plant operation, very low transport activity)	$0.3*10^{-3}$

Table 12.3 Gas Ignition Probabilities in typical process plant situations

Explosions in vessels, pipes, and buildings

Ignition

For an explosion to occur within a pressure vessel or tank, four criteria must be satisfied.

- there must be fuel vapour or gas
- there must be air present
- air must be present in sufficient quantities so that the vapour is within the flammability range, that is above the lower flammability limit
- there must be a source of ignition.

Within a vessel, the usual ignition causes are reactions and process heat, static electricity, although other causes are possible. If the first three conditions are fulfilled, it is usually only a matter of time before ignition of some kind occurs.

The list of possible ignition sources, even within a tank, is very large. Some of the causes can be quite surprising.

- electrostatic charge
  . splashing
  . charged liquid
  . trash
- electrical sparks
  . switching
  . short circuits in equipment, cables
- friction of moving parts
- welding on the vessel
- adiabatic compression of vapour lighting
- fire outside the vessel
- gas analysers (faulty)
- welding on the vessel
- pyrophoric substances (especially FeS, asphalt)
- chemical reactions
- hot surfaces (especially internal piping)
- escape of vapour from the vessel, ignition, and burn back
- process heat to above the auto-ignition temperature
- stray substances which catalyze combustion.

Table 12.4 Causes of Ignition Within a Vessel

Example

In an oil sludge tank, expanded polystyrene particles from waste packing entered. the tank was not supplied with inert gas, because there were no obvious causes of ignition, no splash filling, and because the sludge contained "only heavy oils". In practice, a small proportion of light oils C5 to C7 was present and saturated the air space in the tank.

The polystyrene pieces charged a level sensor and the resulting spark initiated an explosion.

Example

A large underground oil storage tank was being ventilated prior to maintenance. ventilating the tank unfortunately, during one ventilation, the fan used was set to suck rather than to blow. When flammable vapour reached the fan, it ignited and caused an explosion. This destroyed the tank.

Petrochemical storage tanks are at their most dangerous during filling and emptying. Special safety instructions are required for this stage. a risk analysis of the procedure is almost certainly worthwhile. the most important aspect, though, is to inform maintenance personnel of the safety precautions, the reasons for them, and the consequences of not following them.

Confined explosions

Confined explosions can generate much higher pressures than unconfined.The energy available when an explosion takes place can be calculated approximately from the heats of formation of the fuel and oxygen, and from the heats of formation of the combustion products. This assumes complete burning in the explosion. The heating of the gasses causes an increase in pressure.

The energy available from a complete combustion explosion is given by the Gilbs function values, if the pressure is (relatively) constant and the Helmholtz free energy if the volume is constant.

In practice using heats of formation overestimates the amount of energy available in the explosion, because the explosion products at high temperature are not usually the "complete" combustion products and because some energy goes to expanding the products and some energy to raising the entropy of the new mixture of products.

In general, the combustion products in an explosion do not represent complete combustion. More complex calculations take into account the fact that combustion products may be dissociated at high temperature (Chetah, D.R. Stull, Loss Prevention 7, 1973). Even this kind of program does not provide perfect results, because average data and equilibrium reactions are assumed. Nevertheless, for risk analysis purpose, a high degree of accuracy is seldom needed for the following reasons.

- the gas air mixture which burns or explode will almost never be an ideal stoichiometric mixture.

- the degree of mixing of air and fuel is almost never perfect, and differences in combustion

In view of these uncertainties, the highest degree of accuracy in combustion energy determination is hardly justified. Examples of the different amounts of combustion energy produced are given by Stull.

Once the amount of energy released in gas/air explosion is known, the temperature can be calculated from the specific heats vary though the combustion products. Specific heats vary though with temperature, so that it is generally necessary to use plots of specific heat against temperature. Curves with average specific heat up to a certain temperature are ideally used.

Such specific heat curves, are given by Smith and van Ness, Introduction of Chemical Engineering Thermodynamics). Alternatively more complete data are given by Wagman (1952) and JANAF. These curves are for specific heat at constant pressure. For completely confined explosions, specific heats at constant volume are required- For an ideal gas C_v (per mol) is given by:

$$C_v = C_p - R$$

The temperature of the products of the explosion is given by

$$\Sigma\, n\, C_v\, (t - t_0) = \text{energy released by burning}$$

Here n is the number of moles of product, t is the final temperature, and t_o is the initial temperature.

The procedure for finding the resulting temperature and pressure is then to

1. find the heat released in the reaction
2. guess the resulting temperature t
3. find C_v for the products, as an average value, up to t
4. evaluate $\Sigma n C_v(t-t_0)$ and find whether the result is smaller or larger than the heat available from burning
5. change the temperature, increasing if there is more heat available than required to reach t, reducing it otherwise, and repeat the calculation
6. stop when the temperature reaches a more or less constant value.

Pressure is given by

$$P_2 / P_1 = n_2 T_2 \;/\; n_1 T_1$$

assuming that the combustion products behave like an ideal gas, a reasonable approximation at the temperatures involved.

Example

A stoichiometric mixture of hydrogen with air burns in a confined volume. The initial temperature is 25°C and the initial pressure is atmospheric. What pressure results ?

The reaction is:

$$2H_2 + O_2 + 3.76\ N_2 = 2\ H_2O + 3.76\ N_2\ -115{,}600\ \text{cal/mol}$$

Guessing a temperature of 2000°C gives:

$$C_v\ N_2 = 5.6$$
$$C_v\ H_2O = 7.35$$

then

$$t = 25 + 115600/(2*7.35 + 3.76*5.6)$$
$$= 4000°C$$

The values for specific heat should now be revised. As an approximation, the pressure is given by

$$P = ((273.4 + 4000) / (273.4 + 25))(5.76/6.76)$$
$$= 12.2\ \text{bara}$$

This is in fact much higher than the actual pressure due to incomplete combustion, heat loss etc. Pressures for unvented confined explosions are given in table 12.4

Table 12.5 Pressures due to unconfined combustion

Fuel	**Concentration** max pressure/ stoichiometric	**Pressure ratio**
Acetylene	1.75	11.03
Ethylene	1.3	9.56
Hydrogen	1.17	7.55
Methane	1.07	7.86
Propane	1.19	9.03
n-Hexane	1.12	9.21
i-Octane	1.12	9.64

(after D.G.Lewis, course notes)

Pressure piling

A problem which is quite often met, with interconnected tanks, or particularly, with furnaces, is that there are several interconnect vessels. If an explosion occurs in one vessel pressure builds up, and after a short time, is transferred to the second vessel. A typical example is shown in fig 12.5.

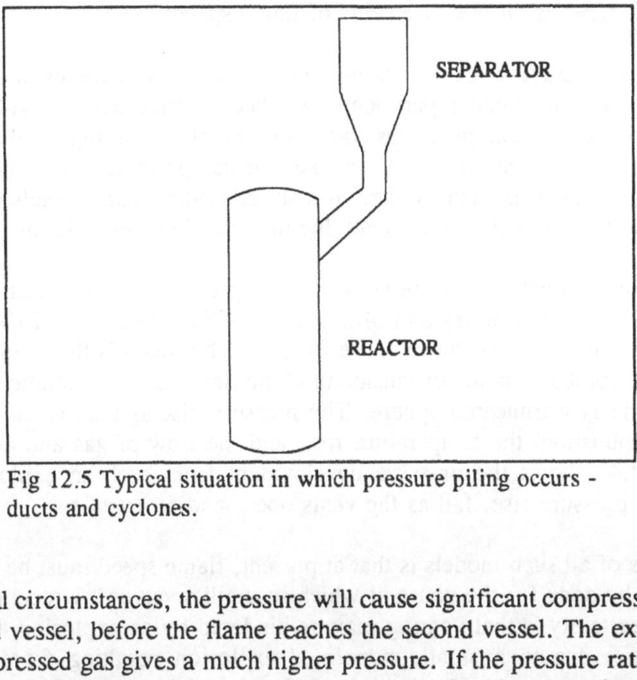

Fig 12.5 Typical situation in which pressure piling occurs - ducts and cyclones.

Under normal circumstances, the pressure will cause significant compression of the gas in the second vessel, before the flame reaches the second vessel. The explosion in the now precompressed gas gives a much higher pressure. If the pressure ratio for a single vessel as calculated above, is x the pressure ratio for the second vessel is, at a maximum x^2. In practice such a pressure would hardly ever be reached since very few vessels are designed for a bursting pressure of 64 times their working pressure.

Explosion venting

One way of reducing the load due to explosions is to provide vents in the vessel. Such vents may be permanently open, or maybe opened by the pressure of the explosion. Typical of such relief possibilities are hinged doors, or thin metal panels which are burst by the explosion pressure. In practice, pressures may approach these valves.

For such vented explosions, the sequence of event sis as follows. First, a small cone of flame builds up. The heat, and resulting expansion causes the pressure in the vessel to rise. This rise in pressure causes the relief panel or door to open. Once the panel has opened, gas begins to flow through the opening. In the first instance this reduces the pressure.

The flow of gas, however, produces several kinds of instabilities, which generate turbulence and more combustion.

The result of these instabilities is that, after the initial fall in pressure on opening of the relief panels, the pressure can rise again. The degree of pressure rise depends on the size of the panel in relation to the size of the vessel.

Because of the complexity of the combustion process in vented vessels, prediction of pressures is based on experimental results (Bradley and Mitcheson have presented an excellent theoretical description in combustion and Flame 32, pp 221, - 236, 1978, but this treatment depends on measurements of flame speed).

Calculation of overpressures for large vented volumes involves a good deal of uncertainty at present. Some experiments have been carried out with volumes as large as 200 m^3, but the actual pressures vary very much according to the presence of turbulence generators, and especially, pockets of gas protected from the main flame. In experiments at the Christian Mikkelsen Institute (1986) flame speeds up to 25 times the normal laminar velocity were found,due to instabilities even in an empty vessel.

Given a reliable estimate for flame velocity, overpressures can be predicted using the dimensioning curves of Bradley and Mitcheson. (1978). The model of Fairweather and Vaisey gives similar results, but is more flexible. This regards the flame as spherical with complete combustion and expansion until the flame hits the volume wall, at which stage the flame is a truncated sphere. The pressure rise is then calculated from the energy of combustion, the temperature rise, and the flow of gas and air through the vent. Fig. 12.4 shows the pressure development from such an explosion with the typical initial pressure rise, fall as the vents open, and renewed rise.

The weakness of all such models is that at present, flame speed must be specified, but this can only be done for the phase at which instability sets in. A recently discovered source of uncertainty is that venting may be hindered by an explosion external to the vented volume, due to a small unconfined explosion resulting from the venting. (A.Evans, in comments to presentation of Evans and Puttock, 1986). The situation seems to arise if there is a cloud of gas outside the vent, or if significant quantities of gas are expelled through the vent.

A major problem with all pressure venting models at present is that they are only verified for vessel volumes up to about 250 m^3 with avery few experiments at volumes of about 1000 m^3. Also the models used for dimensioning do not apply for very elongated rooms. Problems arise in predicting the effects of gas escapes in tunnels, and large gas escapes in oil and gas platform modules and in large industrial buildings. The problems arise because there can be may ways in which flames accelerat in such environments, particularly if there are obstacles which create pockets of gas. The result is that significant overpressures occur, in excess of those described by the standard models. In the absence of more detailed and larger scale experiments, caution is necessary in design. For difficult cases, detailed finite element calculations are available which can a design basis for venting.(Haverstad, T.A., in reports from SINTEF, Trondheim, Norway)

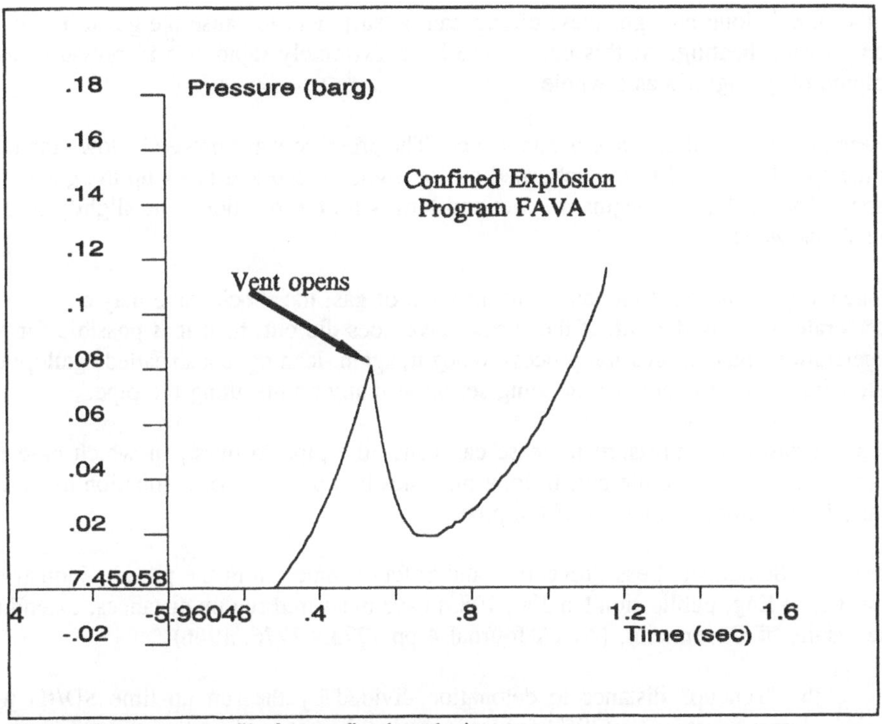

Fig. 12.6 Pressure profile for confined explosion.

Detonation

If a vessel is long or and explosion takes place in a pipe, the calculations for pressure given above do not apply, since the gas may detonate, Detonation occurs when flame speeds accelerate to the local speed of sound, and flame and pressure waves become complete.

After ignition in a long vessel, or pipe, the expansion causes flow of gas. The flow causes turbulence. This turbulence leads to more rapid burning. As a result, experiments show a tendency for flames to propagate along pipe walls. The extent of this effect depends on whether the vessel or pipe has an open end

Another phenomenon which causes acceleration of the flame front is Taylor instability. When a heavy fluid is accelerated into a light fluid, the interface between the fluids is unstable. Waves develop at the fluid interface. The surface of the flame becomes extended. This in turn increases the burning rate with explosions in closed vessels. The photographs show a stroboscopic record of the flames. When the flame reaches the wall of decelerates, and the wave form develops.

The result of these acceleration mechanisms is that the flame travels faster. This in turn causes a more rapid flow in the pipe. This more rapid flow causes more turbulence, and compress the gas ahead of the flame.

If the pipe is long enough, these effects can be sufficient to cause the gas to ignite by compression heating. At this point there is an extremely rapid rise in pressure, as a volume of gas ignites as a whole.

Thereafter a detonation wave is established. The pressure wave proceeds down the tube at the speed of sound in the unburned gas,or higher speed and heats up the gas in the shock. The heated gas begins to react and burn with the reaction zone slightly behind the shock wave.

Depending on the reactivity and concentration of gas, the shock wave may continue to accelerate, or may die out. If the shock wave does die out, then it is possible for the acceleration and compression process to begin, again, leading to a so called "galloping" detonation, with the detonation being set up at many points along the pipe.

In some cases, the pressure increase can cause the pipe to burst, in which case the detonation will tend to die out. In long pipelines it is possible for detonation to lead to many bursts along the length of the pipe.

Steen and Schampel (Loss Prevention and Safety promotion in the process industries, Inst. Chem Eng. publication No 290, 1983) have measured run up distances, extending the results of Oppenheim, (AIAA Journal 4 pp 1773 - 1776, 1966)

- the "run up" distance to detonation divided by the run up time SD/tD was almost constant (±120%) for various pipe diameters, for ethylene and propane air mixture except for large diameters (25 cm).

- the run up distance increases, more or less linearly with diameter

If the pipe is connected to a tank in which combustion starts, the run up distances are reduced by the flow induced by the combustion in the tank, typically by a factor of two.

The pressures in detonating gases are considerably greater than those in deflagration typically a factor two higher than the maximum completely confined explosion pressure for practical cases with hydrocarbon i.e. 16-20 bar if unvented. (The pressure experienced by a bend or blanking plate may be up to twice this, because of pressure wave reflection.

The pressure experienced by a pipe band or blanking plate to the reflected pressure, which is twice the side on pressure. This is not the complete story however. A detonation does not in practice proceed as a simple plane wave. In most cases, many small detonations centers are established and expand spherically in the direction of the wave, overlap die out, and reform continually.

The cell sizes are characteristic for each gas mixture, temperature, and pressure. This is important because a minimum of one cell appears to be required for propagation of a detonation along a pipe. With only one cell, the detonation propagates in spinning fashion, rather than as a simple wave. The minimum diameter for this is given by where d is tube diameter δ is cell size.

The cell size is also important in determining whether explosions can propagate from a pipe into a tank or into free air. A rule of thumb is that thirteen cell diameters are required for such propagation.

Table 12.6 gives cell sizes for some typical gases.

This means in practice that there is a minimum size for propagation of explosions

Table 12.6

Substance	measured cell size
Acetylene oxygen equimolar	4 - 5 mm
Hydrogen air Stoichiometric	1.5 mm

In process plant risk analysis, detonation becomes important for columns into which there has been significant air leakage. It is important for pipelines in which air is present. And it is important for safety value blowdown lines, vent lines, and reactor feed lines, if air and flammable are present.

If detonations do occur in a vessel it is very unlikely that the vessel will be able to withstand the pressure. Detonation rupture of a vessel is especially likely to cause explosion fragments, Detonations in pipes will also, with high probability, cause ruptures. These are especially hazardous if they spread fire between parts of a plant.

Sewers may be subject to explosion, if leakage of gas, volatile solvents or gasoline occurs. Sewer explosions are fairly common, occurring every few years in some plants. Most often, the effect is to raise man hole, covers, sometimes to throw manhole cover in the air, The conditions for detonation are present in such cases, however if the correct mixture of gas and air exists, and such explosions have been known to destroy equipment over a wide area.

Example

The following case illustrates a rather unexpected form of explosion. Oil films which are deposited on pipe walls can be brought to detonate if they are heated.In a compressed air supply system piston ring wear in an "oil free" piston compressor led to oil deposits on piping walls. During a period of high demand, air temperatures rose and ignited the oil film.The resulting detonations burst the air piping at eight places around the plant.

Although the mechanism here may appear strange, explosions of this kind are not uncommon, having occurred in three plants in the authors experience.

Condensed phase explosions

Chemicals in reactors, and in storage, can cause violent explosions through unwanted reactions. These may involve reactions between chemicals, or they may involve decomposition or condensation reactions. Very thorough studies have been made Barton and Nolan (1984) and by B. Rasmussen (1987).

The basic phenomenon involved is a reaction which produces heat, and as a result accelerates. In some cases the reaction is a normal one, which goes out of control, in other cases it is a completely unwanted reaction. Rasmussen gives a (11 page) check list of possible causes. for unwanted reactions, the reaction can occur due to

- incorrect mixing
- impurities or wrong substances added
- omitted substances (e.g. inhibitors)
- accumulation of products or raw materials
- erst while unknown reactions.
- incompatible construction materials
- contamination of vessels
- ingress of air, water etc.
- stray catalysts.
- leakage of impurities
- accumulation of a reactant or product or unwanted concentration of a product
- ageing of a product or raw material
- wrong process conditions
- inadequate mixing

for normal reactions runaway can occur if

- the reaction is exothermic, and cooling fails
- errors in temperature and pressure control
- errors in addition of raw materials, catalysts, inhibitors
- stray catalysts
- failure of mixing
- hot spots on heat exchanger
- fire
- inadequate design

Rasmussen found that most such explosions could be predicted by a good process of hazard identification.

Some reactions may be so violent that and explosion is immediate e.g. decomposition of vinyl acetylene after accumulation in a butadiene unit (Jarvis 1972). In other cases, pressure relief through burst discs is possible. Note that even if an explosion does not take place, a runaway reaction can release toxic material. for this reason, burst discs often vent to blowdown systems, dump tanks, or scrubbers.

Various kinds of reaction calorimetry are available to help in the prediction of runaway reactions, and to support cooling system design. e.g. differential thermal analysis, heat flow calorimeter, accelerating rate calorimetry, and dewar flash tests. for determining the needs for relief of reactors, a complete package of methods has been developed by Fisher et al in the DIERS research program (Fisher 1985).

Overpressures from reactor and storage explosions is generally calculated using the TNT equivalence approach. In this, the energy which may potentially be released is calculated in one of the following ways

- from the energy of formation of reactants and products assuming works case energy release
- using an advanced computation technique such as "Cheetah".
- from reaction calorimetry.

The energy is expressed in terms of the equivalent amount of TNT which would release that energy. Curves are then available bases on empirical studies giving the explosion overpressure as a function of distance. Curves have been given for example by Baker and by Lewis.

Modern practice generally involves use of simple computer calculations to achieve the same result.

Dust Explosions

Mixtures of dust in air can explode if ignited. In practical situations, serious explosions involve a sequence of events. First, a simple deflagration (flashing fire) occurs. This causes sufficient flow of air and smoke to disturb more dust. The disturbed dust then provides the fuel for a dust explosion.

Dust explosions have occurred for example in flour and soya silos, in powdered product silos, in driers, and in coal mines. Such explosions have destroyed and topped silos.

Design data for dust explosion relief has been given by Bartknecht (1981) by Palmer (1973) and by Field (1982). Calculation of the consequences of dust explosions has been described by Baker (1982).

Missiles and Missile damage

Many explosions produce fragments, which can be thrown for considerable distances. Unconfined vapour cloud explosions can, for example, break walls, roofs, and piping. Explosions in buildings can break walls and doors and cause them to be thrown for considerable distances.

Predicting missile formation theoretically is extremely difficult. The number of missiles formed, and the speed and direction of these, are determined by a large number of parameters, each of which has only a limited experimental basis for support. Baker et al (1982) gives a good background for various theoretical aspects. In particular, they point out aerodynamics of fragments can cause them to glide for much longer distances than would be expected from simple ballistic considerations.

Practical missile analyses, for risk analysis work, is empirical. The classic work of this kind is by Clancey who plotted missile sizes and thrown distances, as a function of explosion energy. This was based on experience from bomb raids on Britain 1939-45.

More recently, Holden and his colleagues have reviewed empirical data from 130 pressure vessel and tank ruptures. By far the largest proportion generating missile were BLEVES. Holden and Reeves give:

Fire events	fragments projected	89
Fire events	no fragment projected	24
Non fire events	fragments projected	17

They found typically 1,2,3, and 4 fragments projected for cylindrical vessels with an average of 2.4 to 3 and with very few accidents with more than 4 fragments. By contrast, they found an average of 8.3 fragments for spherical vessels. They found a distribution of fragments by distance which is very similar to that given by Baker et al. The values are given as plots, but have been converted by the author to the following

$$P = .38\ (R/R_{max})^{-.62}$$

where P is the probability density, per meter, for a missile to be found at a given distance R, and R_{max} is the maximum missile range.

Maximum range is determined empirically as

$$R_{max} = (0.4\ M\ H/m)^{0.5}/2g$$

where M is the mass of gas, H is the heat of combustion, m is the missile mass, and g is the acceleration due to gravity.

The directions in which missiles are thrown, is non uniform. Holden and Reeves give a ration of 60:40 for the ration of fragments in two 60^0 sectors along the axis of a cylindrical vessel, compared with those outside these sectors. Fragment distribution from spherical vessels was found to be very of the directional, but the direction not predictable.

The probability of a person being hit, or of a specific process unit being hit by a missile given that an explosion occurs is:

$$P_{hit} = N\ P_d\ A\ /(2\ \pi\ r)$$

where

N is the number of missiles
P_d is the probability density, as above
A is the target area, or the missile area, whichever is largest
r is the distance from explosion ground zero to the target

A is here either the target area, or the missile area, if this is much larger than the target.

Mechanisms by which missiles are cast, are important. The shell of a tank may be broken, and parts thrown out by pressure on its surface. Cylindrical vessels often loose end caps, or split around the circumference. The resulting "end taps" can be propelled like a rocket by ejected liquid. The distances to which such rocketing sections are propelled are generally larger than for other missiles.

Missiles are important, both as a direct threat to persons, and a cause of domino effects. Holden and Reeves cite that most of the 11 persons killed and 10 persons injured among onlookers at the Deer Lake, Pennsylvania LPG tanker BLEVE, were victims of a rocketing tank section.

Missile penetration

Penetration and damage caused by missiles involve extremely complex phenomena. A large number of experimental studies have been performed for military purposes. More relevant for process plant protection are those developed for the design of nuclear plant containment.

The National Defence Research Committee empirical model is described by Kennedy. This describes penetration distance for a flat faced cylindrical missile, end on, into concrete:

for $x/d < 2$ $\quad G=(x/2d)^2$
for $x/d > 2$ $\quad G=(x/d)-1$

where x is the penetration depth in a thick concrete target
d is missile diameter
$G = 27.4*10^{-5}\ D\ d^{0.2}\ \tau^{1.8}\ /\sigma_{c0.5}$
D is "calibre density", $D=M/d^3$
M is missile mass
τ is missile velocity
σ_c is concrete compressive strength

If e is the thickness of the target which can just be penetrated, then

$18 > e/d > 3$ $\quad e/d = 1.32 + 1.24*(x/d)$
$e/d < 3$ $\quad e/d = 3.19*(x/d) - 0.718*(x/d)^2$

Berriaud gives a similar empirical relationship

$$\tau^2 = 1.7\ \sigma_c\ r^{1/3}\ (d\ e^2\ /\ M)^{4/3}$$

here r is concrete density

Barr et al found an agreement with experiments to within 20% for the first formula and 30% for the second.

Hughes reviews the different formulae, and gives modification for shape of the missile. One of the most important of his observations is that penetration and spilling depend not only on the thickness and missile properties, but also on the rigidity of the wall section. A wall which is free to flex suffers less damage than one which is very rigidly supported, at the missile velocities considered. He points out the dangers of using the formulae for small rigidly supported structures.

Penetration of tanks and pressure vessels by missiles is generally more significant than penetration of concrete protection. Impact of missiles on steel plates, such as tank and vessel walls, has been investigated by Nielson (1985) Zaid et al (1973), White et al (1963) Beynett and Plunkett (1971) Calder et al (1971) Miyamoto et al (1979) Woodward (1978) and by Jowett (eg. 1986).

Missiles from BLEVE can travel to over 1 km indicating a missile velocity of about 100m/sec. An empirical equation which have been widely used is the SRI (White) and BRL, both based on energy principles. The SRI correlation is:

$$E = \frac{\sigma d}{10.29}(42.7\, t^2 + wt)$$

where E is the kinetic energy for penetration
σ is the target material ultimate tensile stress
d is the missile diameter
t is the material thickness
w is the plate width
l is missile length
v is missile velocity for penetration
The validity range for the equation is

$0.1 < t/d < 0.6 \quad 0.002 < t/l < 0.05$
$10 < l/d < 50 \quad 5 < w/d < 8$
$8 < w/t < 100 \quad 21 < v < 122$

The equations apply for flat ended cylindrical missiles, sharp nosed missiles have been studied by Miyamoto et al (1979) with an equation for target fracturing as follows

$E = 2.9 * 10^9\, t^{1.5}\, d_e^{1.5}$
$d_e = t\,(1 + 2.9\,(\tan \beta/2)^{2.1})$
$d_e = d$ for cylindrical and spherical missiles, and if $d_e > d$
β is the nose angle of the missile

For both these equations, determining the width of the part of the plate considered to be rigid is a difficult problem. For a plate with close supports (e.g. a reinforced tank) the distance between supports is a reasonable choice. For a simple tank shell, the characteristic deformation diameter seems to be the reasonable choice.

Example

Truck crashes on tanks represent an important group of accidents, both for transport and stationary plant. A chlorine rail tanker has a wall thickness of 9 mm. The equivalent plate breadth is set equal to twice the deformation length for the tank

$w = 2\,s = 2\,\pi\,(R\,t)^{0.5}\,/\,(3\,(1 - v^2))^{0.5}$
≈ 0.5 m

where S is the deformation
t is the plate thickness
R is the plate radius

The structural features of a truck impacting on the tanker, are essentially those of the cab, reinforcement. The equivalent missile is about 5 cm in diameter.

This gives an energy for penetration. according to SRI, of about 10000 Joule. Not all of the truck mass will be effective in the collision, since the truck will deform. If the effective mass is therefore of the order 1000 kg, the penetration speed would be 3.5 m/s. This speed is well below that for validity of the formula. It is also necessary to check the possible energy of deformation of the cab. The calculation nevertheless demonstrates that the truck could readily penetrate the tanker.

References

Strehlow, Luckritz, Adamczyk and Shimpi — The Blast Wave generated by Spherical Flames, Combustion and Flame 1979

Lind and Whitson — Explosion Hazards Associated with Spills of Large Quantities of Hazardous Materials. Report CG-D-85-77, 1977

Knystautas et al — Direct Initiation of Spherical Detonation by a Hot, Turbulent Jet, Seventeenth International Symposium of Combustion, 1979

Van Wijngerden, C.J.M — Experimental Investigation into the Strength of Blast Waves Generated by Vapour Cloud Explosions in Congested Areas, 6th Int Symp Loss Prevention and safety Promotion in the Process Industries, 1989

Lewis, D.J. — Unconfined Vapour Cloud Explosions: Definitions of Sources of Fuel, Prog. Energy. and Combustion Science, Vol. 6. pp121-126

Wiekema — Journal of Hazardous Materials 3 No. 3

Kogarko et.al. — Int. Chem. Eng. 6, 1966

Brinkley and Kirkwood — Determination of Explosion Yields, Loss Prevention 3, 1969

Geiger — Loss Prevention and Safety Promotion in the Process Industries, Swiss Society of Chemical Industries 1980

Auton and Pickles — CERL Laboratory Note RD/1/Nm210/78, 1978)

Baker et al — Explosion Hazards and Evaluation", Elsevier 1983)

Harris, Marshall, and Moppet	The Response of Glass Windows to Explosion Pressures, I.Chem. E Symposium Series no 49
Scilly, N.F. and High, W.G.	The Blast Effects of Explosions(1986) 5th Int.Symp. Loss Prevention and Safety Promotion in the Process industries, Cannes
Clancey, V.J.	The Effects of Explosions, I.Chem. E. Symposium Series No. 25, 1982
Stull, D.R.	Chetah, Loss Prevention 7, 1973
Smith, J.M. and van Ness, H.C.	Introduction of Chemical Engineering Thermodynamics, McG-H 1959
Wagman (1952)	Selected Values of Chemical thermodynamic Properties, Natl Bureau Standards, 1952
Bradley and Mitcheson. (1978)	Combustion and Flame 32, pp 221, - 236, 1978
Lee, J.H et al.	Blast Effects from Vapour Cloud Explosions, Loss Prevention, 1986
Fisher, H.G.	DIERS research program on emergency relief systems, Chemical Engineering Progress, 1985
Bartknecht	Explosions, Springer Verlag 1981
Palmer, K.N.	Dust explosions and fires, Chapman and Hall 1973
Field, P.	Dust Explosions, Elsevier, 1982
Baker	Explosions, Elsevier, 1982
Holden, P.L. and Reeves, A.B.	Fragment Hazards from the Failure of Pressurised Liquefied Gas Vessels, I. Chem. E. Symposium Series No. 93, 1985
Barton and Nolan	Runaway Reactions in Batch Reactors, I. Chem. E Symposium Series No. 85, 1984
B. Rasmussen	Unwanted Chemical reactions in Process Industry, Report Risø-M-2631, Risø National Lab. 1987
Evans, A., and Puttock,J.S.	Experiments on the ignition of dense flammable gas clouds.

	5th Int Symp Loss Prevention and safety Promotion in the Process Industries, 1986
Scilly, N.F, and High, W.G.	The Blast effects of explosions. 5th Int Symp Loss Prevention and safety Promotion in the Process Industries, 1986
Steen and Schampel	Loss Prevention and Safety promotion in the process industries, Inst. Chem Eng. publication No 290, 1983)
Oppenheim	AAA Journal 4 pp 1773 - 1776, 1966
Hughes, G.	Hard Missile Impact on Hard Concrete, Nuclear Engineering and design 77, 1984
Kennedy, R.P.	A Review of the Procedures for the Design of Concrete Structures to Resist Missile Impact Effects. Nuc. Eng. Des. 37, 1976
Barr et al.	Studies of Missile impact with concrete structures, Nuclear Energy Vol. 19, No. 3, 1980
Nielson	Empirical Relation for Perforation of mild steel plates, Int. J. Impact Eng., 3 pp 137-142, 1985
Zaid et al	An Examination of the Perforation of Mild Steel Plate, Int. J. Mech. Sci. 15, 1973
White et al	Containment of fragments from a runaway reactor, SRIA-113, 1963
Beynett and Plunkett	Plate Impact and Plastic Deformation by Missiles, Exp. Mech, 11, 1971
Calder et al (1971)	Projectile impact on an infinite viscosity plate, Int. J. Solid structures, 7, 1971
Miyamoto et al	Experimental study of the strength of steel plates subject to missile impact, 5th SMIRT conf. 1979
Woodward, R.L.	The penetration os metal targets by conical projectiles, Int. J. Mech. Sci. 20, 1978
Houweling, J.	A generalised shock wave model, lost prevention and safety promotion in the process industries, EFCE, Basle, 1980.

13 Organising Risk Analysis Projects

Risk analysis for Process plant design

When organising risk analyses for a new plant, there will generally need to be a number of analysis steps, each specially suited to a different stage in the design process (Fig 13.1). The following set of procedures has been found to be very effective, and is used by a number of companies.

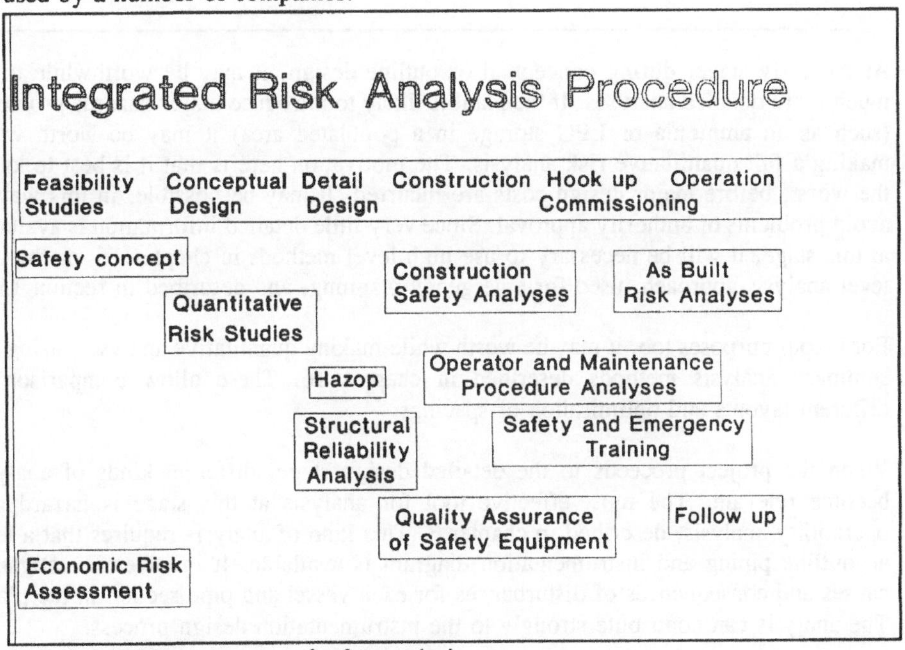

Fig. 13.1 A full programme of safety analysis

At the outset of planning for a new process plant, only very little information is likely to be available, probably just the process description, the flow sheet, and a plant layout or plot plan. Nevertheless, it is at this stage that the greatest benefit can be obtained from analyses, since it is possible to make major changes in design, without incurring large costs or delays. At this stage, a preliminary hazards analysis can be carried out.

This is a rather superficial analysis, intended to identify the main classes of hazards (see chapter 3).It is also possible at this stage to review the process for "inherent safety". Inherent safety is property of some plants, which have very little risk, because the substances or processes they use do not have the physical capability of causing accidents. (Kletz 1984). Table 13.1 gives a short check list which can be used in such a review.

Table 13.1 Some approaches to inherent safety

Reduce inventory
Use low energy low toxicity materials
Use continuous rather than batch reactors
Ideally, use tube reactors
Process dilute mixtures
Use self limiting reactions
Use passive safety barriers
Remote location
Produce hazardous materials where and when they are needed

At the early stage, during conceptual or outline design, it may be worthwhile to do much more detailed analyses. If the plant is likely to introduce major hazards problems (such as an ammonia or LPG storage in a populated area) it may be worth while making a full quantitative risk analysis. The motivation here is that it is best to know the worst, before major design costs are incurred. It may be possible, in this way to avoid problems of authority approval. Since very little detailed information is available at this stage, it will be necessary to use high level methods in chapter 16, or the unit level analysis approach, used for emergency planning, and described in section 13.2.

For layout purposes too, it may be worth while making quantitative analyses, using the summary analysis methods described in chapter 16. These allow comparison of different layouts and optimisation of spacings.

When the project proceeds to the detailed design stage, different kinds of analysis become relevant. The most effective tool for analysis at this stage is hazard and operability analysis, described in chapter 3. This kind of analysis requires that a least an outline piping and instrumentation diagram is available. It involves investigating causes and consequences of disturbances for each vessel and pipe section in the plant. The analysis can contribute strongly to the instrumentation design process.

Quantitative risk analysis can be used, for example to compare two designs, to decide whether a single shutdown valve or doubles valves are required etc. Such analyses are generally very limited in scope. Full scale quantitative analyses may be carried out too, though this in unusual for on shore plant. For North Sea oil and gas production platforms, such analyses are required for authority approval. Choice of approach here is often a question of how much is at state. An error in design for a large oil production platform, requiring later rework in order to meet authority requirements, can be extremely expensive.

Once designs are complete, operating procedures may be developed, and software for monitoring and control may be configured. At these stages, human error and software analyses <u>should</u> be performed. The largest numbers of design errors and problems are generally found here.

During commissioning, or prior to initial plant start up, it is very worth while to apply the safety audit procedures, as described in Chapter 17.

The results from a study of risk analysis during the design of a process plant are shown in fig. 13.2. As the design proceeds, an increasing number of problems are found, but even with the best analysis techniques, some problems will remain, to cause accidents during operation.

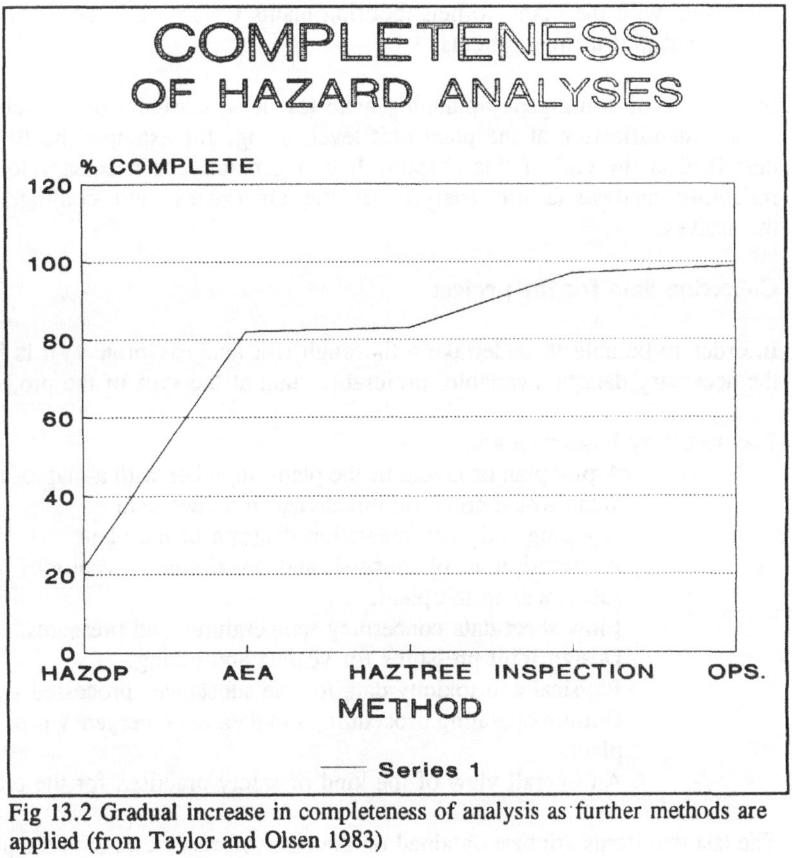

Fig 13.2 Gradual increase in completeness of analysis as further methods are applied (from Taylor and Olsen 1983)

Risk analysis for emergency planning

Risk analysis provides an ideal first step in emergency and contingency planning.The scope of the analysis will depend to some extent on the objective, but even more on the risks involved. The ideal should be to analyse just those risks which eventually turn out to be significant, and no more.(Of course this is only roughly possible, and then only with a considerable amount of judgement.) Limitation of scope is necessary, not only to reduce costs, but also to avoid drowning the analysis in detail.

Generally, the largest plant units with the highest inventory of hazardous substances should be analyses. It is a good idea to make a few trial consequence calculations, particularly of lesser hazards, in order to define the limits of interest. However it may be necessary to add units with smaller inventory, if they present new types of hazard, or if the probability of accidents in these units is especially high.

In addition to considering the plant itself, attention should always be paid to transport associated with the plant, to neighbouring plants which might be involved in domino effects,and to drainage systems.

In the case of contingency planning analyses, it will usually be sufficient to make a hazard identification at the plant unit level, using, for example the PHA procedure, described at the end of this chapter. It will generally be necessary to make a good reliability analysis/failure analysis of the emergency and contingency measures themselves.

Collecting data for the project

In order to be able to undertake a thorough risk analysis project, it is important that the necessary data be available, preferably right at the start of the project.

The necessary basis data are:

- A plot plan or layout of the plant, together with a map of the population areas which could be threatened in an accident.
- A piping and instrumentation diagram of the plant.
- A description of normal and maximum inventories of hazardous substances in the plant.
- Flow sheet data concerning temperatures and pressures.
- Design limit pressures for vessels and piping.
- Physical and toxicity data for the substances processed and produced.
- Outline operating procedure, and detailed emergency procedures for the plant.
- An overall view of the kind of safety practices for the plant

The last two items are best obtained on the basis of interviews with the plant operating staff.

Hazard Identification

For a contingency planning project, the preliminary Hazard Analysis procedure, PHA, as described at the end of this chapter, is generally sufficient to identify the starting points for the worst risks. Alternatively a hazard classification tree analysis, as described in Ch. 3, may be carried out.

For tasks requiring more detailed work, hazard and operability analyses should be carried out. The methods described in Ch 3 are suitable. For large projects, various computer aids are available. Computer methods should not be used however until the manual methods have been mastered. and can produce safety barrier diagrams as a result. For existing plant, action error analyses for critical operations are desirable.

For existing plants too, an onsite inspection of the plant should be made as part of the hazard identification procedure.

Selection and description of scenarios

In order to be able to understand an accident, it is necessary to describe all of the significant steps, from initial cause through to final emergency treatment, and clean up.

The hazard identification provides a basis for knowing the kinds of releases, explosions, fires, etc which may occur. However, if each of these is analyses in detail, the amount of work involved is very large. It is usual to select the most important releases and to group releases.

For example, it will generally be necessary to include an analysis of rupture and major leak from each of the major vessels in a plant. Only if these give similar results will it be possible to group them. For leaks and ruptures of piping, grouping will be necessary. Pipes with similar operating pressure and contents should be grouped. For long pipes, leaks at just two or three places will be chosen.

It is necessary to define the size of each release. There are some standardized sizes defined in terms of small sizes which are convenient to use, because failure rate data have been collected for these. These diameters are

	diameter inches	diameter mm
very small	0 - 0.15"	0 - 3 mm
small	0.15 - 1"	3 - 20 mm
medium	1" - 3"	20 - 75 mm
large	3 inches and over	75 mm and over or 20% of diameter
rupture		full diameter

For each of the releases, it will be necessary to draw an event tree, or, preferably, a consequence diagram. For emergency and contingency planning purposes this should include at least:

- the release itself
- automatic and manual alarms
- manual and automatic shutdown
- automatic fire protection, sprays and deluge systems, steam and water curtain operation.
- mustering of staff
- general alarm, shut off of ventilation
- rescue
- emergency action, fire fighting
- remote shut down, accident control
- cordoning the area
- evacuation and hospitalization
- clean up.

A picture of a simplified contingency plan consequence diagram is given in fig. 13.3

Consequence calculations

For each of the scenarios, a complete set of consequence calculations will be needed, taking into account the safety and contingency measures provided, and also the possibility of failure of these measures.

In making the calculations, it will be necessary also to take into account different wind speeds and stability conditions.

The possibility for confusion is very large. In order to avoid confusion it will be necessary to make the case names carefully. For example

Tank 101 small leak 2m/s N

where the last two terms refer to wind speed and stability conditions . For the consequence calculations it will be necessary to obtain substance data for all of the substances involved (from the data base for example), to obtain concentration limit data, such as LC_{10} and LEL values (see below).

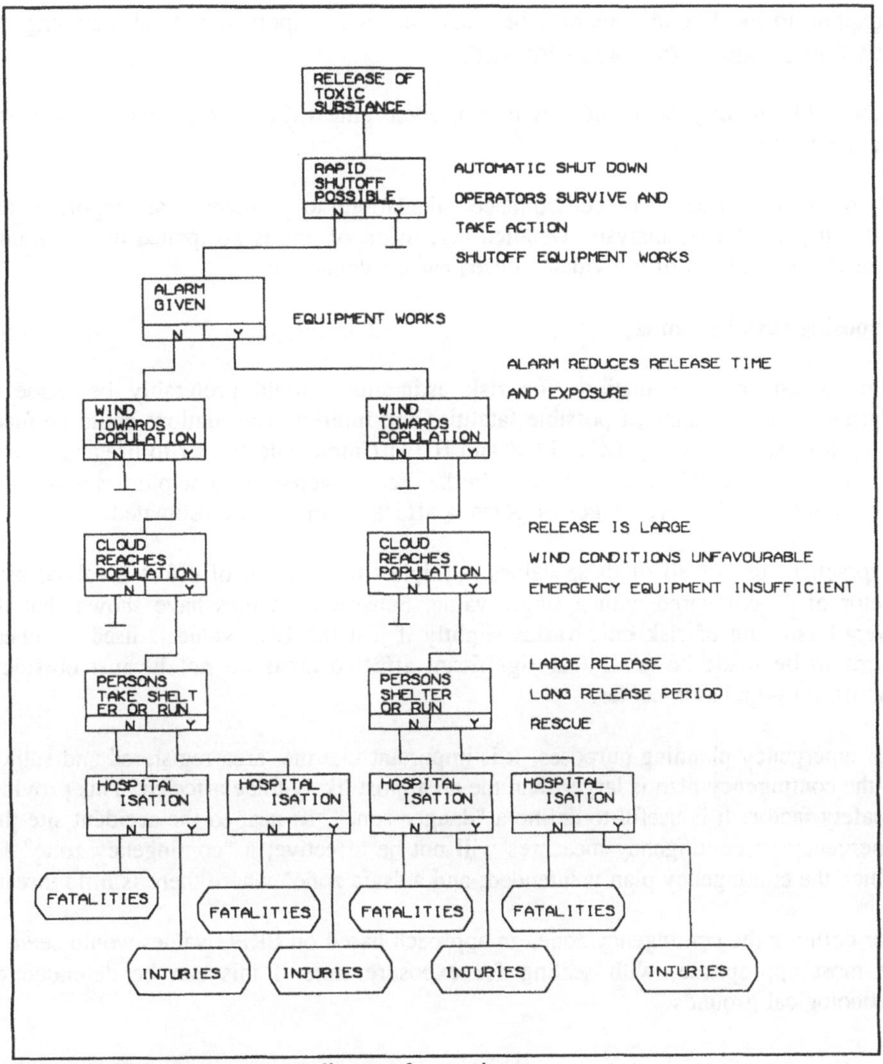

Fig 13.3 Typical consequence diagram for contingency

It will be necessary to chose wind speed data and atmospheric stability conditions for gas cloud and plume dispersion calculations. Typically three sets of values will be chosen eg.

Low wind speed areas. m/s	High wind speed areas M/S
2	2
4	5
6	8

Adaption to local conditions may be made, but it is important for risk mapping that only a few standardized values are used.

U,N and S stability calculations will be needed, unless the consequences of accidents are clear without this.

Choosing the sequence of consequence calculations to perform is an important and difficult part of risk analysis. Detailed discussion of this is postponed however until after the description of individual consequence calculations.

Choosing toxicity limits

The consequence calculations for risk estimation should preferably be made to represent the full range of possible fatalities and injuries. Calculations could be made using for example, LC_{50}, LC_{10}, LC_1, and IDLH (Immediate danger to life and health) values, or the ERPG values (Ch. 10). In the areas covered by toxic plumes with these concentrations, the percentages of persons affected can then be estimated.

In practice, use of all of these values multiplies the amount of effort involved by a factor of 4, compared with a single value. Sensitivity studies have shown that the overall estimate of risk only varies slightly if just the LC_{10} value is used. A check needs to be made however that significant affected areas do not lie just outside a significant population center.

For emergency planning purposes, it is important that the area registered and subject to the contingency plan is larger than the area most likely to be affected. This provides a safety factor. It is useful to define a "danger zone", so near to the accident site that emergency or contingency measures will not be effective; a "contingency zone" for which the contingency plan is intended; and a "safe zone" where there is little threat.

For defining the contingency zone, an approach based on ERPG values would seem to be most appropriate, with scaling for exposure time if this can be defended on toxicological grounds.

Assigning probabilities and frequencies

In order to be able to evaluate risk in a balanced way, it is necessary to calculate both probability and consequence.

To obtain frequency values, for the initial release at the unit level the data given in table 13.2 can be used. More detailed calculations can be made using the data given in Ch. 5. For hazop and action error analyses, it is necessary to make the calculations in detail.

For emergency and contingency planning, placing probabilities on the scenario event trees is a very important step. Probabilities here can in part be determined from the failure rate data given in Ch. 5, with the appropriate assumptions about repair times,

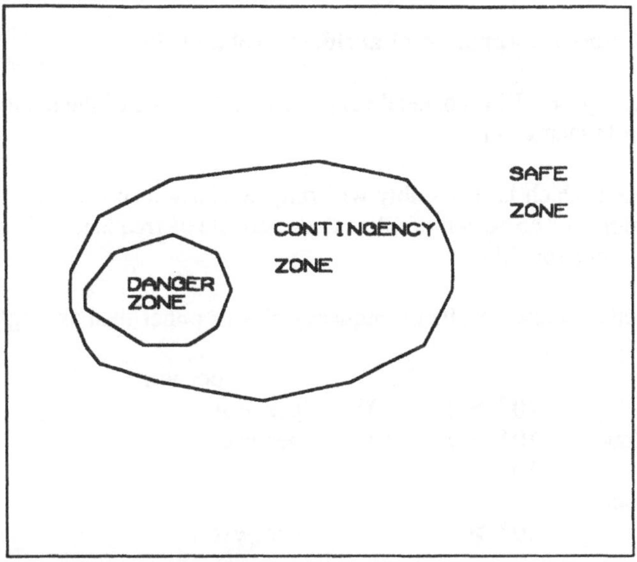

Fig 13.4 Accident zones

and from the human error chapter. For ignition probabilities a table is given in chapter 12 (explosions).

Note that unconfined explosion probabilities will depend on the amount of gas present. For the following amounts of gas in a plume or jet, explosion appears to be possible.

Gas type	Amount
Hydrogen, acetylene	small amounts
ethylene, propylene	200 kg.
pentane, hexane vapour	500 kg.
ethane, propane, butane	1000 kg.
methane	5000 kg
gas with oil mist	500 kg.

For clouds containing these amounts, or more, there is a possibility of explosion. However this possibility does not is not always actualized. On a historical basis, only about 10% of clouds which could explode actually do so (the percentage is higher for hydrogen and acetylene).

TABLE 13.2 Rough estimation of accident probabilities.

1. Accident probabilities depend very much on the state of the plant, management and on human error.

2. For plant which is reasonably will run, regularly inspected, and has no specia operational or corrosion problems, an estimate of frequency of accidents can b given, very roughly.

3. It is better to give results in frequency classes rather than as single numbers eg

high	$F > 10^{-2}$	per year
low	$10^{-2} > F > 10^{-4}$	per year
very low	$10^{-4} > F > 10^{-6}$	per year
rare	$10^{-6} > F > 10^{-8}$	per year
extremely rare	$10^{-8} > F$	per year

4. Poor management can increase accident frequencies by a factor of 30. Very good management can reduce frequencies by a factor of 10

	Equipment type	Accident frequency per 10^4 years
A	Pump, large seal leak	31
B	Pump, rupture (=large leak)	4
C	Compressor, large leak	76
D	Compressor, rupture (=large leak)	8
E	Knock out drum, storage drum, including	
	large leak	2
	valves and piping rupture	0.1
F	Pressure vessel, including valves and piping	
	large leak	0.1
	rupture	0.01
G	Column, including valves and piping	
	large leak	3
	rupture	0.1
H	Heat exchanger	
	large leak	1
	rupture	0.1
I	Storage tank, Cone roof	
	Rupture	0.1
	fire	1
	overflow	100
J	Storage tank floating roof	
	fire	10

	Equipment type		Accident frequency per 10^4 years
K	Batch reactor		
	burst disc rupture,*		100
	overflow		100
	large leak		4
	explosion		0.1
L	Continuous exothermic reactor explosion		0.5
M	Pipe line, buried,	8"	30 per km
	large leak	26"	5 per km
N	Pipe line, land near road		
O	Transport truck, crash release, pr km.		0.0026
P	Pipe bridge, rupture due to crash		10

Sources of data.

1	Assessment, based on API data for one year, 90 refineries	A,B,C,D,E,F,G,H,I,J,L
2	J.R. Taylor, estimate based on operating experience for three chemical companies over 20 years.	K
3	de la Mare RF, and Andersen, Ø Pipeline Reliability, report No. 80-0572, det Norske Veritas, 1980	M,P
4	Estimate based on crash frequency calculations for 4 roads in Venezuela, J R Taylor 1987	N
5	US Data, Hazards of gasoline delivery by truck, Battelle Inst.	O

TABLE 13.3 Probability of failure of emergency measures (rough estimates)

	Probability of failures
Pr. incorrect human action in emergency without training.	0.3
Correct functioning of alarm, periodically maintained (1 month inspection)	0.01 to 0.05
Correct functioning of ESD (typical)	0.03 to 0.05
Correct functioning of deluge	0.01 to 0.05
Incorrect human action, in emergency, with training	0.03 to 0.1
Arrival and operation of fire brigade, with minimum response time N min.	N min. 1.0 N + 2 min. 0.8 N + 5 min. 0.3 N + 10 min. 0.1 N + 20 min 0.001
taking mechanical and access problems into account.	
Ignition in general plant area, without special ignition sources.	.1
Ignition in area with furnace or flare.	1.0
Ignition in zone 1 area (taking failure of equipment into account).	.01
Ignition from large seal leak	0.1
Failure of untrained person to follow emergency plan	0.5

Additionally, it will often be necessary to take situation geometry into account. Wind direction probabilities for risk mapping are automatically taken into account by risk mapping programs (see Ch. 16.). If sizes are very dependent on wind speed, then the wind speed the wind speed probabilities need to be taken into account when calculating event tree probabilities.

Jet fire probabilities and radiation patterns need to be taken into account on a geometrical bases e.g.

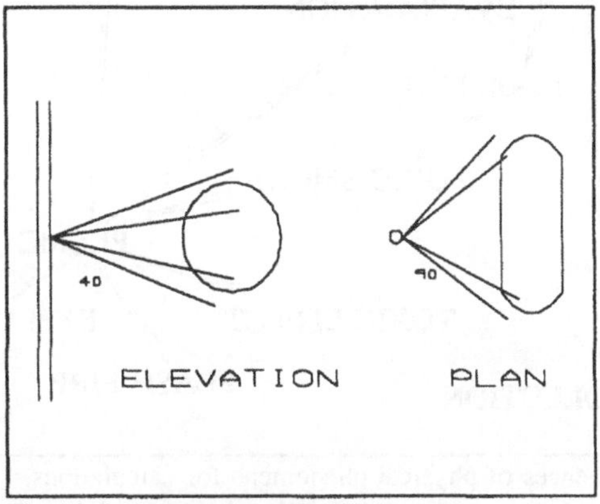

Fig 13.5 Geometric probabilities

$$\text{Geometric probability} = 98^0/360^0 * 40^0/360^0 = 0.05$$

Small sketches may be needed to obtain a reasonable geometrical estimate, taking both jet dimensions and radiation patterns into account.

Planning the Consequence Calculations

One of the most difficult aspects of consequence calculation for risk analysis is to match all of the various phases of the calculation to each other. The problems are particularly difficult if the substance which escapes in an accident is a liquefied gas, since the way in which the liquid turns into gas will determine the concentration, density, and temperature of the release, and therefor the area affected. A general view of the way in which the individual calculations interconnect is shown in fig 2.1.

Seven different cases accident cases are described here, and the sequence of calculations is given for each. These cases cover the most common accident situations in petrochemical plant.

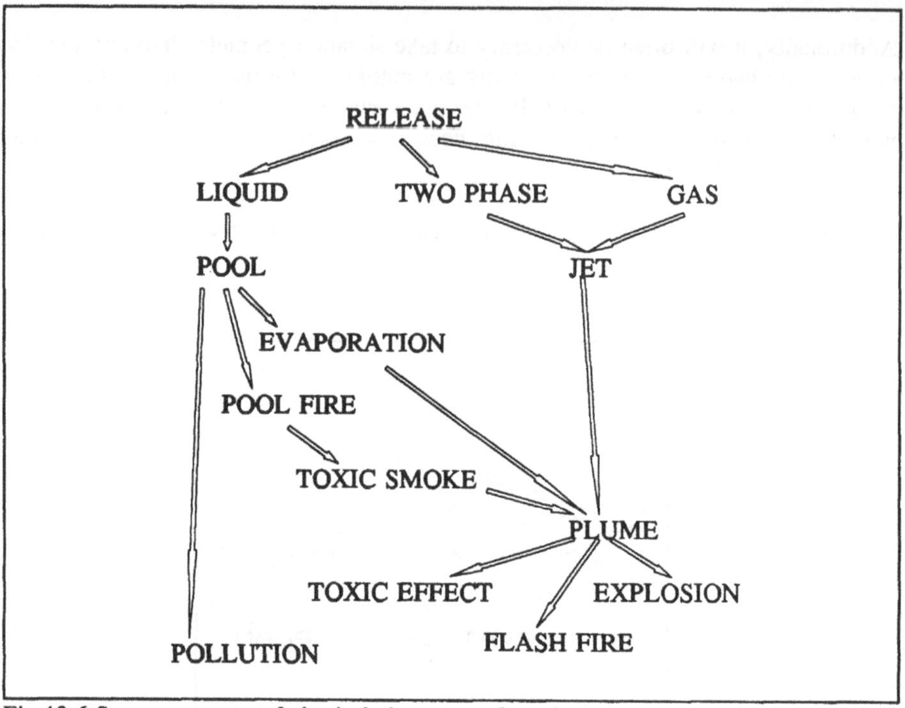

Fig 13.6 Some sequences of physical phenomena for calculations.

1. Rupture of a vessel containing a liquefied gas under pressure and at ambient temperature, or at a temperature significantly above the boiling point (more than 8 deg C)(see fig 13.7)

2. A hole in the gas phase of a vessel containing liquefied gas under pressure.

3. A leak from a vessel or pipe containing compressed gas.

4. A leak in the liquid phase (below the liquid surface) from a vessel containing liquefied gas above its boiling point. (see fig 13.8)

5. A leak in a tank or vessel containing a liquefied gas at cryogenic temperatures, i.e. at or below its boiling point.(see fig 13.9)

6. A leak in a tank, pipe or vessel containing a liquid at or above its boiling point.

7. A release of liquid which is above its boiling point, from a pressurized vessel.

8. A release of liquid (e.g. oil) from a pipeline.

9. A release of liquefied gas from a pipe.

10. A release of gas from a pipe, or from a gas well.

1. Rupture of a vessel containing liquefied gas

When a rupture or a large hole (larger than about 1/10 the vessel diameter) is created in the wall of the vessel, the pressure in the vessel falls and the liquid inside boils. If the liquid is more than about 6°C above its boiling point the boiling is very violent. The liquid is thrown out as a spray of droplets and gas, and mixes violently with air.

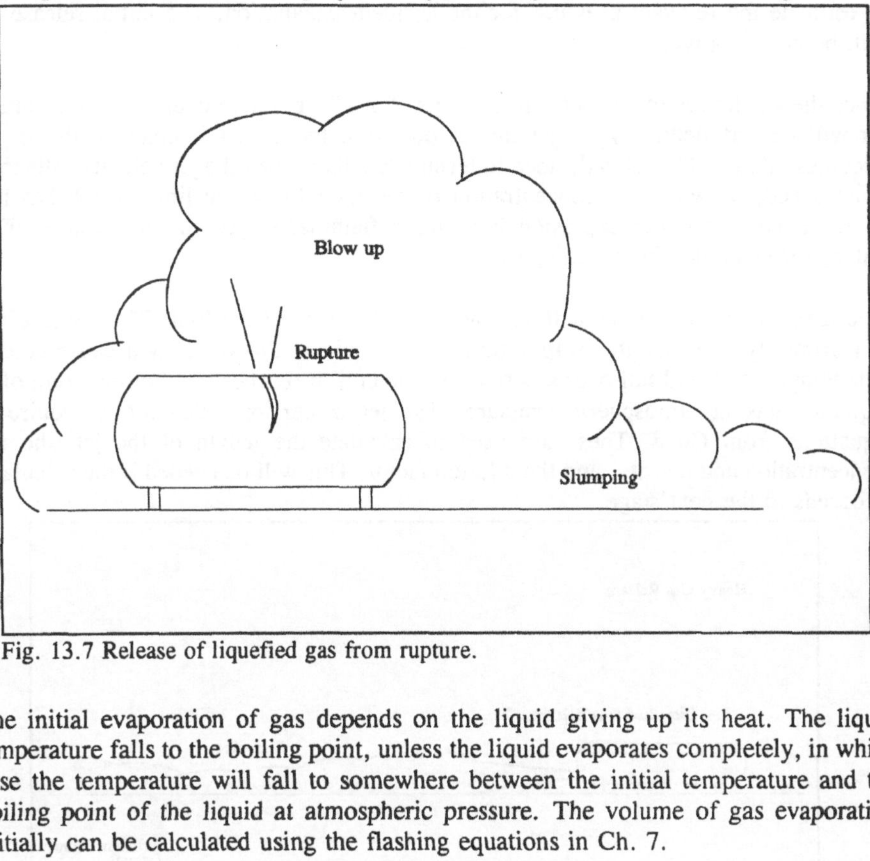

Fig. 13.7 Release of liquefied gas from rupture.

The initial evaporation of gas depends on the liquid giving up its heat. The liquid temperature falls to the boiling point, unless the liquid evaporates completely, in which case the temperature will fall to somewhere between the initial temperature and the boiling point of the liquid at atmospheric pressure. The volume of gas evaporating initially can be calculated using the flashing equations in Ch. 7.

The mixing of gas with air causes the air to cool, as it gives up heat to the unevaporated liquid. The result is a cold mixture of air, gas, and mist. The amount of air mixed into the cloud is about 10 times the initial volume of evaporated gas.

The resulting mixture of gas and air will be heavier than the surrounding air, in almost all cases. This applies even for a gas which is generally less dense than air, since the resulting cloud is primarily a mixture of cold air and gas.

2. Hole in the gas phase of a vessel containing liquefied gas

A hole in the gas phase of a vessel containing liquefied gas under pressure will behave differently from the rupture case described above, especially if the hole has a size less than about 1/20 of the cross section of the vessel, or if the hole is in a long pipe attached to the vessel. Generally the leak will be pure gas.

For the initial rate of release the Gas release equations from Ch 7. Note that as gas is released the liquid in the vessel will evaporate to replace the lost gas, and will cool as a result. The pressure in the vessel will therefore drop. Judgement will be required to estimate the relevant pressure for the accident calculation. The initial release rate will be conservative.

Once the gas leaves the vessel it forms a jet. For all pressures over about 1.2 bara the jet will be turbulent. As the jet moves out from the vessel it mixes with air, and becomes diluted. The jet is deemed to terminate either when the jet velocity falls to the wind speed, or when the concentration of the jet falls to the limit which has been given, either the lower explosion limit for a flammable gas, or the LC_{10} or IDLH value, for example, for a toxic gas.

The temperature of the gas will be low, partly because of the Joule-Thompson effect, but primarily, because the evaporation of the liquid in the vessel will cause cooling. The temperature will fall over a period of time until it reaches the boiling point of the liquefied gas at atmospheric pressure. The jet dispersion calculations require the equations from Ch 8. These are used to calculate the length of the jet, the final concentration and density, and the dilution factor. This will be needed if the calculation proceeds to the next stage.

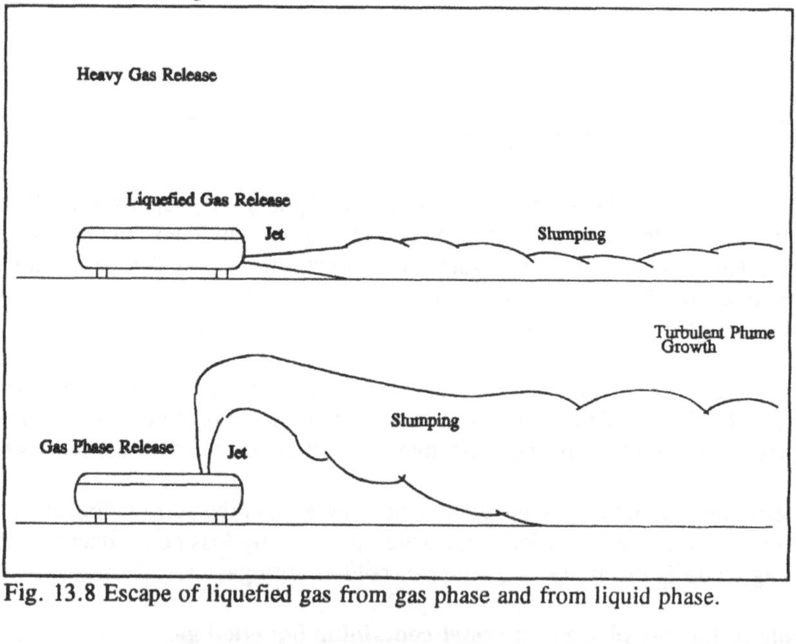

Fig. 13.8 Escape of liquefied gas from gas phase and from liquid phase.

If the end of the jet is reached before the concentration of gas has fallen to the given concentration limit, the gas will disperse as a plume. The form of the dispersion will depend on the gas involved. For light gases (notably ammonia, methane, and hydrogen) the light gas dispersion calculations from Ch.10 provide a suitable calculation. For gases which are heavier than air, taking temperature and the inmixed air into account, one of the heavy gas plume dispersion programs will be required.

3. Escape of compressed gas

The sequence of physical processes here is similar to that for case 2.
However, it is more likely that the gas will be lighter than air, since the cooling effect of evaporation is absent. The sequence of calculations is:

gas release
jet dispersion
light plume dispersion for a gas lighter than air at the actual temperature
heavy plume dispersion for a gas air mixture heavier than air at the actual temperature

As gas is released the pressure will drop in the vessel. It may be necessary to repeat the calculations for several pressures if the pressure falls quickly.

4. Leak from the liquid phase of a vessel containing liquefied gas

The immediate effect of the leak is a flow of liquid from the vessel. This is well represented by using liquid release equations, if the leak is directly from the vessel, or the two phase pipe flow equations if the leak is from a pipe. The flow rate is determined from the pressure at the surface of the liquid (the vapour pressure of the liquid at the actual pressure) the head of liquid between the liquid surface and the hole.

After leaving the pipe the liquid will flash, some of the liquid turning to vapour, the remainder becoming colder. reaching the boiling point of the liquid at atmospheric pressure, unless all of the liquid evaporates. The flash equations calculate the proportion of liquid which evaporates.

Once the liquid has flashed, the velocity will increase significantly, by comparison with the initial liquid velocity (the momentum remains constant however, so the liquid velocity can be used to calculate the momentum jet dispersion).

If the jet has a high velocity, and the amount of liquid compared with the amount of gas is low, the jet will disperse the remaining liquid as a mist. With more than about 4 times as much liquid as gas, then there will be a tendency for the liquid to "rain out". The amount of mist, and the gas air mixture density is then reduced by the amount corresponding to the amount of liquid. The factor 4 times the amount of gas is somewhat high, compared with observations, and is therefore considered to be conservative.

After leaving the pipe the gas liquid mixture will mix with air, and will fall in concentration. The length of the jet and degree of mixing may be calculated using the jet dispersion calculation, and subsequent dispersion calculated using a heavy gas plume dispersion program, as for case 2.

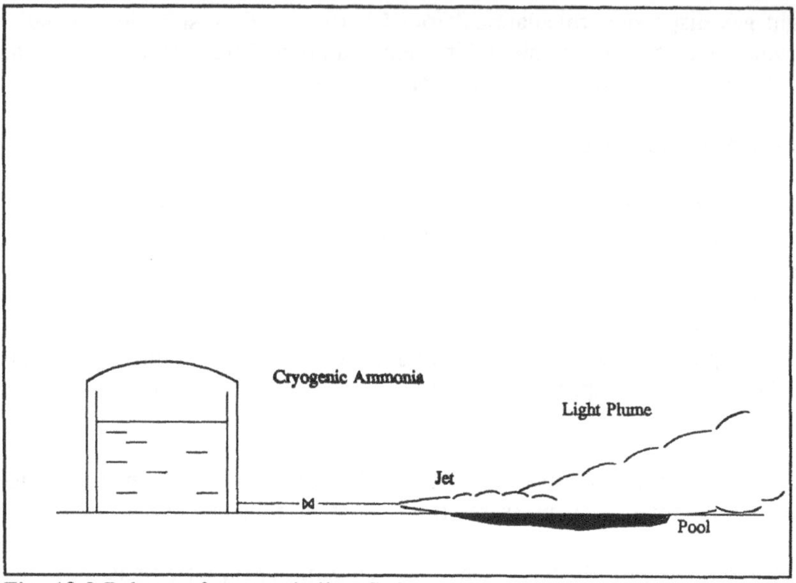

Fig. 13.9 Release of cryogenic liquefied gas.

5. Escape of cryogenic liquefied gas

If the liquefied gas which is released is at or below its boiling point, then there will be no immediate flashing when the liquid is released. The liquid will spread as a pool on the ground, and will boil as it absorbs heat from the ground.

The release rate for the liquid can be calculated using the equations for liquid release from a pipe. The evaporation rate may be calculated using the cryogenic evaporation equations in Ch 8. It will be necessary to determine the maximum pool size, if any, depending on the presence of any walls, bunds, drains etc., since the rate of evaporation depends on the area of ground covered. At a very late stage in the release (tens of minutes or hours), the conduction from the ground ceases to be significant, heat transfer from the wind or by radiation from the sun becoming more important, as the ground cools.

The dispersion of gas can then be calculated using the equations for light gases (ammonia,methane, hydrogen) and a heavy gas plume dispersion program for gases heavier than air. Note that there is very little initial mixing with air at the surface of the cryogenic pool, so that the starting point for the plume calculations is an undiluted gas release.

6. Release of volatile liquid

When a volatile liquid is released, it spreads over the ground, until it meets a barrier such as a ditch, wall, drain or unevenness. The liquid will evaporate in proportion to the heat transferred from the atmosphere or by solar radiation.

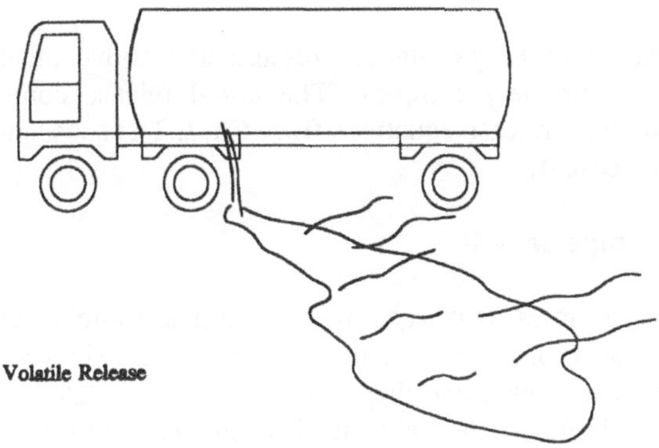

Fig 13.10 Release of volatile liquid.

The rate of release can be calculated using the liquid release equations. The spread of the liquid must generally be determined by judgement of the nature of the surface, but for smooth surfaces the equations for flow over a surface given in Ch. 8 or Ch. 12 may be used. The rate of evaporation may be calculated using the equations for volatile evaporation in Ch.8, and the dispersion using a heavy gas plume dispersion program.

7. Release of liquid above its boiling point

Liquids such as hexane, gasoline which are above their boiling point (for example during distillation or extraction) will behave in a very similar way to liquefied gases. The sequence of calculations corresponding to cases 1 and 4 may be used, although the fact that temperature is reduced on mixing with air needs to be taken into account.

8. Release of liquid from a pipeline

For releases of liquid from a pipeline, the initial flow rate depends on hole size. For a small hole, the flow rate may be calculated from the pipeline pressure. For a rupture, the release rate is given by the pumping rate until the pumping is stopped. After this the flow rate depends on the height difference between the local high point on the pipeline and the point of rupture. The release rate can be calculated using the methods for pipeline release in Ch. 7.

The later steps in the accident calculation can then be calculated as for case 7.

9. Liquefied gas pipelines

For pipelines containing liquefied gas under pressure and above the liquid boiling point, the calculation is extremely complex. The initial release conditions can be calculated using the two phase release equations from Ch.7. The subsequent stages can then be calculated as for case 4.

10. Release from a gas pipe or well

For calculation of release rates from gas wells requires finite element or finite difference programs has been provided. This allows two casing sizes to be taken into account. The subsequent gas release and dispersion may be calculated using turbulent jet dispersion equations (Ch. 9) and the light gas dispersion equations or a heavy plume dispersion program, depending on whether the gas is light or heavy.

11. Fire and explosions

The previous calculations may be use to determine the sizes of gas cloud, plumes, and liquid pools. The results from these may be used more or less directly in the fire and explosion calculations. For clouds and plumes it is necessary to know the amount of gas in the cloud above the lower explosion limit. This is calculate from the length of the plume, wind speed and release rate for plumes (continuous releases), and is the complete inventory of gas, for the clouds (instantaneous releases).

For pool fires the pool size will need to be estimated. Fire size and radiation areas are then determined using the equations in Ch. 11.

Using the risk analysis results in emergency planning

The results of the risk analyses described above can be used fairly directly in emergency planning. The most obvious use is to determine the size of the areas affected, and the number of persons at risk, in each type of accident. The risk analyses can be used for much more than this however. In particular the progression of the accident, and the timing, can be plotted alongside the timing of emergency actions. This allows the needs for early mobilisation, public audible alarms, cordoning(blocking roadways) and the needs for rapid hospitalisation to be determined. It is also feasible to use the methods to determine the effectiveness of emergency measures.

The procedure described below has, at the time of writing, been applied to a full range of hazard scenarios for three cities, and is in use in planning for two more. The results have so far identified several areas where large improvements in emergency plans could be made, within the existing resources. The steps in the procedure are:

Identification of possible releases (as described above)

Evaluation of detection and release <u>times</u>

Evaluation of areas affected by releases (as described above)

Evaluation of effect of emergency actions
- emergency limitation
- warning, rescue, evacuation
- hospitalisation

Timing of emergency actions is important, and can in some cases be critical.The emergency actions taken into account are:

1. Detection of the release
2. Identification of the release
3. Initial attempts to shut off the release
4. Public alarm
5. Cordoning
6. Firefighting
7. Secondary shut off attempts
8. Rescue
9. Hospitalisation
10. Social help
11. Clean up.

It is important to recognise that an emergency action only rarely goes according to a detailed plan. Individual emergency actions may be delayed, or fail, and circumstances may prevent application of the detailed plans. Detailed plans are therefore the basis, or skeleton, which is used as a basis for later improvisation. For this reason, an emergency action is rarely a complete failure, even though individual aspects of the operation may require additional effort.

The evaluation of emergency preparedness is based on the hypothesis of a number of "scenarios". The scenarios describe the sequence of events for each accident type. Additionally, a variety of wind conditions and timing for the individual accident types is important. for the present analysis, the following variants are taken into account:

Windspeed	one low and one normal wind speed
Atmospheric stability	neutral and stable
Wind direction	8 compass directions
Timing	night, day, rush hour

To support the evaluation, cause consequence diagrams have been constructed for each scenario. The two standard diagrams, shown here, can be modified as necessary for the individual cases.(Fig 13.11 and 13.12). (It is useful to mark the main event sequences for each scenario onto the diagrams, and the standard diagrams save a great deal of time, compared with drawing new diagrams for each case.)

For each scenario, the affected area is calculated, and plotted on maps. The effects of a chemical accident depend as much on where it occurs, as it does on timing. In order to assess the number of persons who could be affected, the area was photographed at varying times during the day. This gives a basis for estimating the number of persons exposed. The most effective way of doing this is to use aerial photographs, supplemented by street photographs. (Overflying of the plants themselves is often prohibited, and in many cases where it isn't prohibited, the risk analysis will show that it should be).

Notes to diagram 13.11

The diagrams here are relatively simplified when compared with the actual situation. The diagrams nevertheless adequate for the overall evaluation of effectiveness of emergency measures. However, to judge whether the individual emergency measures will be effective, it is necessary to go into much more detail, see appendix A.

A.
The initial release size for a chemical release depends on the reason and mechanism. Background data for estimating frequencies for this type of event are given in reference 2. Four classes of release size are distinguished

- small
- medium
- large

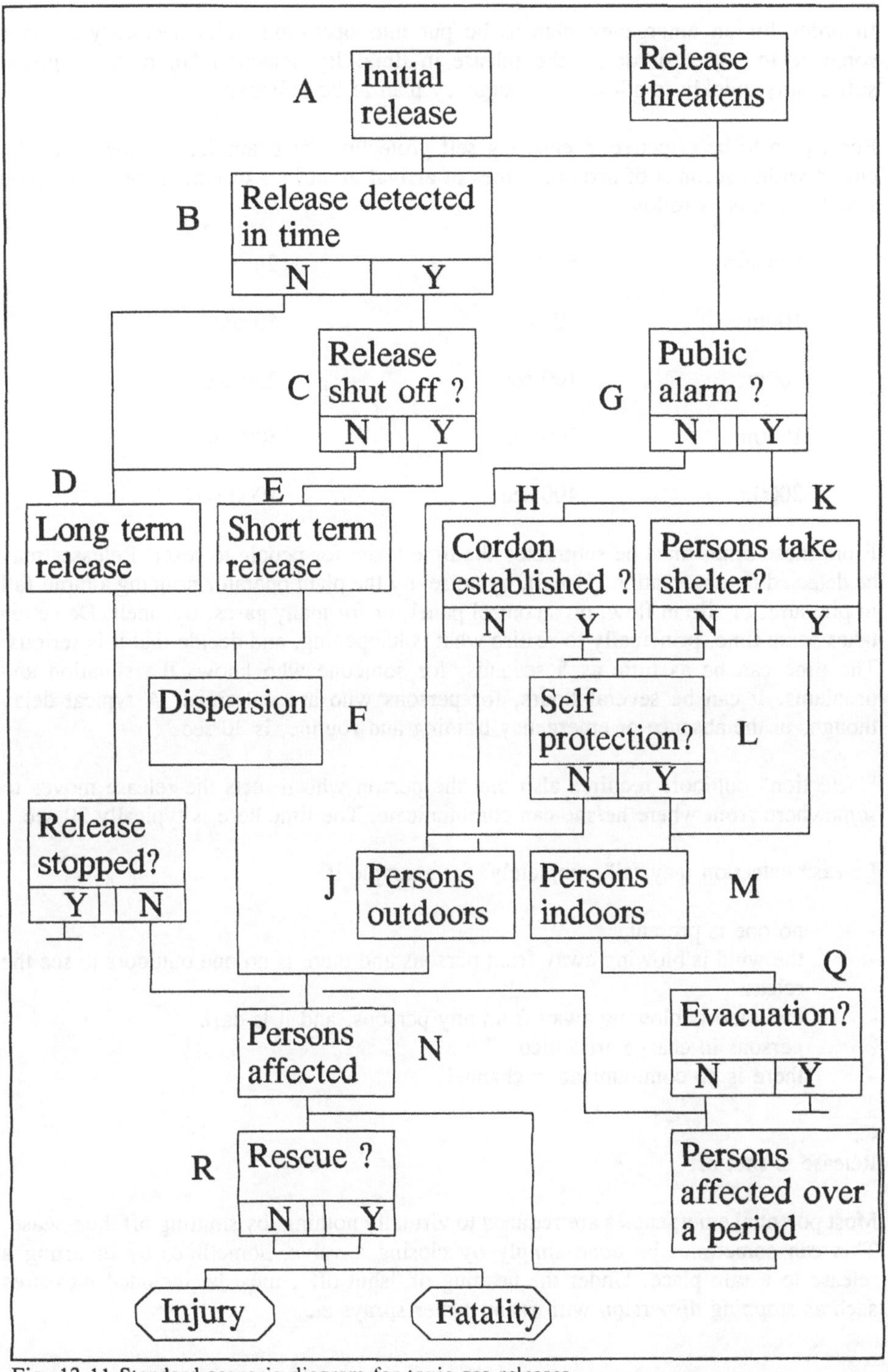

Fig. 13.11 Standard scenario diagram for toxic gas releases.

B.
In order for an emergency plan to be put into operation, it is necessary to have someone to notice or detect the release in time. By detection "in time" is meant sufficiently quickly to allow the emergency plan to be effective.

For a plan to be effective in ensuring self protection for example, windows must be closed within seconds of arrival. Times of arrival at various distances for wind speed 2 and 5 m/s is as follows

Distance	5m/s	2m/s
100m	20 sec	50 sec
500m	100 sec	250 sec
1000m	200 sec	500 sec
2000m	400 sec	1000 sec

From these times must be subtracted the time taken for people to react. Releases may be detected by observation (visual), by noise, by the plant operator noticing a rapid fall in pressure, or rise in flow, on a control panel, or for many gases, by smell. Detection takes some time, principally to realise what is happening, and decide that it is serious. The time can be as little as 5 seconds, for someone who knows the situation and problems. It can be several hours, for persons who are untrained. A typical delay though, in the absence of emergency training and routine, is 30 sec.

"Detection" outdoors requires also that the person who detects the release moves to somewhere from where he/she can communicate. The time here is typically 30 sec.

Release detection may fail completely, for example if:

- no one is present
- the wind is blowing away from persons and there is no one outdoors to see the release
- the wind is blowing away from any persons, and it is dark.
- persons in charge are killed
- there is no communication channel.

C.
Release is shut off

Most potential emergencies are reduced to virtually nothing, by shutting off the release. This can sometimes be done simply by closing a valve, sometimes by diverting a release to a safe place. Under the heading of "shut off", must be included measures such as stopping dispersion with foam, water sprays etc.

D.
A release is "long term" if it cannot be shut off until gas plumes are fully developed, and spread into houses. For houses with open windows, it takes only a few seconds to fill the house. (Typically 2 minutes if windows are ajar, 20 seconds if windows and doors are fully open).

E.
A release is short term, if it can be stopped before becoming fully developed and entering houses. A lignified gas tank rupture will often give a short term, very large release, initially, followed by a long term release, as small pockets of liquid evaporate.

F.
After the release, gas disperses as a cloud (roughly circular initially, but gradually spreading in the direction of the wind) or as a plume.

The "size" of the cloud or plume is determined by the concentration limits which are considered dangerous. This in turn depends on the length of time of the release, on whether persons are protected indoors, and on the ability of the individual to withstand the gas and on the toxic.

G.
It may be possible to protect the public, by getting them to take shelter, by getting them to shut windows, and by preventing them from entering a hazard zone.

The effectiveness of this depends on the speed with which it can be done, compared with the spread of the plume.

The only method of alarm which is effective int he case of an actual accident is an audible alarm/siren. This can be activated typically within 30 seconds of a release. People take a minimum of an additional 1 min. to take shelter. There is always a problem of range of the alarm.

Touring on area in cars with loud haulers cannot be instituted fast enough to have any effect for seeking shelter, no matter what the wind speed, if the accident has already occurred. It may be very valuable though, if a release threatens (e.g. a gradually rising pressure, a fire which is developing and releasing fumes, or threatening a gas storage, or a leak which is getting worse.

Public alarm has an effect later, if the release is long term, in preparing persons for evacuation.

H. Cordon established.

One of the most effective measures for protection is to prevent persons entering an area which is hazardous or which may become hazardous.

This can be done by setting up barriers, or stopping traffic with a police can post, if there is time. There may be sufficient time for this if there are just a few routes, and the release is long term.

A technique which has been used is to set up traffic lights on a roadway, to stop traffic when the release occurs (examples are Moron ammonia plant and Puerto Cabello Refinery).

Persons take shelter by going indoors and shutting windows, in some cases by going upstairs. If the plume has small width and is visible, people can take shelter by walking away from it.

J.
Persons outdoors. The number of these may be estimated by taking photographic surveys of an area, at different times of the day and year.

L.
Self protection, by going indoors or walking away, is possible even when there is no public alarm, provided the gas can be seen or smelled.

M.
Persons outdoors. The number of persons can be estimated by taking the average number of inhabitants, in houses, and counting the houses.

P.
Evaluation may be relevant if a release is long term, and concentrations indoors are likely to build up. An evacuation takes an absolute minimum of 20 minutes to organise, and for a large area 2 hours is considered a good performance. This makes evacuation most relevant for long term releases which cannot be stopped, particularly in areas where people may become ill over a period. since for the most part, long term release rates are generally relatively small, and the areas are consequently relatively limited.

R.
Rescue is relevant for persons who have become ill. Rescue generally requires protective equipment, unless the release has stopped. There are several difficulties, particularly in locating persons.

S.
Hospitalisation will be necessary for all of the persons who are badly affected by gas. Additionally, a much larger number will need to be treated for suspected poisoning.

For large releases, treatment of hundreds of persons within the first few hours is very problematic in most areas.

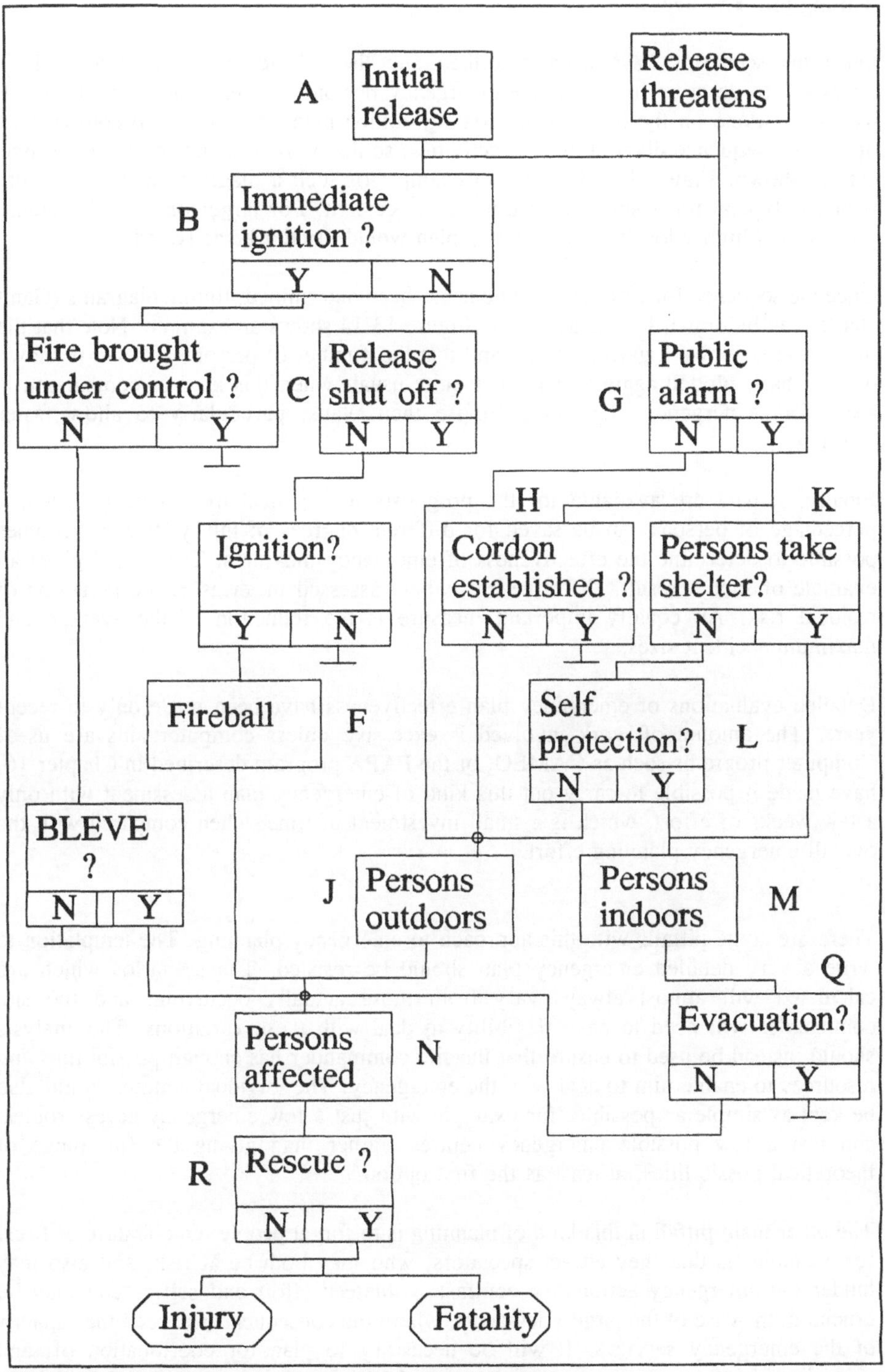

Fig 13.12 Standard scenario diagram for flammable releases.

Timing and resources

Once the scenarios have been described and the affected areas calculated, it is necessary to describe in outline the emergency response for each of them. This will usually be done on the basis of an existing outline plan. It is useful to construct an operations sequence diagram for the activities, so that sequence and resource conflicts can be shown. Figure 13.13 shows an example of such a diagram, with a structure which is typical for a small to medium size accident. For larger ones, civil defense groups, and higher levels of emergency plan would usually be activated.

Once the sequence for emergency actions has been determined, timing diagrams (Gantt charts) can be drawn for the activities. Figure 13.14 shows an example. Note that the time scale is shown logarithmically, and that the number of persons affected is shown on the chart, plotted against time. It is quite usual, when this kind of comparison is seen, for emergency planners to revise their plans, particularly to allow faster warning.

Finally, if data are available for the prognosis for medical treatments (i.e. which percentage of persons can be saved for different degrees of injury) then it becomes possible to determine the effectiveness of emergency measures. Table 13.4 gives an example of such a result. Effectiveness is here assessed in terms of the reduction of residual risk. An equally important measure is the reduction of the average and maximum accident sizes.

Detailed evaluations of emergency plan effectiveness have been made only in recent years. The amount of work involved is excessive unless computer aids are used. Computer programs such as CAMEO, or the PAPA program described in Chapter 16, have made it possible to carry out this kind of emergency plan assessment with only a few weeks of effort, which is a small investment of time, when compared with the overall emergency planning effort.

There are some pitfalls with this approach to emergency planning. The temptation to write a very detailed emergency plan should be resisted. The scenarios which are calculated will almost always vary from those actually occurring, and the site commander will need to have flexibility to deal with these variations. The analysis should instead be used to ensure that the site commander has enough possibilities and resources to enable him to deal with the emergency. The idividual actions should also be kept as simple as possible, for example with just a few emergency access routes, and just a few possible emergency centres, rather than giving the full range of theoretical possibilities, at least as the first option.

The other main pitfall in this kind of planning is to forget people. One feature of fires, for example, is that they attract spectators, who may both be at risk, and also may hinder the emergency action. By contrast, volunteer effort and self rescue may be crucial in the case of the largest accidents, where the consequences exceed the capacity of the emergency services. It will be necessary to plan for coordination of, and provision of information to, such volunteers. Provision of information about who has been rescued, and where they are being treated, avoids a good deal of confusion, and helps to reduce the effects of shock.

Risk source / Emergency measure	Effect	Risk pr million years	Reduct-ion due to EP	Resulting risk
Truck crash BLEVE	Flash fire, radiation burns explosion	90		
Self protection/ sheltering indoors	Possible, removes persons to shelter		2	
Cordoning to prevent ignition of flammable gas.	Possible, removes persons to shelter		2	
Cordoning to prevent exposure of passers by.	Possible if EF arrives within ca. 5 minutes		2	15
Cordoning to keep spectators away.	Possible if EF arrives within ca. 5 minutes		10	9
Evacuation.	May be possible for a few houses		3	
Rescue and hospital-isation			3	8
Cooling with water			3	2
Total		90	18	2

Table 13.4 Effectiveness of emergency measures to deal with transport accidents for LPG (example from a complete city emmergency plan analysis)

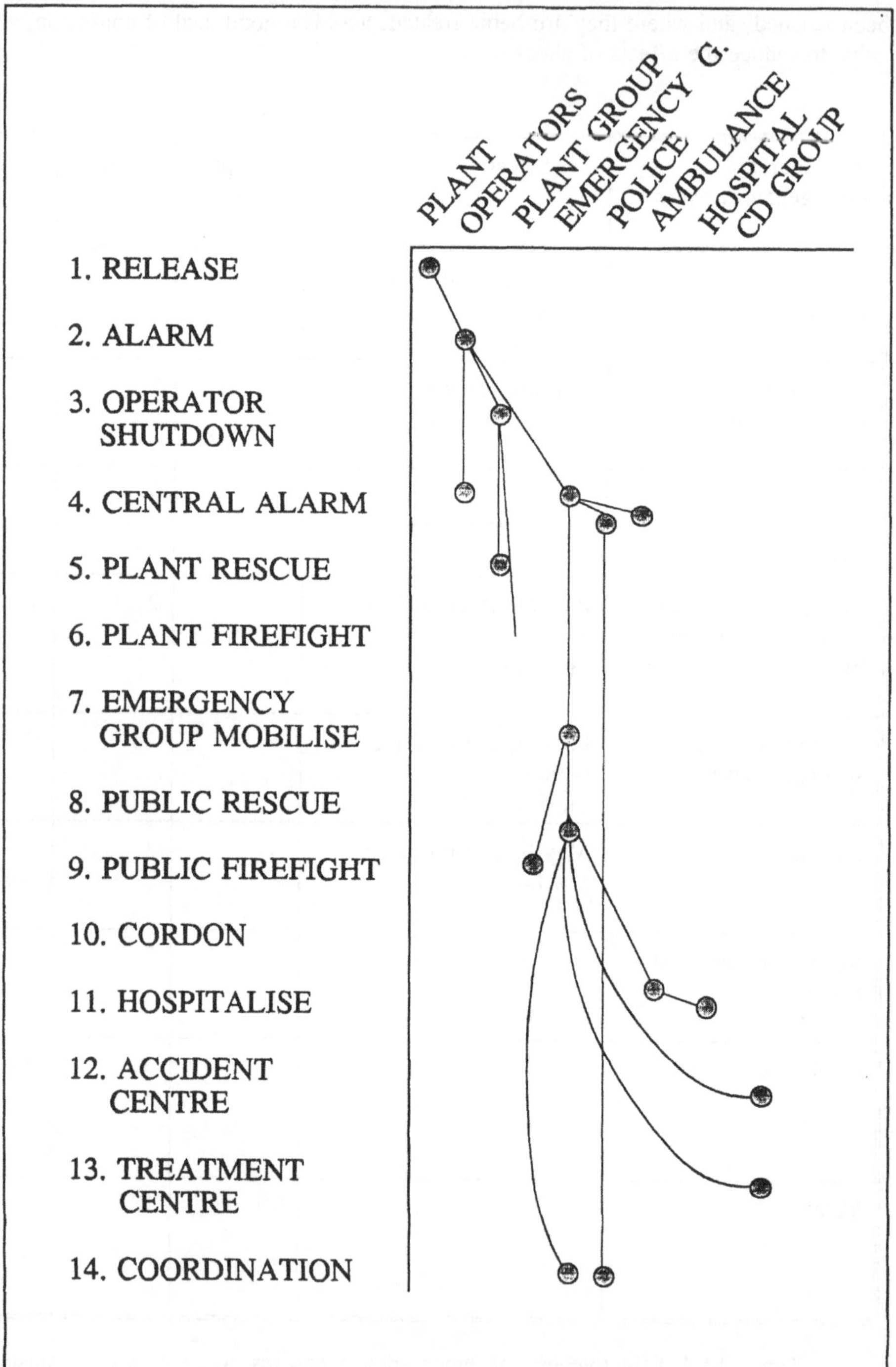

Figure 13.13 Operations sequence diagram for emergency action (example)

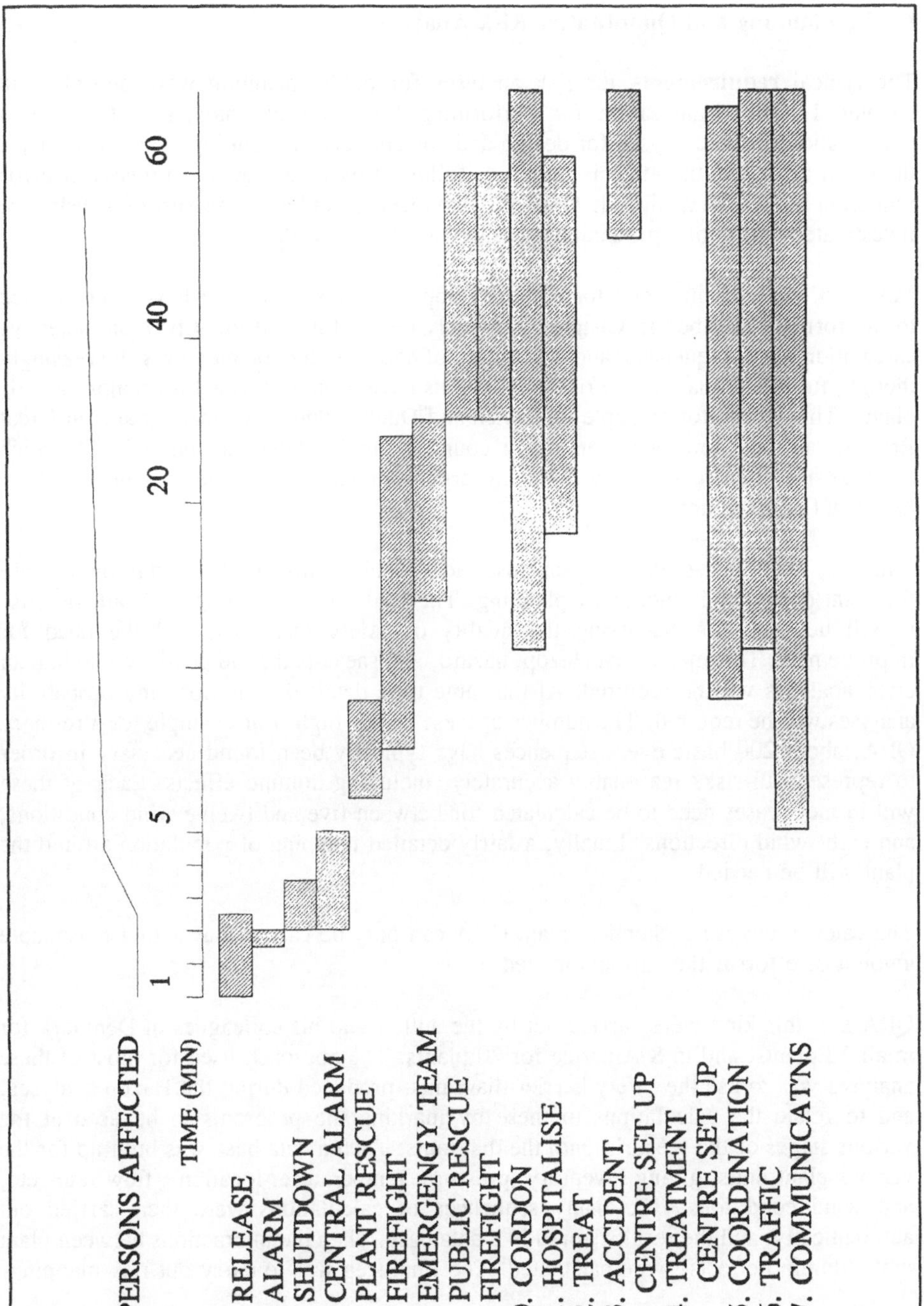

Figure 13.14 Emergency action timing diagram

Public Planning and Quantitative Risk Analysis

The typical **requirements** for risk analyses for public planning were described in chapter 1. The organisation for performing this kind of analysis is typically a combination of the analyses for design and for emergency planning. That is, firstly a thorough error identification is required, followed by as thorough a process of error reduction as can be achieved. Once this has been completed, the **residual risks** are investigated and public protection measures are bases on these.

Most risk analyses intended for authority approval or surveillance have been limited to the format described in Chapter 1, namely detailed hazard identification, selective calculation of consequences, and description of hazard reduction measures. Increasingly though, full scale quantitative risk calculations have been performed for major hazards plants. This is true for example for Dutch and Danish plants, for many plants in Great Britain, and for new plants in many countries in S. America and in USA. Such analyses are often required by financing organisations before funding is provided for new plant construction.

Full scale quantitative risk analyses poses additional requirements, in addition to those for plant design and emergency planning. The analysis needs to be in detail, because it will be used for discussing the quality of safety measures, and the need for improvement. This means that Hazop, hazard, and sneak path, and possibly also human error analyses will be required. At the same time detailed scenarios, and probability analyses will be required. The number of these can be high. For example for a refinery QRA, about 200 basic event sequences have typically been found necessary in order to represent all risks reasonably accurately, including domino effects. Each of these will in most cases need to be calculated for between five and twelve wind conditions, and eight wind directions. Finally, a fairly detailed mapping of population around the plant will be needed.

The calculations for a chemical plant QRA can only be carried out with a reasonable amount of effort if they are automated.

QRA's of this kind were carried out by the author and his colleagues in Denmark for in all 18 plants, and in S America for 20 plants. The approach used for most of these analyses was to use the safety barrier diagrams, produced during the Hazop analyses, and to relate the calculations to these by marking the programs to be used at the various stages of the scenario onto the diagrams. Then a data base was built up for the various plant units, giving inventory, pressure, temperature, location , flow rates etc, and wind conditions. The many sequences of calculations were then carried out automatically, and the results stored in a data base. Then the interactions between plant units (domino effects) were calculated. The final step was to carry out risk mapping.

The advantage of this approach is that the final quantitative results are traceable all the way from the details of hazard and operability analysis to risk maps, so that the effect of detailed safety improvements can be seen. The calculations themselves were found to take only a few hours of personal computer calculation time for each plant, once the calculations had been set up.

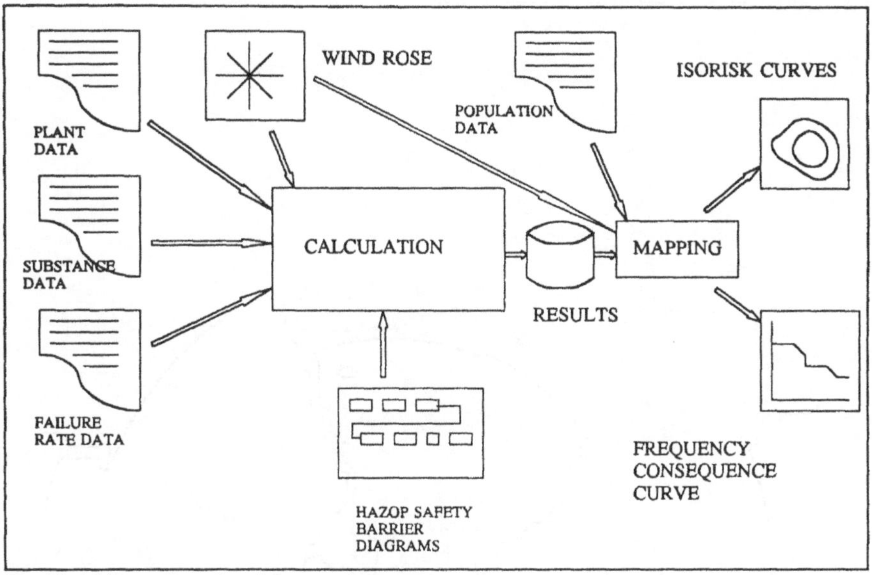

Fig 13.15 Data flow for risk mapping.

The data flow for the calculations is shown in figure 13.15. A typical risk map result is shown in figure 13.16, and the corresponding frequency consequence curve is shown in figure 13.17. Such maps and curves may be compared with the criteria described in chapter 2.

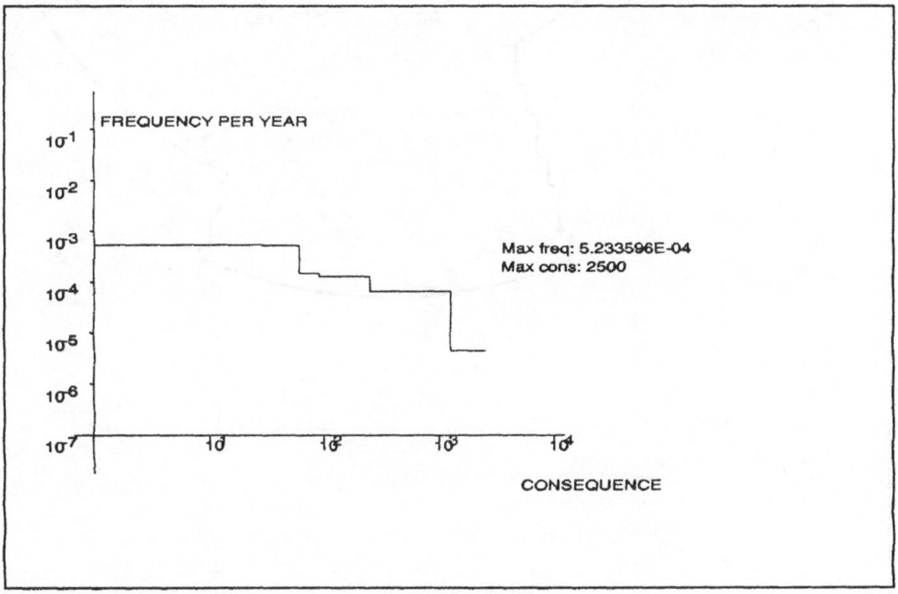

Fig 13.16 Frequency/consequence curve for the plant

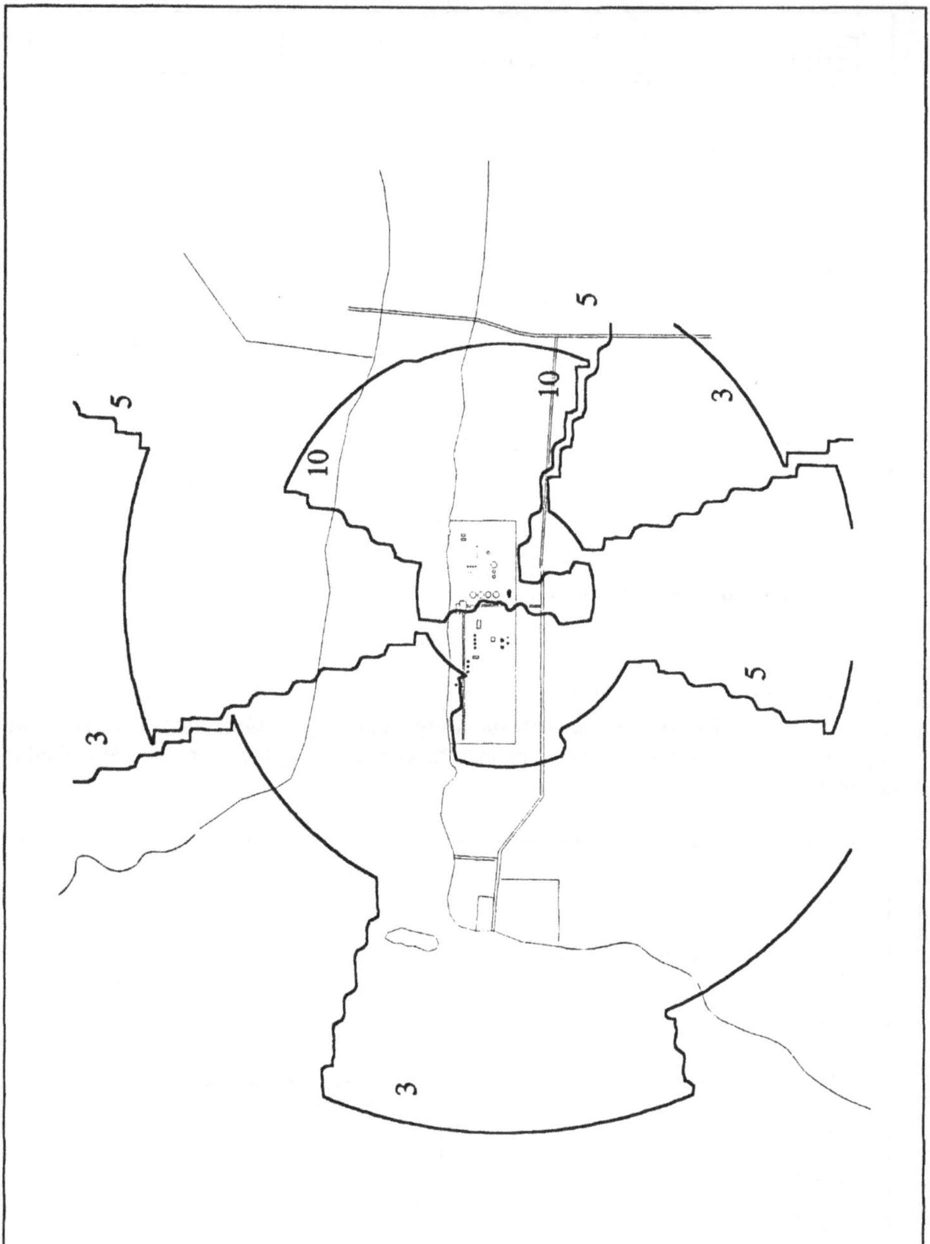

Fig 13.17 Some isorisk curves for example plant. Values are fatalities per 10^6 yr.

The population mapping for this kind of analysis requires identification of population groups around the plant. Sensitivity studies indicate that about 20 to 40 groups need to be identified to give reasonable accuracy for a refinery, rather more for large storage plants for toxic gases such as chlorine or ammonia, with the areas for counting being closer together nearer the plant. Both indoor and outdoor populations need to be identified. To allow this, the easiest approach has been found to be to photograph the area from the air if possible, and also while walking the roads and paths. In this way, a good profile of the population can be built up with just a few days effort. Care should be taken to observe in the morning, afternoon, and evening, because the number of persons outdoors varies. Note also schools, hospitals, shopping areas, and sports fields, since these often figure significantly as threatened groups. Note also how many windows are open for the different housing and building types. This will give an idea of the effectiveness of public alarms, in reducing risk. Finally, try to note seasonal differences, with repeat visits, and note special exposure situations, such as rush hour, or large sports events.

The most useful approach to interpreting these calculations has been found to be comparison between different plants, when these are analysed in the same way. Figure 13.17 shows this kind of comparison, used for twelve plants in Denmark (before risk reduction efforts). For comparison purposes the area covered by the 10^{-6} and 10^{-4} per year fatality risk have been found to be suitable comparison indicators, supplemented by a count of the number of persons living within each of the curves. Used in this way, a"scale" of risk can be built up, against which new plants can be measured.

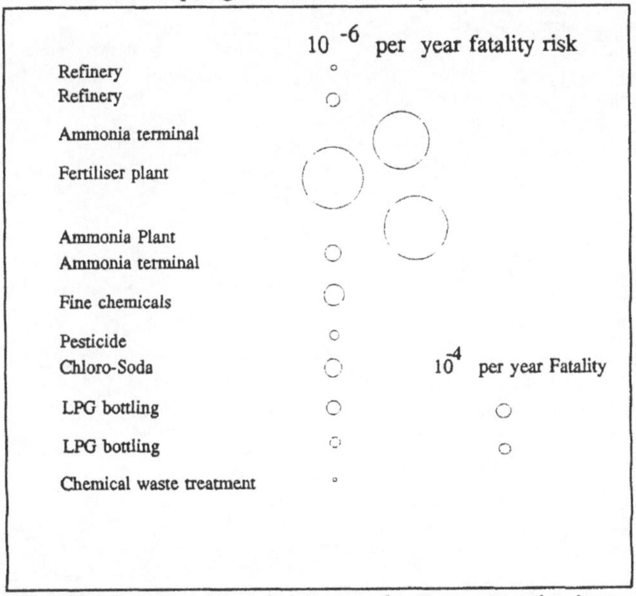

Fig 13.18 Relative sizes of risk areas for some example plants

References

Garcia Tavel, N. and Taylor, J.R. Major Hazards Assessment for Contingency Planning, 6th Int Symposium Loss Prevention and Safety promotion in the Process Industries, Oslo 1989

Taylor, J.R. and Platz, O. Comparison of Risk levels and Safety Measures for some Industrial Plants, Report ITSA 90-6-1, 1991. (In Danish)

14 Environmental Hazards Analysis

A release of a toxic substance will often be able to affect the environment, polluting surface water, ground water, and rivers, and polluting soil so that it cannot be used. Environmental hazards analysis involves

1. determining which substances can be released, and where.
2. determining the rate of release, and total release amount.
3. determining the extent of dispersion and spread, and any hold up which might occur.
4. determining the mode of transfer to humans, animals and plants, and any indirect pathway.
5. determining the toxicity of substances to humans, animals, and plants.

Environmental accidents can often have effects over a very long period, since et is not always possible to carry out a perfect clean up operation.Emergency planning for environmental accidents involves choosing the best techniques at each stage for limiting environmental consequences. In the first stage of an environmental accident, releases may be limited by shutdown, by bunding and retaining walls, by hold up ponds. In the third stage, when dispersion has begun, various clean up methods can be used to limit the risk, such as removal of soil, boom towing and skimming, absorption, and pumping.

At the fourth stage, the effects of damage limitation methods, such as special ventilation in housing, water purification methods etc. need to be taken into account.

An environmental hazards analysis needs to take into account the various routes by which pollutants reach the environment. These are primarily:

- directly to the ground, and from there to surface water.
- directly to the ground, and from there to ground water, which is then used for drinking or irrigation
- directly to the ground, with later spread as contaminated dust.
- directly to the ground, with later spread as evaporated vapour. (This last is especially important near housing).
- directly to the ground, and from there to plants which are used as food, or which die.
- directly to waterway and the sea. Here it is necessary to distinguish between substances which are dissolved, substances which float on water and substances which sink.

Considering the transfer to humans animals, and plants, it is necessary to consider the different pathways through which the pollutant can travel, and the concentration which can be reached at each stage.

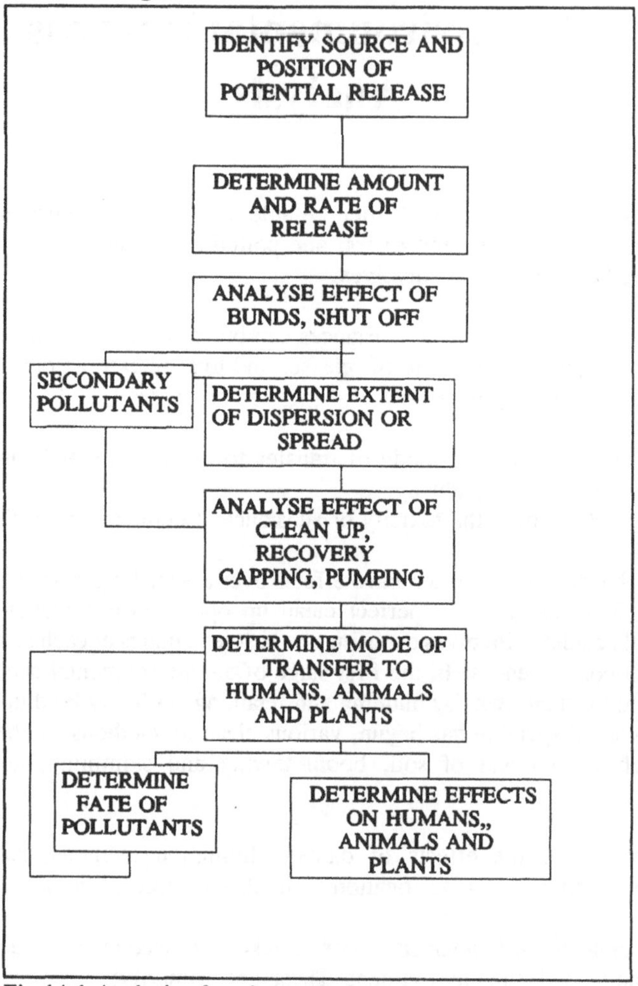

Fig 14.1 Analysis of environmental consequences

These pathways can for example involve, solution in water, absorption on detritus, consumption by plankton, consumption of plankton by fish, and consumption of fish by humans. Along the pathways too, it is necessary to take account of concentration, decomposition, and metabolisation.

In the ground, pathways can involve transfer to plants, which are then eaten by animals or people, transfer to children while playing (eating earth, pica, in the case of very young children), transfer to air by evaporation, and there by homes, and transfer to drinking water.

Along each of these routes, hold up of the pollutant, hydrolysis and oxidation, metabolisation in plants and animals with the possibility of conversion to other toxic products, can occur.

Analysis of all of these aspects is very complex, and only in a few cases are there sufficient data to make a detailed analysis possible. It will often be necessary to make some measurements in the field, and some laboratory studies, before a proper analysis can be made.

There are also considerable difficulties in determining toxicity of the substances once transferred to plants, animals and humans, since the rate of transfer to the organism of metabolisation, and excretion determines the concentration in the organism. Here, data are though often available, because of testing for foodstuff protection.

Analysis techniques.

Because of the difficulties of acquiring data on ecological effects, a complete analysis is often very complicated. To the models described above must be added the possibility that animals and humans will avoid polluted areas, water and foodstuffs. There are several cases of importance however, which can be calculated with reasonable accuracy. Example are:

- spread of a toxic substance on the ground, and evaporation from the surface of the ground, and toxic effects of this route.
- spread of substances to ground water and spread in ground water.
- spread of substances via waterways, when the waterways are used as drinking water sources.
- spread of liquids which float on water in waterways lakes and sea areas.

Each of these types of calculation can be used in a hazard assessment only insofar as the toxic properties of the substances are known. However, these properties are known for many practical cases. It is also possible to calculate upper limits on risks for far more cases than can be calculated exactly.

Spread of pollution on and in the ground

The initial area covered by a release of toxic liquid is very dependent of the topology of the area, slopes, drainage bunding etc. For this reason, it is necessary to take local topology into account. For shallow releases, the nature of the ground surface is important, since liquids will be absorbed into the ground. Spreading of a shallow pool can be described by Chezy's formula, or the similar Manning formula (Ch. 8), which can most readily be derived using energy principles. Solving the energy equation for the liquid velocity U gives:

$$U^2 = (H_1 - H_2 + (Y_1 - Y_2)^* 0.5) - 0.5(U_2^2 - U_1^2)$$

An equation of this type can be used as the basis for finite difference or finite element evaluations of affected areas, taking height variations into account.

Spread of liquids in the ground occurs either by simple diffusion, or by absorption in ground water streams, which then carry the pollution to distant areas. There will be a big difference in rates of spread, depending on whether the pollution is above or below the water table. Movement of pollution is in the pores between the earth grains. If the pollution is above the water table, it moves under gravity, and by being washed into the ground by precipitation. Once the pollution reaches the water table, the pollution is spread by the flow of the ground water.

The rate of spread of pollution depends also on the degree of solution of the pollutant. The pollutant may be insoluble, or have limited solubility, and may be absorbed or desorbed on earth, clay etc. The substance may be destroyed by hydrolysis, oxidation, bacterial action etc., and new toxics may be formed as a result of reactions.

Calculation of spread of toxic substances in the ground requires that the water table level and water flow rates be determined. This may be estimated from knowledge of soil and rock types, but careful work requires hydrological studies in the field. If there are water extraction wells in the area, water flow rates can be inferred from water table level reduction. Otherwise it may be necessary to drill wells (this would normally be done only after an accident, or when planning a new development which is likely to cause pollution, such as a tank farm).

The rate of flow of ground water, or of pollution above the water table, is governed by the D'Arcy equation:

$$u = K\frac{\partial H}{\partial x}$$

$$v = K\frac{\partial H}{\partial y}$$

where u and v are the liquid horizontal velocities, H is the water table head, and K is the permeability. Determination of permeability requires either obtaining rock or soil samples in an undisturbed state, or back calculation from tracer substances inserted into the ground water flow, or observation of well heads and flows, and corresponding modelling.

The rate of dispersion of polluting substances in the soil, that is the way in which substances are picked up by the ground water flow, and spread, must be known too. Figure 14.2 shows the kind of process involved.

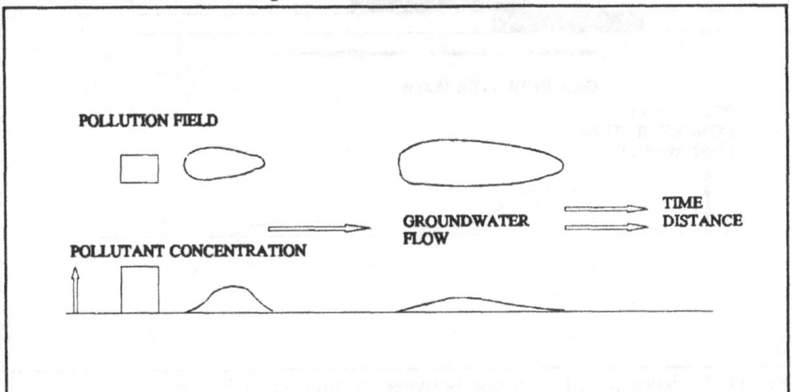

Fig 14.2 Pollutant dispersion in soil

The spread of pollutant can be calculated using D'Arcy's equation, assuming no dispersion in the first instance. For a complete solution, the Bear equation applies:

$$\frac{\partial(c n_e m)}{\partial t} + \nabla \cdot (n_e m (D_m + D) \nabla c) - \sigma m - q c_{in} = 0$$

Where c is pollutant concentration, n_e is the effective porosity, m is the thickness of the flow field,t is time, D_m is the molecular diffusion coefficient, D is the dispersion coefficient, σ is the pollutant source rate, and q is the discharge rate pr unit area.

Dispersion values may to some extent be determined from laboratory studies, by taking soil samples, and observing the dispersion of pollutant through a column of soil. Fig. 12.1 shows the way dispersion appears in the laboratory, the initial pollution being spread out over a distance, some pollutant travelling rapidly, other substances more slowly. The actual dispersion in the ground shows a much more exaggerated effect, due to the inhomogeneity of the ground layers. Pollutant flows most rapidly along the layers with the highest diffusion rate, so that some of the pollutant travels rapidly, some slowly. Field measurements are needed to enable this kind of behaviour to be calculated.

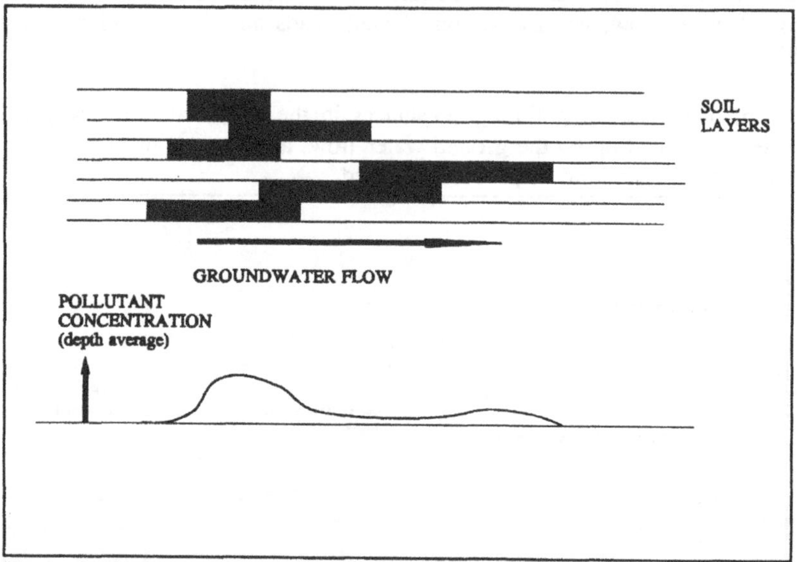

Fig 14.3 Dispersion of pollution between ground water layers.

In practice the equations must be solved numerically, using finite difference or finite element programs, and for accurate solutions, data on absorption, desorption, destruction through hydrolysis or oxidation, as well as permeability and dispersivity data are needed. For risk analysis purposes, the data are rarely needed accurately however. The main deed is to give an indication of the degree of pollution which could arise, its rate of spread, and the concentrations which could be present at water wells. If risks are found to be very high, or if a pollution incident has already occurred, then there may be justification for more detailed data collection and more accurate calculation.

River pollution

By contrast with ground water pollution, it is generally quite easy to calculate the rate of pollution of rivers, and the rate or dispersion, at least to the accuracy needed for risk analysis. White, Lee, and Belcher give a simple empirical model for average velocity of flow:

$$v = a \ln b\, Q$$

where a and b are constants, with values of 1.1 to 1.9, and 0.09 to 1.5 respectively, and Q is the volumetric rate of discharge. This velocity gives the rate at which pollution can spread from a release. The distance travelled will be $x = vt$, where t is time, and the concentration around this point is then given by:

$$c = \frac{mt^{-1/2}}{\alpha} \exp\left(-\frac{(x-vt)^2}{4Kt}\right)$$

Here c is concentration, m is pollutant mass, α is a constant, and K is the longitudinal dispersion coefficient. Values of K of from 10 to 30 m^2/s were found for rivers in S England.

Models for spreading and dispersion at sea.

Models for pollution spread can be divided into these intended to treat

- products which float on water
- products which dissolve or are mixable with water.
- products which sink.
- liquefied gases

For products which float on water, very much theory is available, and a fairly large number experiments supporting the models. The main spreading parameters are surface spread, the effects of wind currents, the effect of tidal currents and the effect of steady currents. Dispersion parameters concern evaporation, dispersion into the water column, and agglomeration tendency, (formation of tar balls, windows etc.). These aspects have been fairly well reviewed in the literature (see eg Audunson), though with considerable remaining uncertainty concerning water column dispersion.

A widely accepted set of models for spread of oil on the water surface is due to Fay:

Initial spread

gravity inertia regime $l = (\delta g\ V\ t^2)^{1/4}$

Intermediate spread

gravity viscous regime $l = (\delta g\ V^2\ t^{3/2} / \mu^{1/2})^{1/6}$

Final spread

surface tension/viscous regime $l = (\sigma^2\ t^3 / \Theta^2\ \mu)^{1/4}$

where l = diameter of the slick

δg = acceleration due to gravity times the relative density of oil v. water

V = the total volume of oil

t = time

μ = kinetic viscosity of water

σ = surface tension of oil

Θ = density of oil

The transition between regimes occurs when the depth of oil falls sufficiently for the next regime to dominate.

Most oils, and many water insoluble chemicals, evaporate on the surface of the water. For crude oil, for example, the light fractions evaporate rapidly for all but the coldest conditions. Several surface evaporation models have been proposed. One of the simplest (Audunson) is as follows:

$$dQ_i/dt = (7.4*10^{-3} + 1.87*10^{-3}W)\ P_i\ M / R\ T\ \delta p$$

Where Q_i = fraction of oil component i

P_i = partial vapour pressure of component i

M = molecular weight of the oil

δp = thickness of the oil slick

R = universal gas constant

T = water temperature

W = wind speed

Oil also disperses into the water column as droplets, caused by wave action. A simple correlation for this is:

$$X = X_0 (W/W_0)^2$$

where X = fraction lost per day to dispersion
X_0 = 0.12 per day
W = wind speed
W_0 = 8 m/s

Practical application of these models requires computer programs. For risk analysis purposes, a full spectrum of releases, wind, current and tidal cases needs to be covered. Such calculations can be made systematically so that all cases are properly covered, or can be performed using Monte Carlo methods. The time taken for calculations however, virtually dictates the use of a Monte Carlo approach. A typical systematic calculation will follow the progress of ten or twenty releases, over a period ranging from a few days to several tens of days, plotting at hourly intervals, for a range of perhaps five different wind speeds, eight wind directions, and a few hundred tidal and current conditions. This gives about 80000 simulations of dispersion, and will be repeated for each grounding, foundering and collision position.

Some special points need to be taken into account in carrying out such calculations. Firstly, use of simple wind rose data to represent wind conditions is not sufficient, since the wind constancy/changeability is an important factor. A much better approach is to represent process, with a constant probability of change for each time interval. This is particularly important in monsoon or fçn areas, where wind direction may be constant for very long periods, and wind current transport can dominate over tidal and background current.

Tidal data are easily obtained for some areas, and extremely difficult to obtain for others, depending on the extent of existing modelling and the amount of observational work which has been performed. In the absence of such data, analysis is virtually impossible. The quality of data required for risk analysis however is much lower than that required for navigational or flood warning purposes however, provided that average tidal current speeds are reasonably accurately represented. In the summaries made in the risk analysis, errors tend to average out.

For substances which dissolve in water and for substances which disperse from the surface, the state of the modelling art is much less developed than for surface dispersion. On the other hand, the dispersion rate is generally very much higher than for substances which float, so that much less detail is required for analysis. It is usually sufficient to take into account a few tidal current and background current conditions, even for critical cases of highly toxic pesticides.

Heavy liquids (heavier than water) and packaged or drummed substances present the biggest challenge for dispersion calculations at present. Even where surface tidal currents are known, it is rare that bottom currents are equally well known. Models of tidal currents present the cheapest approach here, though still requiring considerable effort. The state of the art today is certainly adequate to give reasonable accuracy in risk analysis. The main problem here is the long period over which movement of packaged toxic substances on move, and the relatively low rate of release.

Gas dispersion models have been developed extensively since the first developments by the US coastguard. Both simplified models which can be applied without much effort, and advanced models which can be made very accurate, but which at the same time require considerable expertise to use are available (see e.g. Havens)

Environmental and Commercial consequences.

The simplest calculations of commercial consequences involve coastal pollution. Here, the consequence can in the first instance be assessed as the cost of beach cleaning, and the loss of affected fish farms, shrimp and oyster farms etc.

For large releases, which are beyond the capacity for complete cleaning (supertanker groundings, for example) the losses to the tourist industry may be considerably higher than the clean up costs. A reasonable approximation for the purposes of risk analysis, is to assume that the sea coast tourist industry will be lost for the affected coast for the first year, and at least halved in the second year. (Based os experience from the Amoco Cadiz and Torrey Canyon incidents). Although there is presumably some compensatory tourist activity en areas not affected by the spill, this will not always be within the country affected, and can hardly compensate for the local damage. Ecological damage in open sea areas is extremely difficult to assess. Immediate damage to plankton and fish larvae in breeding grounds can be based directly on toxicity data. however, for smaller spills, this is hardly relevant, because of recovery by invasion from other areas. If a fishery stock is marginal (at the limit of stability) any large scale disturbance (from a large spill) could be catastrophic. Recovery could take many years. The secondary effects of pollution in increasing sensitivity to parasites and disease need also to be taken into account.

Wild life sensitivity (birds, sea animals) is an important aspect of the pollution consequence, but is not directly available for quantitative analysis, since it is difficult to set a value on wild life resources. A minimum valve can be set in terms of the money spent for protection of wild life.

For fisheries themselves direct, damage is large only in the case of very large or persistent releases of toxic substances - these represent a continuing hazard. The problems of tainting and continuing toxicity will generally prevent the use of the affected area of the fishery while the pollution persists, and will cause damage to fishing gear which can be important. An extremely important practical factor here is the "political" or market risk. It will often be necessary to suspend fishing over a much wider area than that which is directly affected on order to retain public confidence, and market stability. In practice rapid administrative action the result can be a major change in buying habits for a period of many weeks. However, this is almost totally dependent on the type and quality of publicity received.

References.

Audunson T. et al.
Slick forecast - A simulation program for oil spill emergency tracking and long term contingency planning. (Petromar 1980)

Fay, J.A. The spread of oil slicks on a calm sea. in Oil on the Sea, Plenum Press, New York and London, 1969

Bear, J. Hydraulics of ground water, Mc G-H, 1979.

White, K.E., Lee, P.J., and Belcher, A.S.B. Time of travel and its significance, in River Pollution Control ed M.J.Stiff, Ellis Horwood, 1980.

15 Transport and Pipeline Risk

Road transport

Road transport is used widely for transport of hazardous materials. Some of this is in containers with limited capacity, some in bulk transport tankers. The quantities involved are very large; most of the gasoline product produced in refineries for example are at some stage transported by tank truck. By comparison with most road traffic, the frequency of accidents for these transports is relatively low. Transport nevertheless represents one of the most significant major hazard risks.

There are two main aspects to road transport risk, each requiring a different approach to risk analysis. The first is societal risk. For these calculations, it is possible to use a method based on the average frequency of accidents per truck kilometer travelled per year. This gives an overall average risk over a complete route, for example. There will usually be some particular stretches of route which are particularly hazardous however, such as rail crossings or difficult intersections. For these more precise models of traffic interactions are generally necessary, in order to determine individual risks.

Models for calculation of transport risks must meet two opposing challenges. They must be sophisticated enough to provide a reasonably accurate picture of the risks over the full range of application. And they must be simple enough to be applied efficiently, without placing impossible demands on data collection. For contingency planning, the models must additionally predict not only the frequency of accidents, but also their size. The size (number of injured) is important for comparisons of transport alternatives, and also for dimensioning the contingency measures such as hospital facilities.

The models described here are very simple to apply. It has been found however, that two models are necessary, one for high population density, and one for low population density.

The frequency of accidents involving people, depends on the density of housing, shops, and other institutions along a roadway, and on the density of traffic. Not only the average density is of interest. Also the distributions of housing along a read stretch is important. The way houses are grouped together determines the size of the largest accidents. A high degree of grouping also reduces the frequency of smaller accidents (*more accidents "miss" the people*). Note that only a very small fraction of persons are generally present outside houses, shops, or work places, and along the roadside, except in shopping areas. Models can in practice be based on the density of housing.

For the models described here, each road section is characterised by two parameters, the grouping factor, which gives the number of persons living close to each other, within a distance of comparable with the size of an accident. The grouping factor depends on how many people live in each house, and on how many houses are grouped together, and is the ratio of the density of population in such a grouping, to the average density of population over the route section considered.

$$G = PD_{group} / PD_{section}$$

For sections in which population density is non uniform, at a scale of the typical accident size (100m) the grouping factor can be determined by multiplying the number of persons per house by the number of houses in a typical group of maximum size 100m.

For sections with high population density, such as a village, in which the gaps between houses are much less than 100 m, the grouping factor is irrelevant, and can be set to 1.

The other important factor is the populated length of a road stretch. Stretches of rural road are typically characterised by long stretches with no population, followed by shorter stretches with some population level.

For the models here, traffic rate, transport rate, and grouping factor are assumed constant for a given stretch of road. This determines the number of stretches into which a given route must be divided.

The scenario for a truck accident is assumed to be as follows. The truck crashes (with another vehicle, or runs of the read), overturns, punctures, and ignites (note that this scenario represents only a small fraction of all accidents).

The ignition may be immediate, as a result of crash, or delay, resulting from the spread of vapour to buildings or other places containing ignition sources. Deaths are assumed not to cause injuries for more than a small percentage of the population (actually assumed to be zero) because of the almost complete sheltering by buildings.

Given that persons are in an accident area it is not certain that they will actually be injured. They may escape before ignition, and they may be sheltered by the houses. The probability of a single person not escaping when involved in an area subject to fire is assessed to be 8%, on the basis of statistics collected for US dept. Energy. For larger groups, potentially affected by an accident of maximum size, the actual size may be much less than the maximum. An impression of the maximum size and effects is given by the accident in Germany in 1988, which burned 12 houses in a town, killed 4, and injured over 100. The size was about 100 m along the street. (Most of the houses were empty as a result of holidays).

The number of persons assumed involved is then dependent on roughly the square of the escape diameter. 10% of releases are assumed total, due to massive impact of the truck with obstacles. 30% are assumed to be large, covering 30% of the maximum group at risk. 30% are assumed to be small, covering just 10% of the group at risk, and 30% are assumed to affect only a few individuals (one or two).

The basis for these values is the consideration of accident modes described in the reference, ch. 8. (The values are preliminary, and require further accident mode investigation).

The model for the low population density accident frequency is then as follows.

$$F_A = F_T * T * L * P * A/G$$

where

F_A = Frequency of accidents

F_T = Frequency of truck accidents causing fire per km/yr

T = Trucks per year travelling the route

L = Length of populated section of route

G = Grouping factor

A = Accident size in km

P = Population density per kilometer over route

The maximum number of persons affected in such an accident is then equal to the size of each group (no of persons per house * number of houses in a group).

The actual number of persons killed is then given for high population densities, virtually every truck accident will result in a potential accident for persons in houses. The frequency of accidents involving persons in houses is then independent of the population density. The accident size is dependent on the population density however.

The model for accident frequency for high population density is:

$$F_A = F_T * T * L$$

with variables defined as above.

The number of persons who could be killed, at maximum, is then given by

$$N_{MAX} = A * P * G$$

The distinction of whether a low or high density model should be used depends on the probability that the accident area will contain persons. For low density areas, this probability is given by

Pr (persons in truck accident) = A * P /G

When this value reaches a value of 1.0, the low density model ceases to be relevant and a high density model may be used.

Note that these two models are approximations, and that a combined model may be produced. However, the combined model requires, integration procedures which make it unsuitable for hand or spread sheet calculation.

Similar considerations apply for traffic accidents involving cars, buses etc crashing with, or affected by, transport of major hazards material. It is important to know the frequency of each type of traffic, or rather its density per kilometer, for each type, and the average number of persons per vehicle (ideally, the distribution of number of persons per vehicle should be known but this is difficult to determine.)

The model for low traffic density, is then:

$$F_A = F_T * T * R * C * A$$

Where

R = length of route section.

For high traffic density, the model is

$$F_A = F_T * T * R$$

The number of persons injured, in the high density section is then

$$N_{MAX} = C * A * K$$

where K is the number of persons per car.

The cut over, at which traffic density is considered high, is then:

Pr (accident concerns cars) >= A * C.

For gasoline transports, under US conditions, the frequency of large, ignited spills is assessed (Batelle 1984) to be

$2.6 * 10^{-8}$ per truck kilometer year.

Navigation risk

Navigational risk is important in several contexts. The first, and presumably most important(!), protection of the ship and its crew, has seldom been addressed explicitly. Risk analysis has however, been used routinely to study the safety of bridges, oil and gas production platforms, and problems of oil and chemicals pollution.

Some of the first explicit risk analyses of navigational risk were made in the 1950's by Jensen, in providing answers to the question of risk improvements which could be obtained in building a bridge between Helsingør and Helsingborg.These studies were important, since they showed that many ship to ship encounters are very complex, and that collisions often involve three or four ships. They were also important in introducing the use of radar recordings as a statistical tool.

Risks analysis in the form of average statistics of ship collisions, groundings, and other losses has of course long been used as a basis for assessing insurance risk, and for supporting regulation of shipping. IMO, for example, publishes shipping loss statistics on an annual basis.

The first and still the main approach to navigational risk analysis for engineering purposes was that of Fujii and his colleagues in Japan. They introduced models of grounding and ship collision, which regarded ships (and shipping hazards such as sandbanks) as objects with a length and breadth and an average velocity. They introduced a factor, the probability of mismanoevre, which expresses the frequency with which there is a tendency to accidents. They could then predict the frequency of collisions on the basis of geometry.

These models have been extremely successful in allowing predictions of the frequency of accidents - and the author for example, used the techniques in accessing frequency of strandings and collisions in the Gulf of Thailand. Comparing these results with IMO statistics revealed a correspondence to within a factor of two in assessing frequency. The advantage of using the models in evaluating risks of this kind is that the models tell not only the frequency of accidents, but <u>where</u> these accidents are likely to occur.

The success of the Fujii models rests on their semi - empirical nature. "Mismanoevre" is quite difficult to define, but the frequency of mismanoevre as defined for the calculations can be measured by observing the frequency of collisions, and strandings in a suitable test area. Confidence in the theory can be obtained by repeating observations in several different places. These show a variability in the "probability of mismanoevre" of about an order of magnitude, so that it becomes necessary to select a value which has been observed in an area similar to the one concerned.

Fujii's and similar models, will be termed here "macro models". While they are very successful at predicting frequencies, they do not say anything about <u>why</u> collisions occur. As such, they can be used for engineering problems, by helping to design for resistance to accidents (e.g. strength of oil production platforms, size and location of oil spill clean up facilities). They do not provide very much information on how to design for prevention of accidents through navigational improvement. For this purpose is necessary to study how accidents arise.

Most ship collisions and many strandings are the result of human error, or of a combination of human error and equipment failure. (Other causes of losses of ships are bad weather, beyond the design capacity of the ship, fire, and pure equipment failures such as engine failures, see table 1). In order to predict collision frequencies in detail, it is therefore necessary to study human behaviour, interactions between captains, helmsmen, instruments, and shipping hazards, in order to be able to predict risks.

Some of the earliest work in this direction was carried out by the Norsk Veritas company in 1980. The risk predictions were at that time only partly correlated with experience. Ten years later, it must be said, that studies of this kind are only at the beginning of their development.

In chapter 3, an approach of this kind, micro modelling, is described. It draws on work on systematic human error analysis in navigation. The theory on which this approach is based has, at the time of writing, been validated to some extent in predicting the frequency of collisions. More to the point it has been validated according to the relative frequency of types and causes of accidents. As such, it gives some information concerning the way in which accidents arise.

Statistics for loss of ships

IMO publishes yearly statistics on loss of ships due to a wide range of causes. Table 1 gives a short except of the statistics. Non collisions losses represent $6.8*10^{-8}$ losses per tanker kilometer year. Statistics of this kind represent virtually the only way of estimating non collision losses, short of a very detailed risk analysis for individual ships.

Macro models of collisions

A number of macro models have been proposed (and validated) for ship strandings. Most of these are based on Fujii's original work . A good summary, and some extensions, have been made by Solem.

The probability of a collision or stranding is calculated as a product of the probability of encountering an obstacle G, and the probability of failing to manoeuvre around the obstacle. This last factor K, is the probability of mismanoeuvre.

When a ship is sailing in a cluster of obstacles, the frequency of encounter can be calculated from the density of obstacles, and the collision diameter D. D is generally the effective width of the obstacle, plus half of the beam, B of the ship.

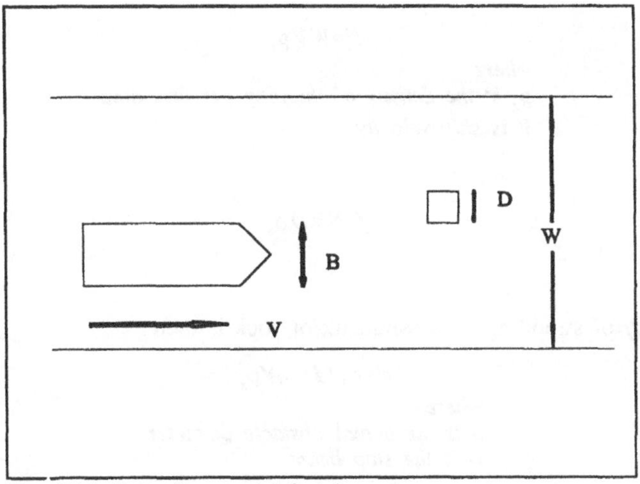

Fig. 15.1 Grounding model for channel

Taking a ship path of length L, and width D, the average number of obstacles encountered will be:

$$n_o = LD\rho$$

where
n_o is the number of objects encountered
L is the length of the channel
D is the diameter of the obstacles
ρ is the obstacle density per unit area

The mean free path between obstacles will be:

$$\lambda = \frac{L}{LD\rho} = \frac{1}{D\rho}$$

a) Probability of encountering a single obstacle in a shipping lane with uniform shipping density.

$$G = \frac{D}{W}$$

where
G is the encounter probability for a ship
W is channel width

$$f = NG$$

where
f is the encounter frequency
N is the frequency of ship passage

Substituting gives

$$N = WV\rho_s$$

where
ρ_s is the density of shipping per unit area
V is ship velocity

$$f = NVD\rho_s$$

The frequency of stranding on a sandbank or rock is then

$$F = K_F(d+B)V\rho_s$$

where
d is the actual obstacle diameter
B is the ship beam

The factor K_F here is the Fujii number, the probability of mismanoeuvre, as observed in a number of Japanese straits. The observed values are between $10^{-3.0}$ and $10^{-4.2}$ with the best values being $10^{-3.7}$. Vestergaard and Taylor used these values in an analysis of shipping collisions and strandings in the Gulf of Thailand, finding an agreement with the $10^{-3.7}$ value of within a factor of 2, which is less than the uncertainty involved in the observations of traffic density.

Note that some authors use B/2 rather than B in the above equation. This applies if only one side of the obstacle is hazardous. Some authors also introduce a factor $1/\pi$. This depends on the way in which mismanoevre is defined, and only affects the observed value of K_F.

When obstacles are very wide, wider than the channel width, Fujii et al give a revised model, which, in the notation used her is

$$F = K_F WV\rho_s$$

b) Clusters of obstacles.

For an area in a channel with many obstacles, the equations above become

$$G = DL\rho$$

as above
and

$$F = K_P \rho \rho_s VWDL$$
$$= K_P \rho \rho_s VDA$$

where A is the obstructed area

c) Ship collisions, crossing lanes.

$$G = \rho_s LD.2\sin(\alpha/2)$$

where α is the angle between shipping lanes

Collision frequency is then

$$F = K_P \rho_{s1} \rho_{s2} LD.2\sin(\alpha/2)$$

Note that this model is very crude for lanes which cross at small angles. Since it does not take effects into account. The model was given by McDuff

d) Ship collision in random traffic.

The collision frequency for random traffic may be found by integrating the above equation, for all angles α. This gives

$$F = K_P \rho_s^2 VDA$$

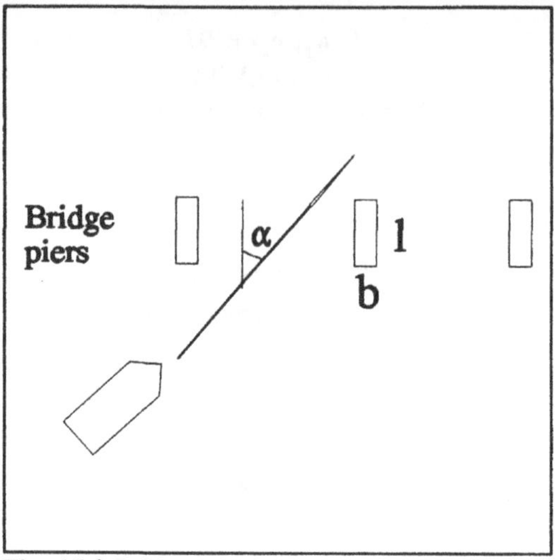

Fig. 15.2 Bridge collision model

e) Collision with bridge piers.

The formula given above for a single obstacle (a) can be used for a bridge pier. The width of the obstacle can readily be derived from geometrical considerations. The effective diameter of the pier is then:

$$D = B + b\sin(\alpha) + l\cos(\alpha)$$

f) Channel bank collisions

For narrow channels, the theory given above for ship and obstacle collisions hardly applies, because of the special manoeuvering difficulties. Most channel groundings appear to occur as a result of the need to avoid ship collisions.

If the frequency of shipping through the channel is N, and the channel width is W, the frequency of encounter is then, as above.

$$f = \rho_s^2 VWDL$$

The probability of a mismanoeuvre causing a bank grounding is then dependent both the probability of mismanoeuvre (eg. sailing to the wrong side) and on the width of channel available to avoid a bank. The distance available for manoeuvreing depends on whether there is a fixed lane system, whether sailing rules are followed. An approximation is:

$$M = W/2 - B$$

or

$$M = W - B$$

The second equation applies if there is no lane discipline.

There will be a probability distribution associated with M. If this is given a uniform distribution, then the probability of the manoeuvering distance being greater than the minimum distance m is given by:

$$P = \frac{m}{W - B}$$

Bank collisions may also occur as a result of straight forward mismanoeuvre, independent of ship collision avoidance. Here, the question is whether there is sufficient channel width to allow a mismanoeuvre to be corrected:

$$F = K_F K c N L \frac{m}{W - B}$$

Where F is the frequency of bank strandings
K_c is the frequency of course corrections per unit distance required for navigating the channel.

Note that there is no guarantee that the value of K_F for a narrow channel will be the same as that for a wide channel. A difference is in fact to be expected. Observations on the Chao Pryah river, indicate a value of K_F of $10^{-3.2}$.

Taking both contributions, ship avoidance and navigation error,into account, this gives:

$$F = (K_F \rho_s^2 VWDL + K_F K_c \rho_s VWL) \frac{m}{W - B}$$

g) bridge pylon collisions

Many bridges are designed with piers or pylons set into, or just projecting from, the channel bank. In these cases, the pylons cannot really be regarded as "obstacles, they are actually part of the bank. The model given for bank strandings can then be used. The "length" of the channel which is significant is then given roughly by the width of the pylon, plus the ships length. In practice grazing collisions will also occur, so that if these are relevant, the length taken for the avoidance manoeuvre needs to be taken into account.

Simulation modelling for ship collisions

As will be apparent from the descriptions of macro models for collisions, the models can be made increasingly sophisticated, and begin to take details of navigation and obstacle avoidance into account. In doing so, they become closer to reality. At the same time it becomes increasingly difficult to obtain empirical data to support the models, because of the increasing numbers of parameters involved.

Solem discusses a simple avoidance manoeuvre. "At a distance s from a projected point of collision, the manoeuvre is initiated. The speed of the vessel is V and the time to complete the manoeuvre is T. If T is greater than s/v, the collision will be avoided.

A study of actual manoeuvering times indicates an exponential distribution so that:

$$K = e^{\frac{-s}{V\tau}}$$

where K is the probability of mismanoevre
and τ is the average manoevre time

Making uses of Fujii's data, with a value of $K_F = 2.10^{-4}$ gives a value of s/v of 8.5. Using Solems own data for channel strandings he finds $K^F = 3*10^{-5}$ corresponding to s/v =10

It is clear that delay in course correction is only one cause of mismanoeuvre. The complete list of errors is much longer, involving at least:

- delay in course correction
- wrong amount of correction
- oversight, not observing an obstacle
- wrong direction of course correction
- communication error, from captain to helmsman
- setting error for compass
- radar alignment error
- radar bearing or distance reading error
- navigation error in reading map

- delay in position taking

- decision error in multiship situations.

With all this complexity, it maybe surprising that simple macro models can be used at all. In practice they give results which are accurate to within a factor 2 or 3 (i.e. they predict this closely the actual collision and stranding rates). The reason for this can only be the power of empirical adjustment. The models work because parameters have been adjusted to make predictions fit facts.

The problem with this approach is that errors can be expected in unusual situations e.g. special obstacles types such as bridges, and that the models do not provide any information on how to reduce risks by improving navigation. This is often needed for example in design of terminal facilities and approaches.

It is quite possible to apply the techniques of human error analysis to this problem allowing the effects of navigation aids, visibility etc to be taken into account. The amount of work involved means in practice that computer programs can take statistics of shipping into account too, so that probability of two and three ship encounters can be incorporated into the models. To do this, however, requires that not only the ships, but also the steering and navigation process, must be taken into account. Monte Carlo modelling, in which ships are simulated on a statistical basis, are generally needed, in order to take the full range of possibilities into account, with a reasonable amount of computing effort.

Fig. 15.3 shows such a model, used by the author for bridge collision and terminal approach collision and stranding calculations. The diagram shows the states for a given ship. The program selects ships from a statistical distribution, and traces their paths along one of a number of sailing routes. When the ship approaches an obstacle or a way point, course corrections are applied according to a set of sailing rules, and the state of the ship is changed. Sailing rules are selected on the basis of proximity, relative speed, course, and sailing state. The sailing rules describe the normal rules of safe navigation, various steering practices and typical response times, and the various errors which can arise in navigation.

Such simulators require statistics of shipping sizes,frequency, and also grouping which is important in determining multi-ship situation frequencies. Ship manoeuvering parameters are needed. Also probabilities are needed for a large number of human error parameters. Virtually the only way of obtaining these is from simulator studies, so that the calculations need to make use of "manual" ship navigation simulation too.

Comparison of results of such "micro models", with the macro models described in the previous section, show good agreement in simple cases. As might be expected, results for narrow channels, and in the presence of special navigation hazards, differ considerably from those for macro models.

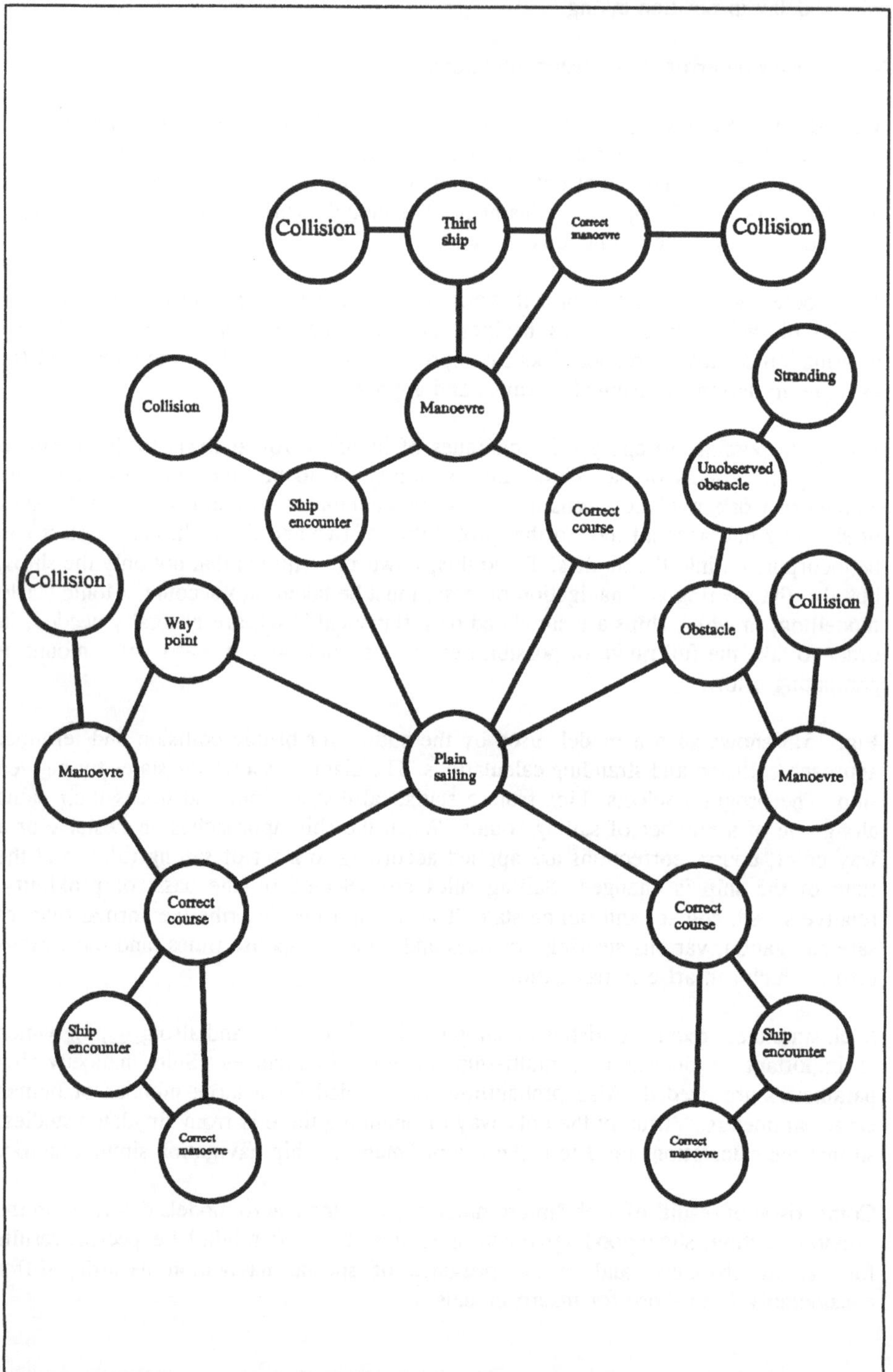

Fig. 15.3 Navigation model for error analysis

Pipeline risk

Pipeline and transport analyses differ from plant analyses in that risk is associated with a line source rather than a series of point sources of risk. Calculation risk involves surveying the route in question and determining.

- The probability of accidents of each size.
- The dimensions of any accidents using the consequence calculation methods.
- The types of population center around the route eg. agricultural, small village, town, city with high rices.
- The average population density per kilometer for each type of population along the route.
- The traffic density along the route, for transport accidents.

These values allow the risk levels to be calculated as follows:

Frequency of accident = F*L*E

where

F	=	Frequency of accidents pr Km yr.
L	=	length of route section
E	=	exposure factor for route section

The number of fatalities for a route section is given by = S*D
where

S	=	size of accident
D	=	Population density for exposed population

This section provides both a description of some typical accidents and incidents involving natural gas pipelines, and statistics of natural gas and oil pipeline releases. While oil pipeline accidents are not directly relevant to the present risk analysis, some of the causes of oil pipeline releases are. Additionally, there is a potential for domino effects between the present proposed gas pipeline, and the existing light crude oil product and LPG pipeline, OSSP and alcohol and diesel, which follows the same route. This potential will depend on distance between the pipelines and on the depth of burial if any.

Systematic records have been kept for oil and gas pipeline accidents, by Concawe in Europe, and the National Safety Transportation board. For the most part, data from both oil and gas pipelines is relevant. (Note that it would be preferable to use data from Brazilian pipelines when this becomes available). In all cases it should be remembered that historical data may be pessimistic, since pipeline construction and inspection standards have improved. Frequency of accidents due to external interference, which represents about 30% of all causes, is unlikely to change, however.

The European data is summarised in the 1972 to 1976 CONCAWE report (there have been only limited follow up studies since that date. The data includes 93 incidents on a total of 86400 km of oil and gas pipeline. Of these incidents, some are not possible on gas pipelines, in particular those due to:

- elevated temperatures
- water drop out corrosion
- external corrosion on pipelines without cathodic protection
- pumping station failures
- one poorly protected pipeline with 6 failures

With these eliminations, a total of 39 failures remains, giving a failure frequency of $4.5 * 10^{-4}$ per km year.

US data gives, in a similar way, values varying between $23*10^{-4}$ per km per year (1968 statistics) to $9.4 * 10^{-4}$ per km per year (1975 statistics).

The causes of failures were summarised by the UK Health and Safety executive, as follows (Turner):

Cause	US	Europe	UK
Corrosion	45	38	28
External interference	28	32	28
Defect	12	23	12
Incorrect operation	2	4	-
Other	13	3	22
Total	100%	100%	100%

Hole sizes were similarly assessed

Small	Less than 20mm	87%
Medium	Between 20 and 80mm	10%
Large	Over 80mm	3%

Note that for natural gas lines, a clean 20 mm hole can only with difficulty support a high pressure jet flame, unless it impinges. Small holes are therefore only in part relevant for major hazards calculations.

More recent European data have been assessed by Blything. (1975-80), and which gives an interesting variation of frequency according to size.

Diameter	No. of failures	failure rate per km per year
6 - 10	35	$1.2*10^{-3}$
12 - 14	12	$7.7*10^{-4}$
16 - 18	10	$5.3*10^{-4}$
20 - 22	6	$4.2*10^{-4}$
24 - 30	3	$2.1*10^{-4}$
> 32	4	$2.4*10^{-4}$

The variation can largely be explained by the increase in wall thickness as diameter increases.

An increase to above 0.5 cm seems to give a reduction of failure rate of about a factor 2.

De la Mare and Anderson provide an assessment bases on gas pipelines alone, giving $6.5 * 10^{-4}$ failures per year. For French pipelines, they found $32.5 * 10^{-4}$ failures per km per year. (Mostly small diameter).

On this basis, values of failure rates of $5 * 10^{-4}$ per km per year seem suitable as a generic failure rate for all failures, with 10% being medium, 3% large or catastrophic and 87% being small in size.

Description of actual accidents are available for a few pipeline releases, with some examples as follows:

A rupture of a line due to an excavation at Emden in 1988 gave a full rupture on a 24" transmission pipeline. A crater about 12 m in diameter was created and a vertically directed jet flame with a length of 250 m as estimated from photographs.

A different kind of accident is the 1973 Bowie, Maryland accident. Gas seeped from a small leak and accumulated in a house, causing an explosion when it ignited.

The accident at Cubatao in 1985, in which large quantities of hydrocarbon liquid were releases due to a pipe rupture, and which causes a large number of deaths, is relevant tot he present study. It illustrates the potential hazards of domino effects from the gas pipelines.

References

A Jensen	The future Collision risk in the Sea Area between Helsingør and Helsingborg. IFORS meeting M.I.T. 1962.
J Wiley and Son	Collision risk in Øresund 1966 Decision and Control.
A Jensen	Traffic Operation Research Futurology North Holland 1980
Macduff, T.	The probability of Vessel Collision. Ocean Industry. (1974)
Perrow.	Normal Accidents (1983).
Toyida, and Fujii.	Marine Traffic Engineering. Journal of Navigation, Vol. 24, No. 1, p.24-34.
Fujii and Tanaki.	Traffic Capacity. Studies in Marine Traffic Engineering. Journal of Navigation, Vol. 24, p. 543-552.
J. A. Fay,	The spread of oil slicks on a calm sea. Oil on the Sea (D.P. Houlst (Ed)), Plenum Press, New York and London (1969)
Fujii, Yamanouchi and Mizuki.	II - The Probability of Stranding. Some Factors Affecting the Frequency of Accidents in Marine Traffic. Journal of Navigation, Vol. 27, No.2,p. 239-243.
N.C. Vestergaard, and J.R.Taylor	Probability of ship collision in Thai Waters. SRE conf, Helsinki 1986
de la Mare, R.F. and Andersen, Ø	Pipeline Reliability, Report 80-0572, Det Norske Veritas, 1980
de la Mare	A Study of Pipeline Reliability, in Offshore Oil and Gas Pipeline Technology, Gulf,1985
CONCAWE	Performance of Oil Cross Country Pipelines in Europe.
Rhodes R.E.	An Assessment of the Risk of Transporting Gasoline by Truck, Battelle Pacific North West Laboratory, 1978
Blything, K.W.	In service reliability data for underground cross country oil pipelines, SRD R 326, UKAEA, 1984
Turner, M.J.	The safety evaluation of cross country pipelines, in I Chem E Symposium series No 58, 1982

16 Computer Software for Risk Analysis

One of the main changes in the practice of risk analysis for process plant during the 1980's was the widespread use of computer programs for risk calculation. In particular the use of personal computers to support risk analysis has made a large difference to the way in which risk analyses are carried out. The analysis process has become more effective, and at the same time, has become much more quantitative.

The first risk analysis programs used for risk analysis of process plant during the early 1970's were fault tree calculation programs, followed, towards the end of the 1970's by various consequence calculation programs. By now, nearly every aspect of risk analysis has specific computer based support. Several integrated sets of computer programs have become available commercially, bringing advanced risk analysis techniques within the capacity of even the smallest safety engineering group. The economics of risk analysis projects generally dictate that at least some types of software are used.

16.1 Programs for Consequence calculation

Consequence calculation programs are the mainstay of chemical plant risk analysis. While other types of program can make risk analysis more effective or convenient, they ar not absolutely necessary. Many consequence calculations cannot be made at all, however, without use of a computer.

There are several styles of computer program for consequence calculation. The simplest take a single aspect of an accident, such as gas release, and make calculations for this alone. Virtually all of the earliest programs were of this kind, and the results of the newest research work are often expressed in this way.

Several integrated packages consisting of separate programs have become available. The most widely used of these WHAZAN (Technica 1985) was developed originally to accompany the World Bank's Guideline on Hazard Assessment . The PAPA programs originally developed by the author to accompany this book, are of this kind. Such sets of programs allow the analyst to study each aspect of an accident in detail, and do make sensitivity studies for each part of the release, dispersion, fire, explosion and toxic effect sequence. They also allow output data to be transferred as input data to the next, allowing an integrated set of calculations to be carried out easily. Generally such programs provide data bases for substance data, and for storage of results.

A third type of program provides calculations for a complete scenario, integrating for example release, dispersion, and toxic effect calculations. some versions, and toxic effect calculations. Some versions of the Degadis program, described in chapter 8, are of this kind. The EAHAP programs provide an integrated set of such calculations with sufficient flexibility included to allow a fairly complete coverage of the usual types of chemical plant accident scenarios.

A fourth kind of program provides "integrated analysis". This kind of program provides both probability and consequence calculations for a complete set of accident scenarios, often treating a whole plant in one calculation. Such programs are described in section 16.4.

Hazard identification programs

Hazard identification programs are a relatively new tool as for as practical applications are concerned. During the 1970's much research was carried out on the development of automated techniques for fault tree construction. This research continues but has never been very widely applied. The reasons appear to that - the programs on expensive data input stage, before analysis can begin, and the fact that chemical plant risk analysis generally involves many simple accident possibilities rather than a few complex ones. The author has nevertheless been able to identify some hazards with these tools, even after very thorough manual analyses have been completed (Taylor 1982), and in a few cases, to avoid serious design errors.

The most practical results from the early fault tree analysis work has been the development of programs which allow fault trees to be constructed by piecing together predrawn fragments of fault trees, in a fairly flexible way. The CAFTAN program provides an examples of these. (Poucet).

More recently, "hazard and operability facilitators", essentially data bases or spread sheet programs, which allow recording and updating of hazard and operability study results have begun to become commercially available. The best of these provide check lists of disturbance types and causes, so facilitating a uniformity of terminology in analyses, and improving consistency. Examples are Hazex (the program written to accompany Ch 3 of this book) and The newest programs use "expert systems" (programs with large amounts of built in knowledge) to support hazard and operability analysis. (Suokas 1989).

The author has found such programs, in the hands of a skilled analyst, to produce consistently more complete analysis results than manual analysis, and an much higher quality of documentation than is usual in manual analysis. Whether the same is true for inexperienced analysts, or those who only occasionally carry out an analysis, remains more questionable.

If such programs are used, care needs to be exercised in setting up the physical arrangements for the analysis, otherwise the important aspects of communication within the hazard and operability analysis team, are lost. The best technical solution which has so far been found to this problem is to provide a large video monitor which can be read by all of the analysis team, but with the main attention of the analyst group focussed on the plant drawings. The hazop team secretary is responsible for performing the recording i.e. for the typing.

Human Error analysis

Very few practical programs to support human error analysis have been written. The main ones are those written by Human Reliability Associates Limited, and the Herrex program, which was written to support the projects on which this book is based.

The two sets of programs are very similar in function and intention, the main differences being in style and detail of the error causes considered.

The Herrex programs implement the action error mode and action error cause procedures described in chapter 7. The programs work in two different ways, either by requiring the analyst to describe error consequences, or by providing a plant simulator to evaluate error consequences.

Probability calculation

Programs for probability calculation have traditionally been those which perform fault tree analysis, and implement the algorithms described in chapter 5.

Several such programs were developed during the 1970's, primarily as a result of the needs for risk analysis of nuclear power plant. Newer programs have improved primarily on the style of earlier developments, by providing graphic facilities for drawing the fault trees, and for improvements in calculation algorithms. Examples or ORCHARD (SRS), Faunet, Relcon , and CARA (Sintef). The last deserves special mention particularly because of its very effective analytic algorithm for fault tree calculation.

The author must admit both to having written fault tree analysis programs, and for having found only limited use for them in the field of chemical plant safety analysis. (Such programs are essential for nuclear plant analyses, and for military and aerospace systems). There are teams which use fault tree analysis as the primary technique for hazard identification in chemical plant, and therefore well placed to make fault tree calculations. Large scale fault tree analysis is however usually an alternative to hazop analysis, which is generally more popular for chemical plant.

Frequency calculations for event trees, for cause consequence diagrams and for safety barrier diagrams exist, but have not been widely circulated or become commercially available. Part of the reason is perhaps, that ordinary commercial spread sheet programs, such as Lotus 1 - 2 - 3 (Lotus Corporation), Excel (Microsoft), etc. provide a very convenient approach to calculation.

A The main advantages of using such programs are the case of modification, ease of sensitivity analysis, and production of tables for incorporation into reports. A typical application of such programs involves tabulating:

- the name of each event sequence
- the frequency of the initiating event
- the name and availability value, for each safety barrier in the event sequence
- the overall frequency for the event sequence (this being the product of the earlier factors,can be produced automatically).
- the size of the consequences for the accident event sequence.

Summation of risks, frequency consequence graphs, and histograms can be produced by these programs using just a few control commands.

Integrated risk calculations

A few computer programs are available which carry out a full sequence of risk analysis calculations, starting with just a plant layout, inventory of hazardous substances, and vessel size, pressure temperature and piping size information. The programs create scenarios, apply accident frequency data, make consequence calculations, and generally produce frequency/consequence curves and risk maps.

Such programs have been criticised for providing a "block box" approach to risk analysis, giving little feedback to understanding and to accident prevention. This criticism is not strictly fair. It is true that such programs are complex, and can be used blindly, but all produce a great deal of detailed information, both at the scenario construction stage and at every step in the total calculation. A combination of such programs together with a good hazard and operability analysis can produce results of very high quality. The weakest point with such programs is that they generally provide only a conventionalised allocation of accident frequencies, and do not take specific features of individual plant equipments and safety measures into account. All such programs provide facilities for modifying accident frequencies however and for input of "hand selected" accident scenarios.

One of the first such programs was SAFETI, written by Technica limited, originally for the Dutch Authorities (Ale). This program remains one of the major achievement in the field of risk analysis programming, providing a wide range of high quality consequence models.

The IFAL technique was developed by the Insurance Technical Bureau in the 1970's, as a manual technique to provide rapid risk analysis, for insurance purposes. This method has been automated, and also gave an impulse to a series of further developments, "Rough and Ready Risk Reckoner" (Taylor and Justesen 1983), OCRA (OC Consultants) and "Summary" (Taylor). These programs use approximate or empirical techniques which are less precise than detailed consequence calculations. As a result complete plant analyses can in some cases be made using personal computers, with just a few hours of computing time. The algorithm for such programs is given in fig. 16.3. One program of this type, which is not generally available, but which nevertheless is important is RISKAT. This program is used by the British Health and Safety executive in order to be able to check safety reports, submitted to the executive by individual companies.

A specific programs set

This book was originally written as a set of teaching notes to support a long sequence of risk analysis projects (Taylor 1989). A set of computer programs was also written to support the projects. These were designed to achieve three objectives:

- To allow each step in the risk analysis to be carried out separately both to provide flexibility, and so that analysts could understand what was going on.

- To integrate detailed hazard identification, probability calculation, consequence calculation, and risk mapping, into one system.

- To provide for large (whole plant) multi scenario analyses in an automatic way.

The design of the program system is known in figure 16.1. Each of the programs in the set can be activated individually, but data are transferred from program to program. The hazard identification programs are used to provide scenario descriptions (in the form of safety barrier diagrams). These are stored in a data base, and can be used to control long sequences of consequence calculations. The results from these calculations are also stored in a data base, from where they can be used for risk mapping and societal risk calculations.

As a result, a program set of this type can be used for simple design support analyses, purely qualitative analyses, or complete plant quantitative risk analyses. The major advantage of this appears when plants are extended or modified. The additional work for incorporating plant extensions typically involves hazop studies and layout studies at the early stage, and full emergency planning studies later. Updating is fairly straightforward in that only the few aspects which have been modified , and any new additions, need be calculated. The analysis can become a "living analysis" where the full consequences of even small changes can be incorporated.

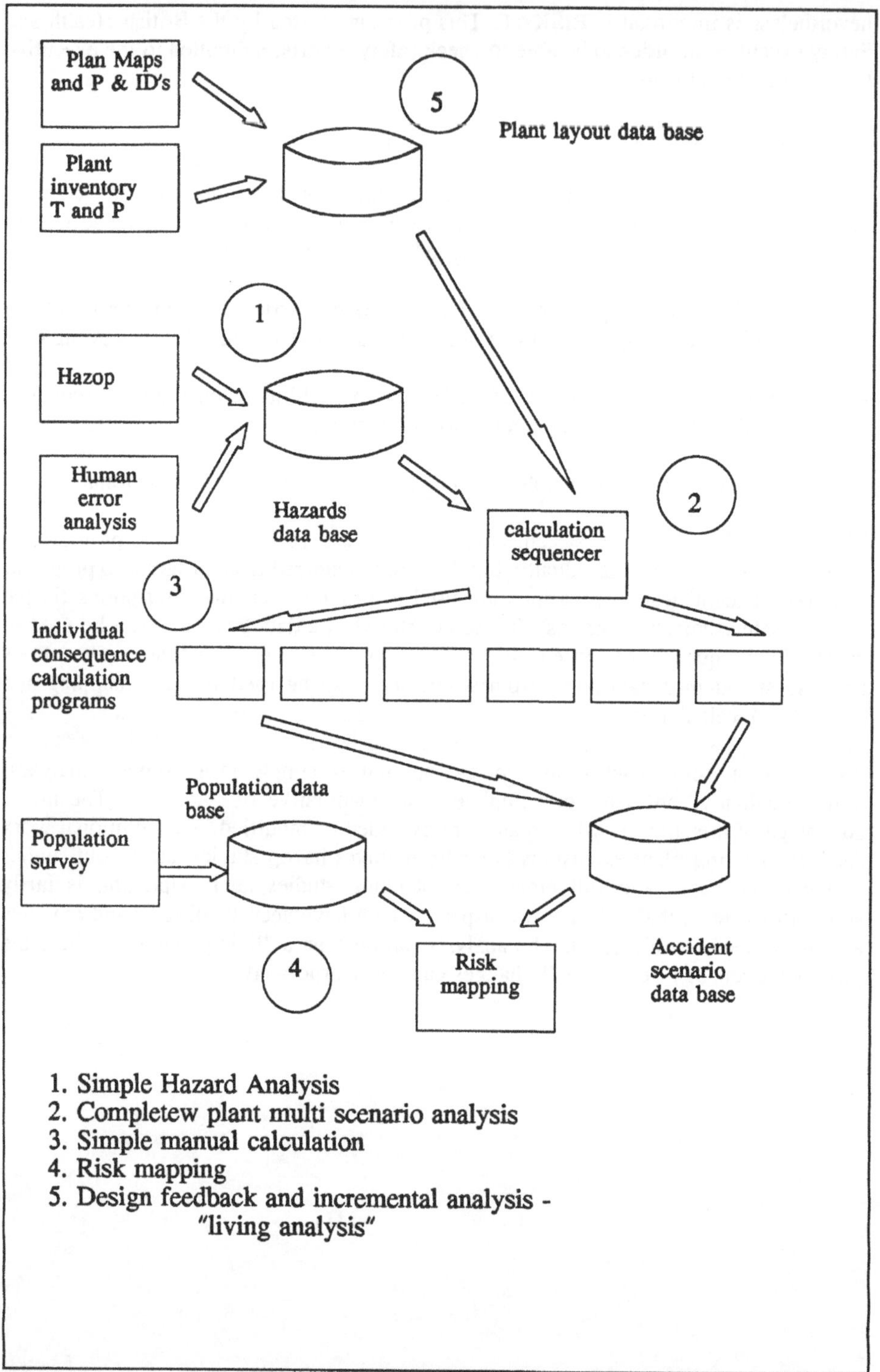

Fig. 16.1 Integrated risk analysis calculation program set.

Using risk analysis programs

Computer programs have become virtually essential to professional risk analysis today. Use of programs is not problem free, however.

All computer programs are based on physical models. These models have generally a limited range of application, in which they are considered to be applicable. They generally have a much smaller range over which they have been validated by comparison with experiment.

It is often tempting to "stretch" programs, using them outside their area of applicability. such use can hardly be expected to produce good results. Good programs limit the range of input data which can be accepted. However, it is virtually impossible to check for extreme combinations of data values. A typical program may have ten or twenty input parameters, and checking combinations of these two at a time requires at least 50 checks.

For this reason it is important to understand the model on which the program is bases, and its limitations. This is even more true for large integrated programs.

Some programs make use of iteration or time steps. Choice of iteration or time step size is almost always a compromise between accuracy and speed. In such cases, trial calculations should always be made, to ensure that results are not unduly sensitive to step size.

Large scale use of computer programs can lead to a great deal of confusion, unless care is taken. All calculations should be named uniquely, and the date and time and author added, to prevent confusion and mistakes at a later date.

References

Note: Development in computer program capabilities is often very rapid. The first published reference for each program is given here. For up to date information it is generally best to contact the program authors.

Ale, B. and Whitehouse, J.J. SAFETI- A computer based System for Risk Analysis of Process Plants, in Heavy Gas and Risk Assessment III, ed S.Hartwig, Reidel, 1986

Munday, G, Philips, H.C.D., Singh, J, and Windebank, C.S. Instantaneous fractional Annual Loss, in Loss Prevention and Safety Promotion in the Process Industries, 3rd Int Symposium, Basle, 1980

Technica Ltd. — The Whazan Program, Technica Ltd, Lynton House, Tavistock Square, London WC1H 9LT, England

The World Bank — Manual of Industrial Hazard Assessment Techniques, Office of Environmental and Scientific Affairs, The World Bank

Heino, P., Suokas, J., and Karvonen, I. — Expert System for Hazop Studies, in Loss Prevention and Safety Promotion in the Process Industries, 6th Int Symposium, Oslo, 1989

TNO — The EFFECTS program, TNO, PO box 45, 2280AA Rijswijk, Holland

Energy Associates — The EAHAP program

J.B.Fussel — Computer Aided Fault Tree Construction, in Reliability and Fault Tree analysis, SIAM, 1975

Taylor, J.R. and Olsen, J — The Rikke Program for automated fault tree and Cause Consequence analysis., Risø National Laboratory, Denmark, and
A comparison of Automatic Fault Tree Construction with Manual Methods of Hazard Analysis, in Loss Prevention and Safety Promotion in the Process Industries. 4th Int Symp, Harrogate, 1983

Lapp, S.A., and Powers, G.J. — The Synthesis of Fault Trees, in Nuclear Systems Reliability Engineering and Risk Analysis, SIAM, 1977

Taylor, J.R. and Hollo,E. — Experience with algorithms for Automatic Failure Analysis, ibid.

Wu, J.S., Salem, S.L. and Apostolakis, G. — The use of decision Tables in the systematic construction of fault trees, ibid.

A. Poucet — Interacting Expert Systems for Plant Safety Analysis, in Probabilistic Safety Assessment for Plant Safety Analysis, Elsevier, 1991 and Joint Research Centre, 21020, ISPRA, Italy

SRD	The Orchard Program, SRS, UKAEA, Culcheth, Warrington, WA3 4NE, England.
Pape, R.P. and Nussey, C.	A Basic Approach for the Evaluation of Risks from major Toxic Hazards, in The Assessment and Control of Major Hazards, I.Chem E. Symposium Series no. 93, 1885 (description of the RSKAT program) HSE, Sheffield, England.
Taylor, J.R.	The PAPA Programs for Process Plant, Pipeline, Platform and Transport Analysis. ITSA 1989 Taylor Associates, Erantisvej 5 , 4171 Glumsø, Denmark.
Sintef	The CARA program, Reliability and Safety Department, Sintef, N- 7034 Trondheim-NTH Norway
Platz, O, and Olsen, J.V	The Faunet Program, Risø National Laboratory, 4000 Roskilde, Denmark.

17 Safety Management and Inspection

Risk analysis has little impact on safety, if the analysis is based on idealised drawings and descriptions of the plant, while the actual plant is constructed in a completely different way, or is operated in an unsafe fashion. An inspection is an essential part of any safety analysis. Ideally, a full review of safety management is desirable, but a detailed study of the actual state of equipment is virtually essential. Similarly, for new plant, an inspection prior to commissioning is important. In this chapter, check lists are given concerning some of the aspects which may be considered.

Safety management

Plants of similar design vary very widely in actual accident rate. Joschek, in a survey of one company's worldwide experience found a variation in accident rate by a factor of 60. The present author has found variations in specific incident rates for similar equipment of a factor of over 100, due to differences in maintenance practices and routine inspection.

Some aspects of what is normal, and what is considered good, are listed below. The information is based on a survey intended as a base for assessing the importance of safety management to risk assessment, made during 1988 to 1990. While risk analyses does not require that safety management is perfect, it is unwise to trust the results if the management is definitely poor.

The list of topics which were considered important in this area were:

- management safety policy, motivation and follow up.
- safety organisation and responsibility.
- quality control
    - during design
    - during construction
    - during operation
    - of maintenance and modifications
- accident investigation and follow up
- training and experience
- housekeeping and safety practice
- operating procedures / safety procedures
- plant documentation
- emergency equipment and training
- emergency planning

Note that this list is far from complete as a general safety management programme, and is given only as a basis for assessing the relevance of risk analysis results.

Management safety policy, motivation and follow up

It is a good practice to have a clear and explicit policy. For example:

> " Safety is of primary importance, and should have the same priority as production or economy."

Having a loosely applied safety policy, or no safety policy but just a few rules, which are applied, means that risk increases considerably. Having no safety policy, or one which is not enforced, giving a poor standard of safety administration, increases risk by at least a factor similar to those presented by Joschek. This can be seen by comparing "best" and "worst" statistics for industrial accidents.

Motivation is the way that management ensures that safety policy is put into practice. Good techniques of motivation are

- Management always follows good safety practices themselves. (Always wear helmets in the plant, do not break rules to achieve production)

- Management show an interest when unexpected safety problems show up, and establish projects to cure the problem.

- Management ensure that safety training hazops, and special subject reviews are made, so that staff become aware of safety issues.

Follow up is also a part of good practice. Management during tours of the plant observe safety problems. Management require a report on safety, for example four times each year, along with production and economy reports. Most important there is a regular audit, once each year, with independent inspectors.

Safety Organisation.

It is difficult to describe precisely the form of safety organisation required for good practice, since this varies according to the kind and size of operation. Some general principles of good practice are clear though.

- There should be one person responsible for safety policy in the company, and for ensuring that it is followed. That person should be at the director level.

- Each of the engineering and operation functions should be staffed by a person who is technically qualified and reasonably experienced. (A management development programme may be necessary to ensure experience.)

- There should be a person responsible for technical aspects of safety, serving as a consultant or staff support to the line organisation. This person (or head of department) should be experienced both in safety engineering and operation.

- There should be a way of involving operating staff in safety (Safety committee, Safety campaign etc.)

- For a major hazard plant, there should be a person responsible for safety and emergencies on 24 hour call, or several persons, on a 24 hour rota. There should always be a competent person on site for emergency responsibility.

- The lines of responsibility in safety should be clear. There should be an organisation structure diagram in which safety responsibility is shown.

The impact of a breach of any of these principles on safety is not clear. What is clear, is that if the performance is inadequate in this area, then safety policy is also inadequate. Moreover, if the emergency preparedness function is not properly led, then credit should not be taken for correct performance of safety procedures.

Quality Control

Not very much can be said about the absolute requirements for quality control in a plant, since this varies so much. Some general principles can be stated, for major hazard plants though, which correspond to good practice.

- There should be a design review or loss prevention review, including hazard assessment during design.
- The standards for construction should be stated, and recognized standards used.
- There should be a quality control programme and a quality assurance activity during plant construction. This should follow a national or international standard.

- There should be a safety review and an inspection prior to commissioning, and a safety report after commissioning.
- There should be a safety follow up after six months of plant operation.

- There should be clear procedures for operation, and a periodic audit to check that they are followed.

- There should be a schedule for testing and inspection. Completion of tests and inspections should be recorded and signed. There should be a periodic audit of these.

- Repairs, and restoration after testing should be inspected independently, for major hazards plant.

- Materials, and the state of equipment should be inspected periodically, and the remaining lifetime of the equipment assessed.

- There should be quality control checks for new materials and products, if these can be significant for safety.

The impact of breaches of these principles on safety can be estimated directly. For example a pressure vessel inspected to the highest standards, and routinely reinspected, will have a probability of failure 100 times lower than for a vessel which is just pressure tested once. Risks during commissioning and initial operation are also 100 times higher than at other times unless special measures are taken.

The other aspects of quality control can be quantified directly by applying human error analysis methods to the most critical stages.

Accident investigation

It is easy to state the principles of good practice in this area.

- After an accident, there should be an investigation by a competent person or group, with some independent person to supply an independent view.

- The investigation should be confidential.

- The objective should be to investigate all possible causes, and to find out as much as possible about ways of preventing recurrence. The investigation should strive to describe more than one cause, and to reach back to ultimate causes. The investigation should avoid apportioning blame.

- The lessons learned from the investigation should be applied, and there should be a programme of work for their application.

An excellent level of practice extends these principles, so that near misses are also investigated. The critical incident technique in which employees are asked to describe the most recent near miss they can remember, is helpful, but requires experience to carry out.

- Disturbances should be reported in the plant unit operating log book reports in a constructive way.

Training and experience

The needs for training and experience vary widely from plant to plant. The variation is one of degree however. Operating a chemical, refinery, or loading plant, is a complex task, and cannot be performed properly without training and experience.

A good standard of practice is

- That there is a training programme for all staff (including maintenance, office, and cleaning staff).
- That the training programme is adapted to needs.
- That there is a periodic follow up for all staff.
- That a person responsible for operation or management of a major hazards plant has experience, and has demonstrated competence, in operating or managing a non major hazards plant of similar complexity.

There is a problem when staffing new or green field plants, that experienced staff may simply not be available. In this case, the best available staff should be obtained, and should be present at commissioning. Contractors, may be required to include training and initial operation in their services.

During periods of sickness, strikes, transport difficulties, manning can be a problem. An excellent practice is to establish minimum manning and experience criteria for the plant, and to shut down or reduce production (if this is possible) if these criteria cannot be met.

It is virtually impossible to quantify the overall effect of training on risk. Swain and Guttman give typically a factor of 10 to 30 increase in reliability for individual operations at the plant operator level, but this is far from the whole story. The main problem here is in quantifying the degree of training.

Housekeeping and Safety practice

The effect of good housekeeping and good safety practice on risk levels is hard to quantify, but corresponds to an important aspect of safety management, so presumably has a similar level of impact.

Good practice includes:

- Storage of materials, tools, equipment etc. only in authorised places, or neatly and without obstruction during abnormal operations.
- Marking escape and transport routes.
- Roping off / fencing areas during abnormal operations.

- Control of contractor staff.
- Tidiness and clear marking of areas during new construction.
- New construction equipment follows safety requirements.
- No ignition sources near hazardous areas.
- Safety instructions for contractor staff.
- Marking equipment for identification.
- Marking and clear identification of safety equipment.
- Good access to the site should be controlled.

If these principles are not followed, ignition probabilities increase towards one, the value of emergency procedures is impaired, and the frequency of fires starting increases.

Operating and safety procedures

Operating and safety procedures are areas where companies are most often found to be sinning - even good companies. It is difficult to keep procedures up to date.
An <u>excellent</u> standard of practice includes therefore:

- ensuring that there are operating procedures for each plant unit.
- ensuring that there are shut down start up procedures.
- ensuring that there is a set of emergency procedures.
- ensuring that these procedures are kept up to date (always updated before new plant modifications are put in operation)
- ensuring that there is always a completely up to date piping and instrumentation diagram for the plant.
- ensuring that instruments which are not used (do not work) are checked for safety, and are marked as not working on the control panel (and written if the failure is permanent)
- ensuring that there is a log book for each plant unit, and that production, disturbances, and repairs are recorded in it.
- ensuring that there is a shift hand over procedure.

- ensuring that procedures have been read and understood, and are followed by the operators.

- ensuring that procedures take account of practical operator experience.

- ensuring that operators understand the reasons for design features and steps in operation.

As stated above, these measures represent the limit of what can be achieved today. The requirements for updating and operator awareness are particularly difficult to fulfill, and almost certainly require a development of new techniques such as more automated updating and the use of simulator training techniques, if the measures are to keep up with the typical frequency of changes in chemical plant. From comparison of the frequency of operator errors, there is usually a potential for reducing the number of accidents in specific areas by a factor of between 2 and 5 even in good plants, by particular attention for this point.

Critical limits for operation

Any plant is subject to operational difficulties from time to time, so that the ideal safety situation is not achieved. Very serious difficulties require plant shutdown. Less serious problems require only that an effort is made to correct the situation. it is good practice to establish the dividing line between serious and less serious disturbances. Examples of areas in which such criteria may be required are -

- low manning levels

- safety systems out of operation

- low fire water reserves, or low reserves of fire fighting foam.

- poor quality of row materials, leading to danger

- lack of documentation, quality checks etc. for row materials.

- bad weather conditions

Maintenance and Safety procedures.

It is good practice to have a full set of maintenance and safety procedures. For the best plants, a complete set of such procedures often fills a complete ring binder. The procedures for tank entry, for example, typically cover one page of clearly written text. Some of the topics to be covered are -

- The list of instruments and apparatus to be routinely tested and test procedures.
- Routine quality control procedures for materials and products.
- Labelling requirements.
- Making ready for repairs and shut off of electrical equipment.
- Hot work procedures and permits.
- Periodic inspection of vessels, and corrosion
- Apparatus routine maintenance procedures.
- Safe methods of sample taking.
- Tank level measurement procedures.
- Use of cranes.
- Use of breathing apparatus and gas masks, and safety clothing.
- Tank filling and emptying for hazardous materials.
- Renewal of hoses, crane slings.
- Procedures for handling of substances.
- Safe storage and warehousing procedures.
- Gasket and seal replacement.
- Special precautions in selection of materials for repairs.
- Repair and modification authorisation procedures.
- Pipe deblocking procedures
- Procedures for safe use of compressed air equipment.
- Marking and clearing of escape routes.
- Carrying flammables, toxics etc.
- First aid treatment of burns, acid burns, poisoning.
- Testing of fire protection equipment, safety showers, and safety alarms.

Safety procedures should be written concisely, and should be kept simple in content. The important thing is that personal know the regulations and follow them.

Plant documentation

Good plant documentation is essential as a basis for planning, for safe modification, and for emergency action. Good practice ensures that this information is kept up to date. In view of the rate at which changes occur in many plants, this is extremely difficult to achieve, and this means that the proper place for updating is in the plant or unit control room. Updating neatly in red pencil on a control room drawing copy, is preferable to falling behind with updates.The problems of updating are lessening at present, because of the introduction of word processing and computer aided drafting. If these techniques are used, however, it is important to use a proper document control, and signing procedure.

The documentation which is needed for safe operation of a chemical plant includes -

- Substance used as raw materials, products, by products and waste products, including substance properties and health dangers.
- Plant layout.
- Storage positions for substances, including typical and maximum storage levels.
- Functional description of plant units.
- Drawings (P and I D's) of plant units.
- Emergency procedures for plant units.
- General safety procedures.
- General emergency procedures.
- List of standards and specifications for unit.
- Periodic testing check list.
- Instrument calibration list / log.
- Maintenance plan and maintenance log.
- Plant log.

Emergency Planning and training

There should be a set of emergency measures which are an essential part of good safety management practice. The impact of poor practice on safety can be estimated by performing a reliability analysis for the most significant accident scenarios.

The list of emergency preparedness measures include:

- A procedure for emergency communications with a phased list of persons and authorities to be informed.

- A clear and agreed authority structure with an experienced person taking direct charge in case of emergency (e.g. the chief fire officer.)

- A protected room or centre for communications in the case of toxic releases, if these are possible.

- General emergency plans, coordinated with the public authority emergency plans.

- Specific emergency plans for special risks such as tanks containing toxic gases of liquified toxic gases.

- Teams trained to deal with emergencies, including firefighting and controlling toxic releases.

- Sufficient equipment to be able to deal with the largest emergency which is likely to occur.

- Sufficient supplies, such as foam and absorbents, gas mask filters, and air cylinders for the largest form of accidents likely to occur.

- Hospitalization arrangements for persons injured in accidents.

- Proper information on medical treatment for poisoning or burning with unusual or special hazardous compounds.

- Equipment for plugging or patching leaks.

- Material for absorbing or impounding leaks

- Proper protective equipment for toxics, acids, refrigerated liquids etc.

- Periodic exercises (between one per month and one every 4 months is usual in large installations, with variation in the type and scope of exercises).

Specific aspects of plant design

In the sections that follow, a number of check lists are given, which supplement the hazard identification methods described in Ch. 3. Note that some of the points covered concern mandatory safety measures, but most are questions of judgement. Irrespective of whether a problem is regarded as too insignificant to be worth correcting, risk analysis should take it into account.

Good safety practice starts with good plant layout. The important aspects are good safety distances, segregation of offices and canteens etc. from production units (and from services, sewers and water treatment), siting of control rooms, and good access in the case of emergencies. Roadways carrying heavy traffic should be away from production units and storage of hazardous materials.

For new plants, this kind of "requirements" can often be fulfilled with little effort. For most plants, development over many years has left a legacy of layout difficulties. Even though it may be impossible to correct layout problems, it is important to recognise them, in order to avoid making matters worse. Also, any such difficulties must be taken into account when risk analyses are made or assessed.

The classic approach to layout problems is to ensure minimum safety distances are maintained. Such distances are generally based on experience. The risk analysis procedures described in Ch.13 and 16 provide a supplement to these layout guides, and can be much more flexible than fixed distance guides.

It is a good idea to make a plan of the plant, and to mark onto it traffic routes, including those of employees going to and from work (both by car and by foot).

During inspection of existing plant, it is a good idea to watch the plant at work for a period, from a high vantage point. This is particularly good for identifying maintenance and contractor access difficulties, and hazards from crane movements.

Fig 17.1 Parking problem

Fig 17.2 Crane and pipe bridge

Layout

1. Are high toxicity gas releases "downwind"?

2. Will toxic/flammable releases gather in hollows?

3. What is the unit separation. Is it adequate?

4. Are there trenches under critical equipment? Under shut off valves? Can gas gather under critical equipment?

5. Are there high level flows of combustibles?
Over operator platforms?
Over fin heaters/coolers?
Over heat exchangers/pressure vessel?

6. Can downhill flows of toxic or flammable substances occur ?

7. Are flammable liquid pumps situated under anything? over anything?

8. Is ventilation intake clear of gas or vapour releases?

9. Are all valves accessible?

10. Are stand pipes/S.V. outlets well clear of stairways, walkways, other equipment?

11. Is there access for firefighting vehicles?
Head room for high vehicles?

Roadways and exits

1. Is there a clear route up to the plant for emergency vehicles?
Clearance?

2. Is there clear headroom over one roadway?

3. Are pipe bridges protected against cranes, high loads?

4. Do all exists, side roads have a clear view?

5. Do swing doors for trucks/trolleys have a clear panel?

6. Are driveways for trucks / trolleys even, without holes?
Are driveways for trolleys level?

Buildings

1. Is explosion damage limitation used where there is an explosion potential?
   - open construction
   - light roof
   - blow out panels
   - blast wall

2. Is there a clear line for explosion relief?

3. Are windows wired? Non shattering?

4. Are lights adequate?
   Are they guarded?
   Are they explosion proof?

5. Are emergency lights needed?
   Are they fitted?
   Are they fitted to emergency exits and passageways?

6. Are there sufficient exits?
   Where?
   Are they marked?
   Are passageways clear?
   Are they marked?

7. Is the control building reinforced (bunker) in places where there is a large explosion potential?

8. Are there tripping hazards uneven floor?
   cables, waste, materials etc?

9. Are floors clean, free of mud, water, oil?
   Non slip surfaces on walkways?

10. Do stairs and platforms have
    - handrails
    - toe rails?
    - clear headroom?

10. Is heating fitted? Adequate?
    Is there temperature controls?
    Are there alarms?
    Are there barriers to protect from burns?

11. Is smoking prohibited?
    Are signs posted?
    Are staff aware of dangers of smoking in a hazardous area?

12. Is ventilation adequate?
    Is it working?
    Is intake area clear?
    Can intake be polluted?
    Can exhaust be ignited?
    Does exhaust threaten other work areas with toxic release?
    Is there excess pressure in control room which can keep out flammable/toxic atmosphere?
    Is there automatic shut off for ventilation in the case of release of gas?

13. Is electrical supply working?
    Are breakers accessible?
    Do staff know where they are and when to use them?
    Are they explosion proof?
    Are they interlocked?
    Are essential supplies marked?
    Do they have a back up?
    Are there signs and interlocks for circuits under repair.

14. Are there adequate waste containers?
    Are they marked?

15. Are washing facilities adequate?
    Toilet facilities?
    Are toilets in high fire risk area?

16. Are supplies for drinking water and process water completely separate?

17. Are there separate drains for process waste and sanitary waste?
    Do drains have seals?

18. Are floor loading limits adequate?
    Are roof loading limits adequate?
    Are loading limits posted? Are snow loads appropriate for risk level? Wind loads?

19. Are surfaces clean?
    Are they free from toxic or irritant materials?

20. Are surfaces painted?
    Is piping colour coded to standard?

21. Are roofs and walls weather tight?

Other problems

Piping

Good piping design requires great care, not only in ensuring that proper materials and pipe thicknesses are used, but also in detailing. Piping should be supported to avoid stress and vibration. Unsupported pipe sections should be short, the distance depending on pipe thickness. Also, the fitting hand work should be well made, with pipes skimmed up on supports, or supported by hangers, which actually take the weight. For critical piping, such as high pressure risers and downcomers, pipe support springs should be checked periodically for adequate tension.

Small guage piping is a problem in many plants, with long stretches inadequately supported. This is particularly a problem with drain lines. Drain lines which can be pressurised with liquefied gas, or flammables at high temperature, should preferably have a diameter of at least 2 inches, and be properly supported to prevent breakage. Heavy instruments should be preferably be supported on steelwork or large diameter pipe, not on 1 inch piping. Otherwise vibration may cause fatigue cracking.

Fig 17.3 Stilling pot for level guage, supported on thin pipe, and vibrating badly

LPG piping is sometimes subject to small leaks, insufficient to cause a fire risk, but sufficient to cool the piping to -20 to -40°C. If this is the case, check whether the piping is rated for such low temperatures.

Two phase flow in pipes can arise because air or gas becomes mixed into liquid flows, or due to flashing or boiling in the pipe. Two phase flow can cause violent vibration, sufficient to crack flange and support welds. The check list contains several questions intended to identify problems of this kind.

Piping

1. Are there splash trap separator on condenser, to prevent flow of gas in the condensate line?

2. is there two phase flow in an uphill or vertical pipe section?

3. Is two phase flow possible under unusual conditions?

4. Are there condensate collection pockets, which can lead to liquid slugs in gas flows?

5. Are there long pipe length following pressure reduction, which can cause excessive vibration ? If so, is the section well supported and damped?

6. Are thermal expansion stresses relieved?

7. Are there screw fittings on flammable liquid or gas service, which could leak if affected by fire, or could vibrate loose ?

8. Are pipe falls in the right direction?
   Can pockets of liquid collect in piping?

Vessels and piping

1. Is design temperature fixed?
   maximum-----------------
   minimum-----------------

2. Is design pressure fixed?
   maximum-----------------
   minimum-----------------

3. What are safety valve settings?

4. Has materials list been made out?
   Are there any high corrosion rate materials or material combinations?
   What raw materials, if added in error will cause high corrosion rates?

5. Are there any pockets where corrodants can be trapped?

6. Are there any galvanic pairs giving high corrosion rates?

7. What is design pressure for flanges?

8. What is gasket material?
   Is it suitable?

9. Are long pipe sections supported? Are supports adequate when piping is full? Do supports prevent vibration? Are hangers tightened?

10. Are there long pipe sections following pumps and safety valves which can cause oscillation, vibration, valve chattering, and cavitation in the pump?

11. Can vessels be drained?
Is there a draining and cleaning procedure?
Has the corrosion inspection interval been set?

12. On low pressure tanks, is there a vacuum breaker?
Is it large enough to equalise maximum flow? (an open outlet pipe at lowest point in system.)
Can the vacuum breaker be blocked?

13. Are pipes, vessels, and valves labelled?

14. Are vessel and hopper supports adequate? Also when vessel is filled with water? Also when filled with heavy solid?

15. Is glass piping/vessels screened?

16. Are supply tanks ventilated, or fitted with pressure balance pipe?
Is venting adequate, even with ice and snow?

17. Is there a safety or relief valve on all vessels and pumps?

18. Can the vessel tolerate a vacuum?

19. For cryogenic liquids, are all pipe sections protected by safety valves.

Reactions

1. Is heat balance calculated?
   - exothermic
   - endothermic

2. Materials properties tabulated?
   - raw materials
   - products
   - bi products

4. Are materials
   - flammable
   - corrosive
   - toxic

5. Is reaction matrix filled out?

6. Are all raw materials clearly identified?

7. Is there any runaway reactions? Is cooling adequate?
   Can materials be dumped?
   Do reactions accelerate with high temperature, pressure?

9. Are side reactions possible?
   Wrong reactions?
   Under wrong conditions, temperature pressure, catalyst, concentration?

10. Are cleaning procedures specified?

11. Are general plant safety instructions made available?

12. Are criteria for safe operation fixed?
    Maximum plant parameter values?
    Minimum levels of spares, emergency equipment?
    Minimum manning levels?
    Minimum supplies?

13. Is agitation required? Can agitation fail?
    Can failure of agitation be detected?

14. Does material deteriorate on standing?
    If left at high temperature?
    Is deteriorated material dangerous?
    Is a warning included in procedures?

15. Can two batches/substances be mixed accidentally?
    Can hot oil and water?

Ignition

(Whenever there are flammable liquids or gases present or potentially present)

1. Has an area classification been made? In accordance with standard?

2. Is all equipment earthed, including support steelwork? Are earth links marked? Have earth links been painted? Are they still good?

3. Are lightning conductors provided?

4. Are balls in ball valves earthed?

5. Long pipelines entering the plant and railway lines insulated at plant entry?

6. Are there power cables/telephone lines parallel to long stretches of piping?

7. Are there power cables and telephone lines through hazardous areas

8. Are manhole lids earthed? Are filter screens earthed?

10. Do filling hoses have earth clips, and are they in good condition and in use.

11. Is it possible for flammable gasses to enter switching or instrumentation rooms?

12. Are there hot surfaces or steam leaks?

Fig 17.4 Broken light fitting, an ignition hazard

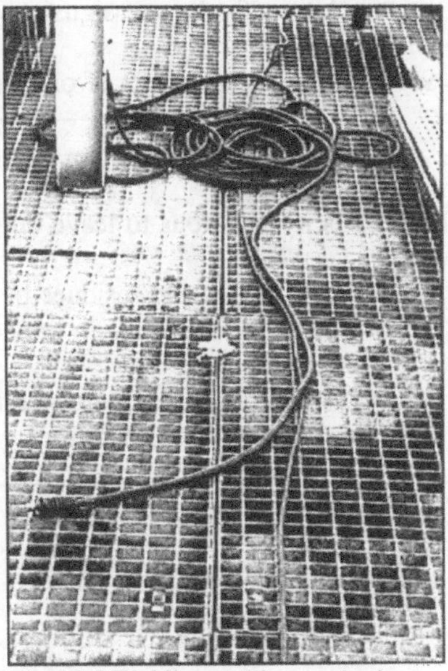

Fig 17.5 Welding cable, a tripping and ignition hazard

Heat exchangers

1. Can the low pressure (shell?) side take the full flow from the high pressure side with one tube sheared, without rupturing?

2. Has the design allowed for rapid heating/expansion/heatshock due to errors or failures?

3. Are correct gaskets fitted?
   Is there a set of reserve gaskets of the right type?

4. Can electrical current arise in the heat exchanger?

5. What is the expected corrosion rate?
   what inspection interval is planned?
   Is it entered in maintenance schedule?
   Is there lower inspection interval for tube sheets?

6. Are flows through parallels exchangers balanced? How?

7. Are flow rates high enough to move tramp metal?
   Is system cleared of tramp metal?
   Is there a filter?

8. Can the exchanger be shut in by block valves?
   Are safety valves fitted? To both sides of the exchanger ?
   Alternatively, are signs posted on valves, and locks provided?

9. What is the effect of one flow stopping?
   Hot side ---------------
   Cold side ----------------

10. Is it dangerous to leave out a reactant?

11. Can wastes and residues be removed? Are they toxic?
    Can residues lodge in crevices?

12. Is tubing well supported so that excessive vibration and fretting will not occur?
    Under all flow conditions ?

Electrical equipment

1. Are all high voltage connections/cables/busses protected? By insulation, cabinets, or guards?

2. Do cabinets have warning for high voltages?

3. Are circuits clearly identified? Is identification correct and voltage marked? Colour coding correct?

4. Are high voltages separated from low? Minimum separation distance.

5. Are there redundant power supplies?
Are they separated?

6. Are there lamps to indicate live circuits?
With circuit test, or double lamps?

7. Are cabinets and equipment grounded?
Is ground resistance tested?

8. Are there high voltage cables in fire hazard areas ?

9. Are wiring diagrams available? Up to date?

10. Is over current protection fitted?
Is response fast enough to prevent damage?

11. Is over voltage protection fitted?

12. Is dust and moisture protection fitted?

13. Are arcing parts enclosed?

14. Are there energisation circuits for capacitors?

15. Are there splices in cable tray bundles?.

16. Is there a safe procedure for de energisation of circuits under repair?
Proper marking for failed circuits, and circuits under repair?
Key system for locking circuits under repair?
Is back feeding from the load possible for circuits under repair?
From a redundant supply or load driven motor?

17. Are safety devices proof against grounding?

18. Power tools grounded/double insulated?

19. Cables in good repair? Do they run over sharp edges?

20. What is the earth voltage pattern from the main power supply? Can it induce currents in the plant equipment?

21. Are terminal boxes sealed ?
Are cables tight within the seals? Can water enter and cause shorting ? Can gas enter with a potential for ignition ?

22. Are there loose cables?
Signal or power cables?
Do cables run over sharp edges?
Can cables be broken by traffic?
By an operator when operating, or as a result of tripping?

23. Are high resistivity liquids present?
Are splashing liquids present?
With earth network?
Are there lengths of plastic piping?
Is there an internal earth/mesh in long piping?

Safety or relief valves, Burst discs

1. What is the design pressure of the vessel or pump to be protected?

2. What is the relief pressure?

3. Is piping up to the relief valve short?

4. Is the flow rate adequate?
Is it adequate to fire?
Even when the vessel is weakened by fire?

5. Can the blowdown line be closed by valves? Are valves locked open?
Can the blowdown line dog?
Weepholes drilled?

6. Can process liquid freeze or evaporate leaving a crust. Should a burst disc be fitted? Is burst disc metal (should not be graphite before a safety valve). Is the space between disc and valve relieved or monitored?

8. Is there a written test procedure?
Are blanking plates removed after maintenance?

9. Is the safety valve adequately supported against vibration?
against jet reactions?

10. Is the relief vent well clear of other equipment?

11. Can reverse flow occur from blow down headers? Is the flow rate adequate even if several SV's open at the same time?

12. Does the blowdown piping have uniform diameter? (narrowing is against standard requirements)

13. Are burst discs approved, and labelled with burst pressure? Is the pressure correct? Are they fitted the right way round?

Control and instrumentation.

1. Are all control signals measurable at test points?

2. Has control pattern been fixed?
   Is it checked for cross coupling?

3. Are loop gains known roughly?
   Have they been checked on the plant?

5. Are failure directions correct (fail open, fail closed, fail fixed)?

6. Are setpoints adjustments locked?
   Should they be?

7. Are all manual switches working (on, off)?

8. Do all interlocks work? Prevent action if false?
   Allow action if true?

9. Do all alarms function? At what setpoint?

10. Do all trip functions work? Are trip levels recorded?
    Do trips occur at specified levels?
    Is there a table of measured trip levels?

11. Is the calibration source checked?

12. Are all instruments regularly tested and calibrated? Is there a check list? Is it complete?

13. Do all instruments measure the correct variable? Is there a compensation factor? Is it known? Does it vary?

14. Are the power supplies for redundant instruments independent? Are the instrument displays independent?
    Do redundant instruments use the same cable trays?

15. Do instruments measure at the correct point? Is there a long time delay for response? Do pressure pulses, vibration, heat tracing. cable induction affect measurements? Is there a long delay on start up before instruments measure stably (analysers etc.).

16. Do instruments and switches give a steady status signal, or a pulse on change of status?

Valves

1. Is packing suitable?
   Type?

2. Is valve the right way round?

3. Can the stem move freely?
   Lubrication procedure provided, maintenance proc. known?

4. Is the right trim fitted?
   Which type?

5. Are material compatible?

   - body ------------------
   - trim ------------------
   - seats ------------------

6. Is pressure and temperature duty correct?

7. Should valve open or close on failure?
   Does it?

8. Is valve right way up for stem to fall if stem breaks?

9. Is valve accessible?

Flanges

1. Is packing material suited to the flange duty?
   - Hard or soft?
   - Pressure rating?
   - Material compatibility?

2. Is flange type appropriate?
   - raised face
   - recessed

3. If flange is intended for frequent disconnection, is it accessible?

4. Are there appropriate procedures for fitting packings? Are packings properly identified?

5. Are there enough bolts on flange for the required duty? Right type - same temperature expansion as flange? Are bolts fitted and tightened?

6. Are bolts and screws of the right length, and screwed right through the nut?

7. Is it possible to fit blinds? Adequate access and clearance?

Control Room

1. Is it built to withstand explosion?

2. Are windows small/wired?

3. Emergency lighting?

4. Air intake well clear of process building?

5. Is a small overpressure provided to keep out gas and vapour?

6. Fire doors fitted, and kept closed?

7. Hand extinguishers for electrical fires?

8. Gas masks? For all types of gas?

9. Are floor drains hazardous? Sumps under building? Cellar?

11. Is there a escape route away from the process plant?

Drainage

1. Are all leaks, releases, collected in bund/drain system?

2. Collection capacity? ----------
3. How is release gathered?
   - sump
   - drain

4. Are drains sealed?

5. Will bund overflow if water is used for fire fighting?

6. Run off and contaminated drains separated?

7. Drainage basin size ----------- 200m?

8. Paving or gravel?

9. Is drainage adequate for fire fighting?
   Are fire water drains free and unblocked?

Pumps

1. Is pump dimensioned for operating temperature?
   Pressure?

2. What is maximum load?
   Is it higher than plant design pressure?
   Is there safety valve/bypass?

3. Can pump operate with no flow?
   Temperature and or flow trip?
   Bypass recirculation for low flow?

4. Is there a non return valve following the pump?
   If not, why not? -----------

5. Can pump run empty?
   Are bearings and seals cooled?
   Trips if cooling fails?

6. Is there adequate NPSH? Also at high flow?

7. Is inlet filter fitted? can it stop pump flow, will air intake through seals be dangerous? How often is filter changed?

8. Is pump on falling part of head/flow curve? Is there a long delivery line?

9. Is material pumped toxic or flammable?
   Double seals fitted?

10. If non return valve fails stuck open, will pump be driven as a turbine?

11. Is there a vortex breaker on pump tank inlet?

12. Is there adequate net positive suction head under all flow conditions ?

13. Is a low NPSH (Net positive suction head) trip fitted?.

Personnel

1. Are staff assigned?
   Is line authority established?
   Is individual responsibility clear?
   Have all staff been informed of their responsibilities?

2. Is responsibility for safety made clear?
   Are safety regulations available to those who need them?
   Have they been read?

3. Have staff received formal training?
   Are staff familiar with plant?
   - with shut down procedure
   - with start up procedure?
   - with emergency procedures?
   - with equipment?
   - with evacuation and alarm rules?
   - with fire procedures?

4. Is protective equipment available?
   In use?
   - eye protection
   - face protection
   - gas masks
   - shower

Procedures

1. Is flow sheet available?
   Is it up to date?

2. Is there a shut down procedure?
   Is it easily understood?
   Is it simple?

3. Is there a start up procedure?

Is it easily understood?

4. Is there a full range of emergency procedures?
   - fire
   - overpressure
   - runaway
   - leak
   - other

5. Are chemical data sheets available?
   Are there spill handling procedures?

6. Are evacuation procedures written up?
   Simple instruction chart prepared?
   Alarm and evacuation instructions?
   Are emergency procedures posted?

7. Are sampling procedures fixed?

Maintenance

1. Are shutdown, cleaning procedures specified?

2. Is marking of shutdown equipment specified? Locking procedures for power supplies?

3. Are maintenance intervals for equipment specified?

4. Are inspection procedures for piping, vessels, specified?

5. Are calibration and test procedure foe instrumentation, alarms, and trips specified?

6. Procedures for shutting off, locking valves, fitting blinds for pipes.

7. Procedures for tank entry prepared and made available to staff? Have they been read?

8. Restart procedures prepared?
   Check list for tools, blinds to be removed?

Further reading

Just a few books are mentioned here, which the author has found valuable in risk analysis work.

Loss Prevention and Safety Promotion in the Process Industries F. P. Lees, Butterworths 1982.

This book is a very important reference book with a very complete list of references up to 1982.

Explosion Hazards and Evaluation, Baker et al, Elsevier 1983.

This book give a thorough treatment of explosion phenomena, including handbook style calculations.

Loss Prevention and Safety Promotion in the Process Industries. International conferences in 1974, 1977, 1980, 1983, 1986, 1989, and 1992. The conference proceedings give an unbroken and virtually complete record of process plant risk analysis developments over 20 years. European Federation of Chemical Engineers.

Index